palgrave advances in
international environmental politics

Palgrave Advances

Titles include:

Jeffrey Haynes (*editor*)
DEVELOPMENT STUDIES

Michele M. Betsill, Kathryn Hochstetler and Dimitris Stevis (*editors*)
INTERNATIONAL ENVIRONMENTAL POLITICS

Terrell Carver and James Martin (*editors*)
CONTINENTAL POLITICAL THOUGHT

Michelle Cini and Angela K. Bourne (*editors*)
EUROPEAN UNION STUDIES

Palgrave Advances
Series Standing Order ISBN 1–4039–3512–2 (Hardback) 1–4039–3513–0 (Paperback)
(*outside North America only*)

You can receive future titles in this series as they are published by placing a standing order. Please contact your bookseller or, in the case of difficulty, write to us at the address below with your name and address, the title of the series and the ISBN quoted above.

Customer Services Department, Macmillan Distribution Ltd, Houndmills, Basingstoke, Hampshire RG21 6XS, England

palgrave advances
in international
environmental politics

edited by
michele m. betsill, kathryn hochstetler
and dimitris stevis
colorado state university

© edited by Michele M. Betsill, Kathryn Hochstetler and Dimitris Stevis 2006

All rights reserved. No reproduction, copy or transmission of this publication may be made without written permission.

No paragraph of this publication may be reproduced, copied or transmitted save with written permission or in accordance with the provisions of the Copyright, Designs and Patents Act 1988, or under the terms of any licence permitting limited copying issued by the Copyright Licensing Agency, 90 Tottenham Court Road, London W1T 4LP.

Any person who does any unauthorised act in relation to this publication may be liable to criminal prosecution and civil claims for damages.

The authors have asserted their right to be identified as the authors of this work in accordance with the Copyright, Designs and Patents Act 1988.

First published 2006 by
PALGRAVE MACMILLAN
Houndmills, Basingstoke, Hampshire RG21 6XS and
175 Fifth Avenue, New York, N.Y. 10010
Companies and representatives throughout the world

PALGRAVE MACMILLAN is the global academic imprint of the Palgrave Macmillan division of St Martin's Press LLC and of Palgrave Macmillan Ltd.
Macmillan® is a registered trademark in the United States, United Kingdom and other countries. Palgrave is a registered trademark in the European Union and other countries.

ISBN-13 978–1–4039–2106–2 hardback
ISBN-10 1–4039–2106–7 hardback
ISBN-13 978–1–4039–2107–9 paperback
ISBN-10 1–4039–2107–5 paperback

This book is printed on paper suitable for recycling and made from fully managed and sustained forest sources. Logging, pulping and manufacturing processes are expected to conform to the environmental regulations of the country of origin.

A catalogue record for this book is available from the British Library.

A catalogue record for this book is available from the Library of Congress

Printed and bound in Great Britain by
CPI Antony Rowe, Chippenham and Eastbourne

contents

list of abbreviations and acronyms vii
notes on contributors xi
1 general introduction 1
michele m. betsill, kathryn hochstetler and dimitris stevis

part i the context of the study of international environmental politics

2 the trajectory of the study of
international environmental politics 13
dimitris stevis

3 theoretical perspectives on
international environmental politics 54
matthew paterson

4 methods in international environmental politics 82
kathryn hochstetler and melinda laituri

part ii the forces that shape international environmental politics

5 the environment as a global issue 113
gabriela kütting and sandra rose

6 international political economy and the environment 142
jennifer clapp

7 transnational actors in international environmental politics 172
michele m. betsill

8 environmental security   203
larry a. swatuk

9 global governance and the environment   237
frank biermann

part iii   normative frameworks for evaluating international environmental politics

10 sustainable development: the institutionalization of a contested policy concept   265
hans bruyninckx

11 the effectiveness of environmental policies   299
jørgen wettestad

12 environmental and ecological justice   329
bradley c. parks and j. timmons roberts

13 general conclusion   361
michele m. betsill, kathryn hochstetler and dimitris stevis

index   371

list of abbreviations and acronyms

AOSIS	Alliance of Small Island States
APEC	Asia Pacific Economic Cooperation
BASD	Business Action for Sustainable Development
CAN	Climate Action Network
CBNRM	community-based natural resource management
CEC	Commission on Environmental Cooperation
CIA	Central Intelligence Agency (US)
CITES	Convention on International Trade in Endangered Species of Wild Fauna and Flora
CLRTAP	Convention on Long-Range Transboundary Air Pollution
CO_2	carbon dioxide
COW	crude oil washing
CTE	Committee on Trade and Environment (WTO)
EB	Executive Body (CLRTAP)
ECA	export credit agency
ECGD	Export Credit Guarantee Department (UK)
ECLAC	Economic Commission for Latin America and the Caribbean
ECSP	Environmental Change and Security Project (of the Woodrow Wilson Center)
EFIC	Export Finance and Insurance Corporation (Australia)
EJ	environmental justice
EKC	Environmental Kuznets Curve
EMEP	Cooperative Programme for Monitoring and Evaluation of Long-Range Transmissions of Air Pollutants in Europe
ENGO	environmental non-governmental organization
ESS	Environmental Studies Section (of the ISA)

EU	European Union
ETUC	European Trade Union Conference
FDI	foreign direct investment
FoE	Friends of the Earth
FSC	Forest Stewardship Council
GATT	General Agreement on Tariffs and Trade
GDP	gross domestic product
GEF	Global Environment Facility
GHG	greenhouse gas
GIS	geographic information system
GIT	geospatial information technologies
GKG	Gaza/Kruger/Gonarezhou transfrontier park
GNP	gross national product
GPE	global political economy
GPS	global positioning system
HIPC	Highly Indebted Poor Countries
IC	Implementation Committee (CLRTAP)
ICJ	International Court of Justice
ICLEI	International Council for Local Environmental Initiatives
ICSU	International Council of Scientific Unions (International Council for Science)
IDGEC	Institutional Dimensions of Global Environmental Change
IEP	international environmental politics
IGBP	International Geosphere-Biosphere Programme
IGO	intergovernmental organization
IHDP	International Human Dimensions Programme on Global Environmental Change
IIASA	International Institute for Applied Systems Analysis
IIED	International Institute for Environment and Development
ILO	International Labour Organization
IMF	International Monetary Fund
INGO	international non-governmental organization
IPCC	Intergovernmental Panel on Climate Change
IPE	international political economy
IR	international relations
ISO	International Organization for Standardization
IUCN	International Union for the Conservation of Nature and Natural Resources (World Conservation Union)
IUPN	International Union for the Protection of Nature
JBIC	Japan Bank for International Cooperation
LDC	less developed country

MARPOL	International Convention for the Prevention of Pollution from Ships
MEA	multilateral environmental agreement
MNC	multinational corporation
NAACP	National Association for the Advancement of Colored People (US)
NAAEC	North American Agreement on Environmental Cooperation
NAFTA	North American Free Trade Agreement
NATO	North Atlantic Treaty Organization
NEC	National Emissions Ceilings (Directive) (EU)
NEPAD	New Partnership for Africa's Development
NGO	non-governmental organization
NH_3	ammonium
NIEO	New International Economic Order
NOx	nitrogen oxides
OECD	Organization for Economic Cooperation and Development
OPEC	Organization of Petroleum Exporting Countries
OPIC	Overseas Private Investment Corporation
PCBs	polychlorinated biphenyls
PD	Prisoner's Dilemma
POPs	persistent organic pollutants
PPGIS	public participation global information system
PPMs	production and processing methods
RENAMO	Mozambique National Resistance Movement
RS	remotely sensed data
SADC	Southern African Development Community
SAP	structural adjustment programme
SBT	segregated ballast tanks
SDI	spatial development initiative
SO_2	sulphur dioxide
TBNRM	transboundary natural resource management
TBNRMA	transboundary natural resource management area
TERI	Tata Energy Research Institute
TFACT	Trust Fund for Assistance to Countries in Transition (CLRTAP)
TFCA	transfrontier conservation area
TNC	transnational corporation
TOMA	Tropospheric Ozone Management Area
UK	United Kingdom
UN	United Nations
UNCCD	United Nations Convention to Combat Desertification

UNCED	United Nations Conference on Environment and Development
UNCSD	United Nations Commission on Sustainable Development
UNCTAD	United Nations Conference on Trade and Development
UNDP	United Nations Development Programme
UNECE	United Nations Economic Commission for Europe
UNEP	United Nations Environment Programme
UNESCO	United Nations Educational, Scientific and Cultural Organization
URA	Unitary Rational Actor
USAID	United States Agency for International Development
VOCs	volatile organic compounds
WCED	World Commission on Environment and Development
WEF	World Economic Forum
WEO	World Environment Organization
WGS	Working Group on Strategies (CLRTAP)
WHO	World Health Organization
WMO	World Meteorological Organization
WOMP	World Order Models Project
WSSD	World Summit on Sustainable Development
WTO	World Trade Organization
WWF	World Wide Fund for Nature

notes on contributors

Michele M. Betsill is Assistant Professor of Political Science at Colorado State University and Affiliate Scientist with the Institute for the Study of Society and the Environment at the National Center for Atmospheric Research. Her research focuses on issues of global environmental governance, particularly related to climate change, and transnational environmental politics. She has published numerous articles and book chapters and is co-author (with Harriet Bulkeley) of *Cities and Climate Change: Urban Sustainability and Global Environmental Governance* (Routledge, 2003).

Frank Biermann is professor of Political Science and Environmental Policy Sciences at the Vrije Universiteit Amsterdam. He is head of the Department of Environmental Policy Analysis at the university's Institute for Environmental Studies (IVM) and director of the Global Governance Project GloGov.Org, a multidisciplinary research programme of leading Dutch and German academic institutions. Most of his research has addressed questions of global environmental politics. He has authored, co-authored or edited six books, the last of which is an edited volume on the debate about a world environment organization.

Hans Bruyninckx is currently associate professor of international environmental politics and sustainable development at the Catholic University of Leuven (Belgium) and Associate Professor of environmental politics at the Wageningen University (Netherlands). Between 1997 and 2002 he was founder and research coordinator of the research group on Sustainable Development of the Catholic University of Leuven. He has been a member of the Belgian Federal Council for Sustainable Development and the Flemish Environmental Council. Recent research projects address topics on the institutionalization of sustainable development in Belgian and EU policies and environmental governance in Russia.

Jennifer Clapp is associate professor of international development studies and environmental and resource studies at Trent University in Canada. Her work focuses on the interface between the global economy and the natural environment. She is author of numerous articles as well as several books, including *Adjustment and Agriculture in Africa: Farmers, the State and the World Bank in Guinea* (Macmillan, 1997) and *Toxic Exports: The Transfer of Hazardous Wastes from Rich to Poor Countries* (Cornell, 2001). Her most recent book, co-authored with Peter Dauvergne, is *Paths to a Green World: The Political Economy of the Global Environment* (MIT Press, 2005).

Kathryn Hochstetler is Associate Professor of Political Science at Colorado State University. In 2003–04, she was Research Fellow in Politics at the Centre for Brazilian Studies, Oxford University. She recently published the co-authored book *Sovereignty, Democracy, and Global Civil Society: State–Society Relations at UN World Conferences* (SUNY Press, 2005). She has also published numerous articles and book chapters on environmental politics and civil society in Brazil, Mercosur and United Nations conferences. She is currently completing a co-authored book manuscript on Brazilian environmental politics since 1972.

Gabriela Kütting is Assistant Professor in the Department of Political Science and Center for Global Change and Governance at Rutgers University. She is also part of the core teaching faculty of MEPIELAN, a UNEP institution training Mediterranean civil servants in environmental law and policy. Her research interests lie in the field of global environmental politics and international/global political economy. She has published widely in the field of global environmental politics and is the author of two books: *Environment, Society and International Relations* (Routledge, 2000) and *Globalization and the Environment: Greening Global Political Economy* (SUNY Press, 2004).

Melinda Laituri is an Associate Professor in the Department of Forest, Rangeland, and Watershed Stewardship at Colorado State University. She conducts research on geographic information systems and their applications in natural and water resource management, indigenous land management and disaster response. Her geographic areas of research include: Puerto Rico, New Zealand, Sweden and South Africa. She has published several articles on public participation GIS (PPGIS).

Bradley C. Parks is a Development Policy Officer at the Millennium Challenge Corporation. He recently completed postgraduate studies at the London School of Economics and Political Science and is co-author of two forthcoming books: *A Climate of Injustice: Global Inequality, North–South Politics, and Climate Policy* (with J. Timmons Roberts) and *Greening Aid: Understanding Environmental Assistance to Developing Countries* (with Michael J. Tierney, J. Timmons Roberts and Robert Hicks).

Matthew Paterson is Associate Professor of Political Science at the University of Ottawa. His research focuses on the intersection between International Political Economy, International Relations theory and global environmental politics. He has published generally in this area, in particular with *Understanding Global Environmental Politics* (2000), and specifically regarding climate change, notably in *Global Warming and Global Politics* (1996). He is currently writing a book on cars and global politics.

J. Timmons Roberts is Professor of Sociology and Director of the Environmental Science and Policy Program at the College of William and Mary. He is author or co-author of 30 articles and three books: *From Modernization to Globalization* (Blackwell, 2000), *Chronicles from the Environmental Justice Frontline* (Cambridge, 2001), and *Trouble in Paradise: Globalization and Environmental Crises in Latin America* (Routledge, 2003).

Sandra Rose is a graduate student in the Political Science Department of the University of Kassel and currently located at the Center for Global Change and Governance at Rutgers. Her research interests lie in the field of critical political economy and the environment.

Dimitris Stevis is professor of international politics at Colorado State University. His research focuses on the social regulation of global and regional integration with an emphasis on environment and labour. He has published several chapters and articles on international environmental and labour politics and co-edited (with Valerie Assetto) *The International Political Economy of the Environment: Critical Perspectives* (Lynne Rienner, 2001). He is currently completing a book entitled *Globalization and Labor: Democratizing Global Governance?* (with Terry Boswell) and is researching the views of labour unions towards global environmental issues.

Larry A. Swatuk is Associate Professor and Head of the Natural Resources Governance Research Unit at the Harry Oppenheimer Okavango Research Centre in Maun, Botswana. Dr Swatuk lectured in the Department of Politics at the University of Botswana from 1996 to 2004. He continues to teach 'Water and Security' in the SADC/WATERNET MSc programme at the University of the Western Cape in South Africa. Among his recent publications are a 2004 co-edited special issue of the journal *Physics and Chemistry of the Earth* entitled 'Water Science, Technology and Policy: Convergence and Action By All'.

Jørgen Wettestad is a Senior Research Fellow and Programme Director at the Fridtjof Nansen Institute (FNI). He is also regularly involved in teaching activities at the University of Oslo. His research interest is regime theory, with specific focus on questions related to the effectiveness and design of international environmental institutions, including the science–politics relationship. Empirically, his work focuses on air pollution and climate change politics within the domestic, EU and global contexts. Recent books include *Clearing the Air: European Advances in Tackling Acid Rain and Atmospheric Pollution* (Ashgate, 2002) and *Environmental Regime Effectiveness: Confronting Theory with Evidence* (with E. L. Miles, A. Underdal, S. Andresen, J. B. Skjærseth and E. M. Carlin) (MIT Press, 2001).

1
general introduction

michele m. betsill, kathryn hochstetler and dimitris stevis

The study of international environmental politics (IEP) has grown in both quantity and quality over the last 30 years, and international relations (IR) scholars have been increasingly more involved, particularly since the late 1980s.[1] From a subdiscipline that attracted mostly American scholars, IEP has now spread throughout much of the world, although rather unevenly. The goal of this volume is to provide a state-of-the-art review of the study of IEP.

Over the years, a number of important volumes have tracked the trajectory of international environmental politics (Caldwell, 1984, 1996; Guha, 2000; McCormick, 1989, 1995; Porter and Brown, 1991; Porter et al., 2000). While these volumes provide important insights into the study of IEP, especially the politics behind it, their primary focus is the practice of international environmental politics. Several other volumes have offered a combination of chapters that examine aspects of the study of IEP along with particular sectors of the international environment (Axelrod et al., 2004; Chasek, 2000; Choucri, 1993; Elliott, 1998; Hurrell and Kingsbury, 1992; Vig and Axelrod, 1999; Vogler and Imber, 1996). Our volume complements these efforts with its systematic attempt to identify the major research issue areas of the field and to provide authoritative accounts of the major concepts, research agendas and debates involved in their study. There have also been a few chapter- and article-length attempts at synthesizing the study of IEP as a whole (Alker and Haas, 1993; Jacobsen, 1996, 1999; Jancar, 1991/92; Mitchell, 2001; Stevis et al., 1989). Our work expands on these projects as there is too much work to be covered by a single article or person, and there has been enough research to require a systematic theoretical review and stock-taking of greater length.

This volume examines the major theoretical approaches and substantive debates in the study of IEP as reflected in a sample of graduate syllabi and texts.[2] We have asked a number of scholars with active research agendas in these areas to provide an account of the past study of that issue area as well as the major questions and debates that characterize it presently.[3] We have also asked them to apply their insights to a case study of their choice in order to illuminate both the theoretical issues that they have addressed as well as to demonstrate how these insights can be employed to better understand specific questions.

As a result the volume is intended to introduce graduate and advanced undergraduate students to the study of IEP, particularly those with some previous exposure to international relations. It can also serve as a complement to the types of volume mentioned above in more introductory courses. Scholars who are embarking on the study of IEP will also find this volume helpful both as a review of the relevant literature and as a guide to how research is being done. Academicians from various disciplines, including other areas of international relations, who are interested in learning more about the study of IEP, either for teaching or in order to initiate a new research project, will find that this volume offers authoritative, accessible and sophisticated accounts of research in IEP.

The contributors to this volume were chosen with an eye towards the increasing globalization of the study of IEP.[4] While we collectively provide an authoritative account of English-language literature, most of the contributors are also familiar with literature published in various other languages and have sought to integrate it where relevant. As a result, this volume will appeal to the above audiences throughout the English-speaking world as well as to anyone who uses English for their research or writing.

The book's chapters discuss a number of themes that are crucial to understanding the theory, method, and substantive content of the field of IEP. Our organizing framework stresses the international politics roots of this field, as the chapters are focused on broad and enduring areas of study in international relations more generally. As Stevis' chapter on the history of the study of IEP shows, such disciplinary frameworks have been important influences on how the field defines its questions and seeks its answers. Specific substantive environmental issues such as biodiversity or water are studied quite differently depending on whether they are framed as, for example, elements of the international political economy or instances of non-state governance.

The chapters are organized into three major sections. The chapters in Part I, 'The Context of the Study of International Environmental Politics', place the later chapters in theoretical and historical context. They review the historical development of international environmental politics as well as the theoretical and methodological approaches used in its study. All three of these chapters stress the diverse perspectives and tools that have been developed over the history of the field. This is a field with few orthodoxies and many debates, as befits a still-emerging and multidisciplinary area of study. The chapters in Part II, 'The Forces that Shape International Environmental Politics', introduce a variety of actors, institutions and structures that have influenced IEP. Each chapter provides an overview of how a particular topic has risen to prominence, discusses the major theoretical views of that topic and identifies lines of future research. In addition, each chapter includes original arguments and evidence in a case study. A similar framework is used in Part III, 'Normative Frameworks for Evaluating International Environmental Politics'. The chapters in this final section discuss the most important standards that have been proposed for evaluating the quality and outcomes of international environmental politics: sustainability, effectiveness and justice.

At the outset of this project, we identified several cross-cutting themes to be addressed throughout the book, as we believed they were central to the study of IEP, regardless of issue area, theoretical perspective or methodological approach. The North–South dimension of international environmental politics is one such prominent theme, emerging in nearly every chapter. It is important in both the study of IEP and in the politics of the international environment as well. While this is a book primarily on international and global environmental politics, we expected that the interface between domestic and higher levels of politics would also be central in many of the chapters, providing links to the comparative politics field within political science. In the conclusion, we discuss how the relatively straightforward treatment of domestic–international linkages in concepts such as 'two-level games' has evolved into discussions of complex interactions across scales captured in ideas like 'multilevel governance'. Such discussions also challenge the state-centrism of many IR theories by tracking the emergence of other types of actors and new forms of governance in IEP. Finally, we anticipated that different research agendas would focus on varying parts of the policy process (for example, agenda-setting, negotiation, implementation), providing a connection to the public policy literature. This reflects our assumption that the field of IEP was converging around liberal institutionalist approaches in which the phases of the policy process are central. The majority of chapters did not find that the phases of the policy process were characteristic or central to the study of the research areas that they covered. We consider this finding in greater detail in the conclusion.

Many of the chapters explicitly position themselves with respect to the extent to which they adopt critical postures of various kinds, illustrating the multivocal nature of the field. Each of the contributors is an accomplished scholar in their own right and individual authors have been encouraged to summarize existing research as well as to stake out their own position. While individual chapters may reflect some perspectives more heavily than others, across the volume as a whole these views are balanced, providing readers with a picture of the rich diversity of approaches used in the study of IEP.

Each of the chapters in Parts II and III includes original arguments and evidence in a case study. The cases are meant to illuminate the theoretical debates and concepts identified in each of the chapters and to provide readers with examples of empirical research conducted by scholars of IEP. The case studies cover a variety of issues including climate change, agricultural trade, desertification, trade in hazardous waste, transboundary resource management, the establishment of a World Environment Organization and transboundary air pollution in several different contexts. The various chapter authors employ a range of methods and approach their subject matter from a diversity of theoretical perspectives. As a result, the case studies reinforce the volume's central aim to introduce readers to the major approaches and debates that characterize the study of IEP.

The volume begins with a presentation of the historical trajectory of the study of IEP. In his chapter, Dimitris Stevis draws on an extensive review of IEP publications, research organizations and programmes as well as interviews with several senior IEP scholars to highlight the ways that international relations scholars have approached the issue and to put IR/IEP scholarship into the context of the broader IEP community. He divides the field's history into four distinct periods and documents how the political geographies of the study of IEP have evolved over time, tracking changes in the substantive issues that have been studied and the voices represented in those studies. He also traces the genealogy of world views on international environmental politics and of the research topics examined in the remainder of the volume. He concludes that the study of IEP has broadened and deepened both in terms of what is being studied and how it is being studied.

Matthew Paterson's chapter introduces the major theoretical approaches used in the study of IEP. He organizes the chapter according to what he sees as six fundamental starting points for enquiry that guide most analyses: international anarchy, knowledge processes, pluralism, structural inequalities, capital accumulation and sustainability. In the process, he examines an array of theories including realism, liberal institutionalism, ecoauthoritarianism, constructivism, pluralism, Marxism, feminism, dependency theory and Green political theory.

In their chapter on methods, Kathryn Hochstetler and Melinda Laituri note that IEP scholars have devoted little attention to the methods they use. Their aim is thus to outline a number of different approaches, discuss how they are used and identify their potential pitfalls. The chapter is oriented around two major categories of methods: positivist (including qualitative, quantitative, rational choice and geospatial approaches) and critical (including qualitative and structural approaches). Given the diversity of the field, they conclude that methodological pluralism is desirable but encourage IEP scholars to pay more attention to their methodological choices in order to avoid unnecessary and unintended weaknesses in their studies.

Gabriela Kütting and Sandra Rose's chapter on the environment as a global issue views the environment as an element of the structural organization of the international/global system. In order to understand this complex and contested concept, Kütting and Rose first take up the historical positioning of the concept. They then separate globalization into its economic, political and sociocultural dimensions and treat the debates about each individually. Such distinctions are inevitably artificial, but prove to be analytically useful as well. The dimensions are then reintegrated in a case study on trade and agriculture.

Jennifer Clapp orients her chapter on international political economy and the environment around three competing evaluations of the relationship: that growth in the global economy is positive for the environment, that the environment is harmed by growth in the global economy, and the third view that either outcome is possible and depends on the presence or absence of global rules that support the possible positive outcomes. These three positions reappear in her discussions of the more specific impacts of global trade, finance and investment flows on the environment and their governance. All of these flows occur in Clapp's case study of the international transfer of hazardous wastes from rich to poor countries.

The following chapter on transnational actors in IEP, by Michele Betsill, begins by pointing out that the issue area lacks a clear consensus on even the nature (or name) of its basic unit of analysis, in part because it has many theoretical roots. Betsill then presents findings on how transnational actors engage in IEP, the effects of their participation, and issues related to their internal dynamics. In this section, she also discusses some of the methodological challenges encountered by scholars of transnational environmental politics. A brief case study of the Climate Action Network, a transnational advocacy network involved in the international politics of climate change, illustrates these points and concepts.

Larry Swatuk's chapter links the study of IEP to one of the central concerns of mainstream IR theory – security. Following a discussion of how environmental concerns have reshaped understandings of security in IR, Swatuk distinguishes between two types of environmental security scholars: those concerned primarily with problem-solving, particularly within a society of self-regarding states, and those taking a more critical and holistic approach to issues of security. He further elaborates the critical perspective in his case study of transboundary natural resource management practices in Southern Africa.

Frank Biermann addresses the question of global environmental governance. He starts by clarifying the main uses of the term and suggests a more empirical approach that distinguishes global governance from international relations at large. He then proceeds to discuss various aspects of global environmental governance, particularly participation by categories of actors other than states, the emergence of private governance and the segmentation of global environmental governance. Drawing upon these insights he elaborates on how Southern participation can be enhanced and advances a proposal to turn the United Nations Environment Programme into a World Environmental Organization, a move that would address segmentation as well as participation.

In the first chapter on possible standards for evaluating international environmental politics, Hans Bruyninckx examines the emergence of sustainable development as a central discourse in international environmental politics and its study. In the first part of his contribution, he traces the emergence of the concept from the early 1970s to the Brundtland Report (1987), the United Nations Conference on Environment and Development (1992), and the World Summit on Sustainable Development (2002). He then examines various debates about the meaning of the concept in policy and academic debates. This is further illustrated by his account of the research on the institutionalization of sustainable development at various levels, from the global to the local. He closes by applying some of the key questions on sustainable development to the politics of the Desertification Convention, arguably one of the most 'Southern' of policy instruments.

Jørgen Wettestad introduces the standard of effectiveness. The chapter begins by discussing three major ways that the concept has been measured in several large projects focused on international environmental regimes. Wettestad goes on to argue that levels of effectiveness can be explained by examining a combination of the characteristics of the problem itself and the institutional capacity available to address it. A case study of the Convention on Long-Range Transboundary Air Pollution serves to illustrate the concepts and arguments of the earlier sections.

Bradley Parks and J. Timmons Roberts examine international environmental justice as a belatedly but increasingly important issue in the study of IEP. After clarifying the emergence and various meanings of the term they suggest that realist and liberal approaches to IR have not addressed the question of environmental justice and, most likely, are prevented by their assumptions from doing so. In their view, world-systems analysis provides the most promising approach for a thorough account of international environmental justice. After clarifying the reasons for that view they apply the insights that follow from this theoretical choice to global climate change by identifying and commenting on ten layers of climate injustice, thus setting an agenda for additional research.

In the final chapter, the editors briefly reflect on the status of the field of IEP as a whole based on the individual chapters in the volume. We conclude that the study of IEP has become broader and deeper over time in terms of research agendas, substantive concerns, theoretical approaches, and the geographical and disciplinary origins of researchers. Consistent with this finding, we note that the field lacks a single normative core. We then make several observations related to the three cross-cutting themes – North–South relations, domestic–international linkages, and phases of the policy process. Looking ahead, we speculate on the future trajectory of substantive, methodological and theoretical debates in the study of IEP. Finally, we discuss the role of IR in the study of IEP and consider how IEP scholars might create bridges to a number of other disciplines.

notes

1. The editors are aware of the debates over the differences of the 'global' and the 'international'. The latter is generally used in a broad heuristic sense to cover both, unless the author explicitly indicates that they are distinguishing between the two concepts.
2. We fully recognize that other scholars might make different choices about the theoretical approaches and substantive debates to include in such a volume.

Some readers may find gaps in the issues presented and/or prefer that a topic addressed within one or more chapters be treated separately. We acknowledge these potential critiques and can only say that the organization of the volume reflects conscious decisions based on our own experiences teaching and researching in the field of IEP, constraints dictated by the publisher and/or the usual challenges of coordinating an edited volume.
3. We gratefully acknowledge support for this project from the International Studies Association, which funded a workshop in 2003, and Colorado State University.
4. Despite our best efforts, the volume does not include contributions from Southern scholars to the extent we would have liked.

bibliography

Alker, Hayward J., and Peter M. Haas (1993) 'The Rise of Global Ecopolitics', in Nazli Choucri (ed.) *Global Accord: Environmental Challenges and International Responses*, Cambridge, Mass.: MIT Press, pp. 133–71.

Axelrod, Regina S., David Downie and Norman J. Vig (eds) (2004) *The Global Environment: Institutions, Law and Policy*, 2nd edn, Washington, DC: CQ Press.

Caldwell, Lynton Keith (1984) *International Environmental Policy: Emergence and Dimensions*, Durham, NC: Duke University Press; 2nd edn 1990.

Caldwell, Lynton Keith (1996) *International Environmental Policy: from the Twentieth to the Twenty-First Century*, Durham, NC: Duke University Press.

Chasek, Pamela S. (ed.) (2000) *The Global Environment in the Twenty-First Century: Prospects for International Cooperation*, Tokyo: United Nations University.

Choucri, Nazli (ed.) (1993) *Global Accord: Environmental Challenges and International Responses*, Cambridge, Mass.: MIT Press.

Elliott, Lorraine (1998) *Global Politics of the Environment*, New York: New York University Press; 2nd edn 2004.

Guha, Ramachandra (2000) *Environmentalism: A Global History*, New York: Longman.

Hurrell, Andrew, and Benedick Kingsbury (eds) (1992) *The International Politics of the Environment*, Oxford: Oxford University Press.

Jacobsen, Susanne (1996) *North–South Relations and Global Environmental Issues: A Review of the Literature*, Copenhagen: Centre for Development Research.

Jacobsen, Susanne (1999) 'International Relations and Global Environmental Change: Review of the Burgeoning Literature on the Environment', *Cooperation and Conflict* 34, 2, pp. 205–36.

Jancar, Barbara (1991/92) 'Environmental Studies: State of the Discipline', *International Studies Notes*, 16/17, pp. 25–31.

McCormick, John (1989) *Reclaiming Paradise: The Global Environmental Movement*, Bloomington: Indiana University Press.

McCormick, John (1995) *The Global Environmental Movement*, New York: Wiley.

Mitchell, Ronald (2001) 'International Environment', in Walter Carlsnaes, Thomas Risse and Beth A. Simmons (eds) *Handbook of International Relations*, London: Sage, pp. 500–16.

Porter, Gareth, and Janet Welsh Brown (1991) *Global Environmental Politics*, Boulder: Westview Press; 2nd edn 1996.

Porter, Gareth, Janet Welsh Brown and Pamela S. Chasek (2000) *Global Environmental Politics*, 3rd edn, Boulder: Westview Press.
Stevis, Dimitris, Valerie J. Assetto and Stephen P. Mumme (1989) 'International Environmental Politics: A Theoretical Review of the Literature', in James P. Lester (ed.) *Environmental Politics and Policy: Theories and Evidence*, Durham: Duke University Press, pp. 289–313.
Vig, Norman J., and Regina S. Axelrod (eds) (1999) *The Global Environment: Institutions, Law and Policy*, Washington, DC: CQ Press.
Vogler, John, and Mark F. Imber (eds) (1996) *The Environment and International Relations*, London: Routledge.

part i
the context of the study of international environmental politics

2
the trajectory of the study of international environmental politics[1]
dimitris stevis

The aim of this chapter is to trace the study of post-World War II international environmental politics (IEP)[2] from the point of view of international relations (IR), primarily as it appears in the English language literature.[3] Over the last fifteen years there has been a proliferation of publications on the subject. A crude counting indicates that the number of books on international environmental issues (including international environmental politics) rose from 92 in 1988 to 198 in 1989 and 325 in 1990.[4] The growth has continued more or less unabated. Thus the central question of this chapter is whether this growth has been associated with a broadening and deepening[5] of the study of IEP or whether the hegemony of certain issues and approaches has led to its narrowing over time. My general answer is that the study of IEP has in fact broadened and deepened over time substantively and theoretically, despite the prominence of specific issues and perspectives and the hegemony of liberal environmentalism (Bernstein, 2002).

The introduction clarifies how I have sought to answer the central question of the chapter and anticipates my findings; what I consider to be within the parameters of IEP; the types of information that I have used; and the rationale for the periodization that I have employed. Following the introduction I examine the trajectory of IEP through four periods while the conclusion identifies some desirable lines of future research.

I address the central question along two dimensions. The first traces the political geography of the study of IEP while the second traces its intellectual genealogy. With respect to political geography I am focusing on two specific aspects that, in my view, capture key dynamics in the study of world environmental politics. First, I trace the scale and types of substantive foci of IEP, which I illuminate with reference to the major environmental issues that received closer scrutiny during the periods chosen.[6] With respect to scale I find that while transboundary and other international issues remain central, there has long been a 'global' component to the framing of environmental issues and thus their study. What has changed over time has been the specific content of the 'global' and the increasing dominance of the globalist discourse. With respect to the type of environmental issues there has been a move towards adding pollution to the extraction of resources and political economy to a 'naturalist' view of the environment.[7]

Second, I trace the geographic origins of the voices represented in the literature.[8] Here I find, along with other analysts, that most of the early research came from the US and the UK, spreading to the rest of the North and to the South over time. Where I may diverge from many analysts is in suggesting that neither the Northern nor the Southern views, in terms of geographical origin, are internally homogeneous. While there are some identifiable patterns, for example, discussions of North–South variabilities are more likely to come from the South, it would be a simplification to allow some patterns to colour our whole understanding.

In general, then, the political geography of the study of IEP suggests a clear broadening of the substantive scope of IEP. The increasing focus on the political economy of the environment and of the growing role of Southern scholars also suggests a deepening of the study of the IEP. However, it is possible for broadening to take place without deepening. Dealing with the intellectual genealogy of the study of IEP seeks to close this gap.

In addressing the genealogy of IEP I ask how the ranges of perspectives or worldviews and of research areas have varied over time.[9] For the purposes of this chapter I distinguish perspectives in terms of the weight they place on the environment – geopolitical, environmental, ecopolitical – and in terms of their emphasis on distribution issues – no emphasis, allocational and redistributive.[10]

It is evident that the same environmental issue, such as climate change or resource depletion, may be approached from a geopolitical or ecopolitical point of view or may be examined in terms of global governance or environmental justice. Similarly, the same research area may be approached from a geopolitical or ecopolitical perspective while there may be various more specific research agendas within research areas. Some liberal analysts, for instance, approach governance from the angle of regimes while others emphasize the role of organizations.

In determining whether the intellectual scope of study of IEP has broadened I asked whether additional theories and research areas joined or disappeared from the mix. As an example of broadening, during the late 1990s, constructivist views became more prominent in IEP while societal politics became an important research area. In terms of deepening I have looked at whether this broadening reflects distinct worldviews and/or the preferences of hitherto excluded stakeholders, particularly the weak. Accordingly, the addition of ecopolitical theories that are sensitive to questions of equity is stronger evidence of a deepening of the field while the addition of liberal constructivism and liberal views of societal politics would be a much weaker indicator.

My view with respect to the genealogy of perspectives and research areas and agendas has changed as a result of the research for this chapter. A prominent reading suggests that the 1960s and 1970s were an era during which the international environment was debated at a more comprehensive theoretical level. On the same view, the 1990s is a period of 'normalization' with systematic research agendas focusing more on the trees and less on the forest. While it is true that the 1960s and 1970s were a period of profound theoretical debates, important points of view were not represented or had not yet emerged, at least with respect to IEP. Murray Bookchin (1962), for instance, had pointed to the broader issues that Rachel Carson eventually made famous, but his brand of social ecology did not find its way into the study of IEP until much later. Questions of environmental justice did not enter the IEP agenda until the late 1980s and into the 1990s. North–South debates in the 1970s were narrowly framed around the environment versus development dilemma. Even though it is also true that there has been a normalization of research since the late 1980s, this has involved a particular subcategory of IEP, what often is placed under the rubric of liberal institutionalism. While this approach has certainly left its imprint on US, Scandinavian and German research on IEP, and has arguably influenced the field more broadly, it has done so precisely at a time when the study of IEP has become profoundly broader and deeper.

But what does IEP from the point of view of IR consist of? While I have used the subdiscipline of IR as my anchor, it is apparent that the framing and study of IEP is not the monopoly of IR scholars, whether we think of IR as a subfield of politics or as a freestanding field. How IR scholars have approached international environmental politics is an important dimension, as their subject matter most directly addresses relations across political jurisdictions. I do not believe, however, that it is possible or desirable to draw narrow and precise lines of demarcation.

It seems to me that at the very least IEP must include work that focuses on the social dynamics of human practices that affect the quality of the environment. Long-standing debates demonstrate that there are deep disagreements over what constitutes a good environment. Yet we can distinguish those who do think about environmental quality from those who are interested in natural resources or pollution as a means to an end, whether military or financial.

Within these general parameters we could further delineate IEP in terms of the people who study it. The narrowest delineation would include only IR scholars who study the international environment. This would unnecessarily leave out many non-IR scholars who employ IR or non-IR theories that do focus on the social dynamics of environmental practices, for example, sociologists, economists, geographers, and so on. While the above heuristic cannot provide us with precise boundaries it does serve two purposes. It forces us to think about IEP in more inclusive disciplinary terms while also placing social dynamics and environmental quality at the centre of the subject matter.

With the above clarifications in mind, a few comments on the information that I have employed are in order. The study of IEP has grown precipitously over the last 30 years. This is manifested both in terms of *research producers* and in terms of *research output*. The category of research producers includes *research organizations, advanced training programmes* and *professional associations*. The category of research output includes *venue*, such as journals and book series, and *research products*, such as books and articles. The frequencies of books and articles are indicative of current research agendas. The launching, location and focus of research organizations, training programmes, professional associations, specialized journals and book series are in themselves evidence of the trajectory of the study of international environmental politics, because they reflect a critical mass of researchers and audiences.

I have also relied on a variety of secondary sources to guide me in writing this chapter. Advice by colleagues, interviews, electronic searches, bibliographies, overviews of the practice of world environmental politics (Bramwell, 1989; Caldwell, 1972; Caldwell and Weiland, 1996; Elliott, 1998, 2004; Guha, 2000; McCormick, 1995; Porter et al., 2000) and the study of IEP (Alker and Haas, 1993; Brenton, 1994; Chasek, 2000; Choucri, 1993; Conca and Dabelko, 2004; Hurrell and Kingsbury, 1992; Jacobsen, 1996, 1999; Jancar, 1991/92; Laferrière and Stoett, 1999; Le Prestre, 1997; Mitchell, 2001; Soroos, 1991; Stevis et al., 1989; Vogler and Imber, 1996) were all helpful. For the era after 1991, in particular, I have also depended extensively on the impressive listing of current publications available in the Environmental Studies Section (ESS) newsletter of the International Studies Association (ISA), the review sections of the journals *Environmental Politics, Global Environmental Change*, and *Global Environmental Politics*, a number of graduate course syllabi, the contributions to this volume, and my own teaching of IEP over the last 15 years.

An account of the practice of international environmental politics organized in terms of periods from one major intergovernmental conference or political development to another is not necessarily the ideal periodization in terms of its study. IR scholars were relative latecomers and, even then, the correspondence between political developments and IEP output seems to be mediated by both external and internal, disciplinary dynamics, examples of which I offer throughout the rest of the chapter. As a result, I discuss the central question in terms of four periods that seem to me to reflect the patterns of IEP research, as these emerge from the frequencies and foci of publications by IR scholars. The periodization could benefit from additional refinement but it serves a useful heuristic and is not without empirical merit. The four periods are from the mid-1940s to the late 1960s; from the late 1960s to the very late 1970s; from the early to the very late 1980s, and from the very late 1980s to the present. I have avoided specific dates to highlight the overlaps and continuities from one period to the next. A few words on the periodization may be useful here. Immediately after World War II there were serious debates over the status of environmental issues – largely resource and population – on the emerging network of global organizations, mostly involving the US and the declining European colonial powers. By the late 1960s, two important changes were apparent: first, an intellectual move towards a more organic view of the globe; second, the rise of the South. While the 1972 Stockholm Conference on the Human Environment is the seminal development, it is only part of a process that was evident a few years earlier, and which is manifested by patterns in the literature. With the Stockholm Conference the quest for reconciling environment and development (or environment and growth, for some) joined questions of resource scarcities, population and pollution as a central issue. While it received increasing attention in the 1970s it was not until the 1980s that the 'sustainable development' synthesis made it one of the two hegemonic discourses in contemporary IEP. During the 1980s, also, there emerged the second grand discourse of 'global environmental change', with a focus on the aggregate rather than the distributive. The Rio Conference played a catalytic role in terms of the study of IEP. So did, however, extensive graduate training during the 1980s, which eventually produced the proliferation of IPE research in the late 1980s and into the 1990s. Finally, the period since the very late 1980s has been one of a proliferation as well as broadening and deepening of IEP research. Clearly, various forces are at play here, particularly since this is also a period of strong pressures towards hegemonic perspectives and research agendas.

from the mid-1940s to the late 1960s: the us ascending

the anglo-american origins of iep

What is worth noting in the early literature, most of it published in the US and the UK, was its decisively global scale, in the sense that the key environmental problems of the day, population and resources, were viewed as global rather than national or regional (Brown, 1954; Kuczynski, 1944; Osborn, 1948, 1953; Thomas, 1956; Vogt, 1948). The move to the global level was reinforced as pollution in the high seas and the impacts of nuclear tests on the atmosphere rose to prominence (Jacobson and Stein, 1966). This is not to say that transboundary and other international issues, such as transboundary waters and regional resources were not significant (see, for example, Bain, 1930; Leith, 1926; Smith, 1949), but, rather, to observe that the global scale was introduced into the study of IEP immediately after World War II, as it was with respect to economic, political and military issues (for historical accounts see Boardman, 1981; Caldwell, 1972; McCormick, 1989; Nicholson, 1970).

Two factors seem to account for this global view of IEP: the global ends and means of American politics and the resource and naturalist legacies of colonial empires. As the new global leader, the US was interested in access to and conservation of global resources and also favoured global organizations and meetings to achieve its hegemony. The declining European colonial empires had already adopted some resource policies in their possessions and they saw the debates over conservation and preservation as another arena over which the post-World War II order was being renegotiated. The differences between the US and Europe were played out in various organizations and meetings, both governmental, such as the United Nations Educational, Scientific and Cultural Organization (UNESCO), and non-governmental, such as the International Union for the Protection of Nature (IUPN) (renamed the International Union for the Conservation of Nature and Natural Resources (IUCN) in 1954 and the World Conservation Union in 1990) and the International Council of Scientific Unions (ICSU). These organizations were behind the growing number of global meetings and projects, starting with the resource conferences of 1949, the International Geophysical Year (1957–58), the International Biological Programme (1964–75) and UNESCO's Biosphere Conference (1968) (for an overview, see di Castri, 1985; Golley, 1993; Mooney, 1999).

The substantive focus of IEP research during this period was on the extraction or use of resources and species, and the implications of population on them (Kuczynski, 1944; Osborn, 1948), with pollution rising in prominence during the 1960s (for example, Jacobson and Stein, 1966).[11] This is not only apparent in the various books and articles published, but also in the formation of major research and policy organizations, such as the Conservation Foundation (US, 1948), the IUCN (global, 1948/1954), Resources for the Future (US, 1952) and the creation of *Natural Resources Journal* (US, 1961), a journal that continues to cover IEP extensively.

By the end of this period there was some evidence of diffusion beyond the US and the UK. Norway played a leading role with the formation of the Fridtjof Nansen Foundation (1958), dealing primarily with ocean resource issues – not surprising given the significance of ocean politics for Norway. The launching of the *Journal of Peace Research* (Norway, 1964) which published a number of articles on the relations between ecology and conflict (for example, Gjessing, 1967) is also an important development, as was the setting of the foundations of the Institute for Environmental Studies at the University of Toronto (1967). In the South, the Argentinian Bariloche Foundation was formed in 1963 and showed an interest in resource issues early on.

The work of international organizations, such as the IUCN and the ICSU, was largely dominated by the North. While the key issues of the day involved the South, its interests and views were hardly considered. By the early 1960s decolonization did force the IUCN to confront issues of environment and development, but it did not achieve a synthesis of its own before the 1972 Stockholm Conference on the Human Environment (Boardman, 1981, ch. 5).

the environment at the door

Most of the IEP research during this period came from non-IR scholars who worked in the natural or physical sciences (Brown, 1954; Ehrlich, 1968; Osborn, 1948; Thomas, 1956; Vogt, 1948). Economists also played an increasingly prominent role (Boulding, 1966; Ward, 1966; for a multifaceted review of international environmental politics during this period see Caldwell, 1972). The conventional IR journals most likely to publish articles on the environment or natural resources were *International Affairs* (UK), *International Organization* (US), *World Politics* (US), the *Journal of Conflict Resolution* (US) and the *Journal of Peace Research* (Norway). A more complete understanding of the origins of IEP research, however, must consider international law journals, such as the *American Journal of International Law* and the *Natural Resources Journal* (US) which were routinely ahead of IR journals in addressing practical and theoretical issues (for a review of IEP based largely on the legal literature before the late 1980s see Stevis et al., 1989).[12]

The most identifiable IEP theoretical perspective was that of geopolitics, a predecessor to the environmental conflict and security research agendas (Bain, 1930; Brown, 1954; Goodrich, 1951; Leith, 1926; Osborn, 1948, 1953; Sprout and Sprout, 1957). Even those analysts who emphasized ecological concepts such as 'carrying capacity' (for example, Vogt, 1948) were influenced by the geopolitical implications of resource scarcities. Populationists (which were largely Malthusian) were also largely concerned with the impact of population on access to resources (Afriat, 1965; Kuczynski, 1944).[13]

By the mid-1960s ecopolitical thinking achieved some autonomy from geopolitics, mostly through the emerging ecological economics (Boulding, 1966; for historical overview see Pearce, 2002). Early on some analysts in this tradition adopted a more organic view of the global, captured by the 'spaceship earth' metaphor (Boulding, 1966; Ward, 1966). Parallel to this writing there was some IR work, interestingly enough from non-US scholars, that was more ecopolitical in the sense that the ecosphere was seen as a factor independent of resource scarcities (Gjessing, 1967; Konigsberg, 1960).

The most prominent of the very few IR scholars of the period who paid attention to the international environment were Harold and Margaret Sprout who came to IEP from political geography (Sprout and Sprout, 1957, 1965, 1971). Starting with their earlier largely geopolitical work, they increasingly integrated an *environmental* dimension (Sprout and Sprout, 1971). Their approach to the environment was colored by its implications for human conflict more than for any intrinsic value assigned to it or any consideration of the impacts of human activity on nature and human well-being, regardless of whether such activities caused conflict. Finally, the IEP/IR scholarship of the period paid limited attention to distributive issues, in general. The limited or non-existent emphasis on distributive issues is also evident in non-IR literature, with few exceptions (for example, Vogt, 1948).

The various research areas that subsequently came to characterize IEP are at most emergent during this period. The limits of science and technology were central to the more ecological accounts (Osborn, 1948, ch. 5). Pollution at sea, nuclear tests and development led to more attention to science and technology (more so than knowledge) by IR scholars (Fox, 1968; Gardner and Millikan, 1968; Jacobson and Stein, 1966). Questions of international political economy were certainly central to the global casting of the resource and population problems but were not addressed systematically (even though social ecologists such as Murray Bookchin as well as urban planners such as Lewis Mumford had addressed them domestically). The North–South dimension was debated within the IUCN (Boardman, 1981, ch. 5) and increasingly the various UN bodies, but I could not find anything in terms of North–South research on the environment – even though development studies were in full swing. The impacts of corporations, economic organizations or of foreign economic policies on the international environment were also not addressed.

There was definitely some discussion about international organizations and their role, particularly at the beginning and end of the period, albeit not always theoretically developed (Gardner and Millikan, 1968; Goodrich, 1951). One of the earliest articles on the subject does point out the innovations of the 1949 United Nations resource conferences and may be worth reviewing by those interested in the subject (Goodrich, 1951). The concept of regime as something that includes both rules and organizations was also used during this era (Jenks, 1956).

What is surprising is the absence of research on non-governmental organizations (NGOs) and the environment in light of the fact that they played such an important international role during this period (Boardman, 1981; McCormick, 1995, ch. 2). Finally, questions of effectiveness and equity were not touched at all, even though the former was identified as an important issue at the 1949 conferences. Issues of method also did not receive any attention by social scientists with the exception of the environmental economists associated with Resources for the Future (on the development of environmental economics see Pearce, 2002).

On balance, then, IEP research during this period was imbued by a geopolitical view of the world that paid limited attention to distributive questions. In this sense, IR scholarship played a very important role in shaping the early study of IEP, including the worldviews of the non-social scientists that predominated during this period.

from the late 1960s to the late 1970s: the south at the door

diffusion of iep to the rest of the north

Global conferences, such as UNESCO's Biosphere Conference (1968) and the 1972 Stockholm Conference on the Human Environment, further legitimated the global approach to the environment (Ward and Dubos, 1972). In addition, the move towards global research programmes continued with the initiation of UNESCO's Man and the Biosphere Programme (1971), the World Climate Research Programme (1980) and the beginnings of the 'global change' research programme, to be examined further in the discussion of the next period, below (see di Castri, 1985).

Ocean politics contributed a great deal to the prominence of the global scale. While the Third United Nations Conference on the Law of the Sea officially lasted from 1974 to 1982, ocean politics has been the subject of almost continuous negotiations since World War II, reflecting the contradictory claims of the US – enclosing of resources without affecting navigation – and the growing competition over ocean resources and uses. Not only had these negotiations been increasingly global in participation and comprehensive in ocean issues but they also introduced the global commons as both an empirical issue and a theoretical approach for IEP scholars (Soroos, 1977; Wijkman, 1982). Journals such as *Ocean Development and International Law*, *Marine Policy* and *Marine Pollution Bulletin* were launched during the first part of the 1970s. Moreover, many IEP researchers cut their teeth in some aspect of ocean politics (for example, Steinar Andresen) or dealt with environmental politics for the first time with respect to some aspect of the oceans (for example, Oran Young).

Increasing global economic integration also played a role, whether it was coming from the continuing significance of resource politics (Kelley, 1977; Laursen, 1982; Moran, 1973) or the more recent argument of the South that many of its environmental problems were due to the structure of the world economy (de Araujo Castro, 1972; Founex Report, 1972). By the early 1970s the relations between environment and development had become a central question, albeit with various answers (see Bruyninckx, this volume, and relevant parts under genealogy of IEP this chapter). An important development at the very end of this period was the 1980 publication of the *World Conservation Strategy* by the IUCN, the United Nations Environment Programme (UNEP), the World Wildlife Fund (as it was known at the time; it is now the World Wide Fund for Nature), and the FAO, which put sustainable development on the agenda. Two years later the United Nations set up the World Commission on Environment and Development (WCED) which produced *Our Common Future* (1987). Even though the global scale coloured IEP research during this period, transboundary and other international issues continued to be important and, in many cases, received more empirical attention than global scale issues. Various marine resource and pollution issues, for instance, were the subject of regional or international politics, rather than global politics. Cross-border air pollution and acid rain were also the subject of research, as were water and transboundary resource issues. The *Natural Resources Journal* and the newly formed *Resources Policy*, *Natural Resources Forum* and *Environmental Conservation* remained quite inclusive in their coverage of both global and subglobal issues.

The US remained in the lead in terms of IEP research producers, particularly with the formation of important think tanks such as Worldwatch (1974) as well as some attention to graduate training in places such as Indiana University and MIT. The formation of a number of Northern producers outside of the US marked the emerging diffusion of IEP research. Among them were the Club of Rome (1968), the Groupe de Recherches sur les Stratégies du Développement [Research Group on Development Strategies] (France, 1973), the International Institute for Environment and Development (IIED) founded in the US in 1971 but moved to the UK soon thereafter, the Institute for Environmental Studies (Netherlands, 1971), and the Beijer Institute (Sweden, 1977). Some of these organizations did not emphasize politics as much as they did the natural sciences or economics, but they all had an important IEP dimension. In addition to national organizations, intergovernmental organizations (IGOs); such as the Organization for Economic Cooperation and Development (OECD), the World Bank and the North Atlantic Treaty Organization (NATO), also took up environmental issues with a decisively Northern bent (see Wilson, 1971, pp. 57–70, for overview).

US-based IR journals, such as *International Organization*, *Journal of Conflict Resolution* and, later, *International Security*, were most likely to publish IEP research, with *International Studies Quarterly* doing so more sporadically. *International Affairs* in the UK and the *Journal of Peace Research* in Norway were also major outlets amongst IR scholars. In addition, IEP research was published in specialized journals, such as those dealing with the oceans or natural resources, as well as law journals, which continued to lead, both in terms of substantive work and in terms of introducing important normative and theoretical questions (see Stevis et al., 1989).

More research came from Europe with the formation of *The Ecologist*, *Ambio*, the *Natural Resources Forum* and *Environmental Conservation*. Also, while American publishers, such as W. H. Freeman, Indiana University and Duke University were pioneers in publishing IEP research, the IIED produced important research, much of it published by Earthscan, the first publisher to specialize in the environment.

During that period there was also increasing research *about* the South (Dahlberg, 1979; Frankel, 1971; Poleszynski, 1977;Sachs, 1974, 1980; Woodhouse, 1972) but surprisingly there was very little on the environment in Development and Area Studies journals (see James, 1978, for review and references). The Research Group on Development Strategies addressed environment and development issues, as did the IIED while the Science Policy Research Unit of the University of Sussex (UK) took up the banner of the South in its constructive critique of the *Limits to Growth* report produced by the Club of Rome (Cole, 1973). North–South relations were also central to the World Order Models Project (WOMP), centred in the US but with strong international connections.

The increased militancy of the South, expressed in the New International Economic Order and its active participation in the Stockholm Conference, served to highlight its stakes in world environmental politics. Research from the South, however, remained limited. One emergent voice was the Indian Tata Energy Research Institute (TERI), formed in 1974. Although its earlier work was technical and limited to energy and India, it did take a moderate 'southern' view. Another important Southern voice was the Argentinian Bariloche Foundation whose most prominent work was its critique of the *Limits to Growth* approach (Herrera, 1976).

Where one can find some research by Southern scholars, however, was in the work of international organizations such as the United Nations Conference on Trade and Development (UNCTAD), UNESCO, UNEP and the IUCN. UNCTAD commissioned a number of studies on trade and the environment that were authored by Southern scholars. The slow but resolute efforts of the IUCN to address environment and development issues were evident in its 1980 *World Conservation Strategy* (IUCN et al., 1980). UNEP played a leading role in launching Infoterra and the World Environment Center (on UNEP see Caldwell and Weiland, 1996, pp. 79–91; McCormick, 1995, ch. 5). There is no doubt that the work of IGOs and, to a lesser degree, that of international scientific organizations did enhance the voice of the South. However, at the end of this period there was still very little in terms of IEP research in Southern countries.

in the shadows of limits and scarcities

In addition to the relative spatial diffusion of IEP research beyond the US and the UK, there is also an emergent arrival of disciplinary IR scholarship. An important development was the formation of the Environmental Studies Section (ESS) of the International Studies Association (ISA) in the mid-1970s by a few scholars, some of whom were not conventional IR scholars but were influenced by ecological thought more broadly. During the 1970s, the ESS was a small section, although the number of panels that it organized and sponsored at the annual ISA conferences rose from one in 1975 to about five per conference by the end of the period. Members of the section were instrumental in establishing the Sprout Award for the most noteworthy book in IEP and produced important IEP research, individually and collectively (for example, Dahlberg, 1979; Orr and Soroos, 1979; Pirages, 1978). It is worth noting, however, that in addition to those scholars that converged around the ESS, many specialists in international organization and international law also published research on IEP (Falk, 1971), much of it appearing in *International Organization* (Kay and Skolnikoff, 1972). Generally, the ESS brought together American and, to a lesser degree, Canadian voices. As of the end of this period the ESS was the only association that explicitly focused on the study of IEP. Worth noting here also is the formation of the Workshop in Political Theory and Policy Analysis at Indiana University (with environment being one of its foci), which played a central role in turning common pool resource research into a global enterprise in the subsequent periods.

There is no doubt that this period was important for the development of the study of IEP. Not only can we trace most of the contemporary research areas to this period, but also this is a period when important broader theoretical debates took place and the various theoretical perspectives took shape (see Paterson in this volume). Within IR it is possible to identify geopolitical (Choucri and North, 1972; Kelley et al., 1976), environmental (Deutch, 1977; Sprout and Sprout, 1971; Young, 1977) and ecopolitical worldviews (Ophuls, 1977; Pirages, 1978, 1983; some of the contributions in Orr and Soroos, 1979). Most of them do not address redistributive issues (Sprout and Sprout, 1971; Young, 1977), but there are important exceptions (for example, Dahlberg, 1979; Falk, 1975; Orr and Soroos, 1979; Pirages, 1978).

While IR scholars finally addressed environmental issues it is worth noting that most of the key works of the period came from natural scientists and economists, and their views cast a long shadow over the study of IEP. The most ecopolitical social analysts were probably those ecological economists advancing 'steady-state' arguments (Daly, 1973; also Daly and Townsend, 1993, for an update that includes the classical texts). The most prominent perspectives, however, were those that drew from the grand narratives of 'limits to growth' and absolute 'scarcities'.[14] This was particularly evident in the work commissioned by the Club of Rome (Meadows et al., 1972; Peccei, 1969; see Onuf, 1983, for a review) as well as the more Malthusian analysts, such as the early Ehrlichs (Ehrlich, 1968; Ehrlich and Ehrlich, 1972), Goldsmith (1972) and Hardin (1968; Hardin and Baden, 1977). Of course, the dire predictions of many of the above analysts attracted stringent detractors (Beckerman, 1974; Simon, 1981).

While questions of scarcity and the commons coloured the study of IEP in the US, an alternative line of thought that eventually rose to great prominence was also taking root during this period. For the most part, environment and development were cast as antagonistic, leading to views that focused on distributive issues at the expense of the environment (for review see James, 1978; also Farvar, 1974). Very early on, however, there were voices that advanced environmentally sound concatenations of environmental and economic policies, whether in the South or the North. While authors such as Schumacher are best known for their advocacy of 'small is beautiful' and its implications for the environment, it is Ignacy Sachs who best characterizes a proactive and positive effort of integrating environmental and ecological priorities into economic policies (Sachs, 1974, 1980; see also Poleczynski, 1977; Woodhouse, 1972). His argument that it is possible to integrate the two not only offers a foundation for sustainable development but also anticipated the presently prominent approach of ecological modernization (on ecological modernization see Christoff, 1996; Mol and Spaargaren, 2000).

In general, then, there is a clear broadening and deepening of the discourse with respect to worldviews during this period, as compared to the previous one. This is also evident with respect to research areas. Research on growth spurred important debates on methods while the role of international organizations received close attention (Hargrove, 1972; Kay and Skolnikoff, 1972). Associated was an increasing discussion of technology and science (Dahlberg, 1979; Farvar and Milton, 1972; Ruggie, 1975). Most of these research areas, however, were not characterized by an emphasis on redistributive issues which were central to the environment–development debates.

As noted in dealing with scale, a globalist discourse was central to this period. The work commissioned by the Club of Rome, for instance, assumed that the various important processes (population, resources, pollution) had to be examined at a global level (Meadows et al., 1972). While the globalist impetus came largely from outside IR, the normative globalists around the World Order Projects Model (WOMP) drew upon international organizations and law (see, for example, Beres and Targ, 1975; Falk, 1971, 1975). One important characteristic of the WOMP project, which did accept some of the assumptions of the Club of Rome and of populationists, was that it placed North–South issues in a central position.

The focus on the global scale and on projections into the future gave rise to major methodological debates over modelling and forecasting (Choucri and Robinson, 1978; Hughes, 1980). On one hand, there were those that criticized modelling and forecasting, particularly that associated with the Club of Rome, on political grounds, either for its disregard for distributive issues (Cole, 1973; Herrera, 1976; for a review see Onuf, 1983) or its pessimism (Beckerman, 1974; Simon, 1981). However, there was also a productive debate over the utility of modelling and forecasting (Cole, 1973; Meadows et al., 1982). Since global change modelling provides much of the background against which contemporary IEP research takes place, it is surprising that such debates are not as central to IR today (but see Boehmer-Christiansen, 1994a, 1994b).

Another 'methodological development' during this period was the use of collective research. Like the Club of Rome, the WOMP was also a collective enterprise. In addition, the American Society for International Law sponsored collaborative projects, whose results were published in the early 1980s, on the effectiveness of international organizations (Kay and Jacobson, 1983), environment and trade (Rubin and Graham, 1982), and nationalism and the uses of the commons (Charney, 1982). These projects presage the teamwork that became prominent in a number of IEP research areas during the 1990s (for example, security, effectiveness and learning).

As the WOMP and the collective projects just noted suggest, the role of international organization and law, precursors of the regime and governance research areas, became central to IEP/IR thinking during this period (Kay and Skolnikoff, 1972; Kennan, 1970). In addition to the more organizational approaches, however, there was also clear evidence of a broader institutional approach emerging around the concept of regimes (Young, 1977). The debates and research on institutional responses can be subdivided into four groups. At a more general level there was a debate about the power and authority of such organizations, with the so-called ecoauthoritarians arguing that the problem required strong and less democratic organizations. Normative reformists, such as those around

WOMP, envisioned more representative and equitable institutional arrangements (Soroos, 1977). International organization and law scholars took a more pragmatic approach to the role of organizations and law (Kay and Jacobson, 1983). Finally, some research on the effectiveness of international agreements was also commissioned by the IIED during the late 1970s.

Related to international organization was the work of Haas (1975) and Ruggie (1975) on technology, knowledge and institutions. Fundamental to their approach was a move towards a more sociological view of science and knowledge. While their work addressed technology one must keep in mind that technology and the environment were quite often taken to be part of the same problematique during this era (Choucri and Bennett, 1972; Dahlberg, 1973; Skolnikoff, 1971). Important critiques of science and knowledge, moreover, also came from analysts tracking the impact of innovations on the South, especially the Green Revolution (Dahlberg, 1979; Frankel, 1971).

The environmental security research theme continued to be present but was not as developed as one would have expected given how central conflict was to 'scarcity' arguments. Two trends were evident here. First, Choucri and North's early work on lateral pressures and conflict was expanded to account for more explicitly environmental factors (Choucri and North, 1975; North, 1977). This is the lineage that has subsequently led to the work by Homer-Dixon (1999), which may be called the environmental conflict approach and was largely published in the *Journal of Conflict Resolution* and *International Security*. A second trend sought to address the impacts of the military on the environment, both in terms of military activities and in terms of the military's use of resources (Hveem, 1979; Juda, 1978). This approach sought to embed the institutions of conflict and war into the broader political economy and can be considered the predecessor of the human security and peace-building approach to environmental conflict.

I have noted that the move towards sustainable development and redistributive concerns became apparent during this period. At one end were developmentalists, mostly politicians, whose primary concern was the possibly adverse impacts of environmental policies on trade (de Araujo Castro, 1972) At the other were those who sought a synthesis of environment and development (Farvar and Milton, 1972; Pirages, 1978; Sachs, 1974). While the question of environment and development occupied the work of various organizations, such as those that cooperated to produce the *Conservation Strategy* (1980), development and area studies journals published very little on the subject, with the exception of the occasional article on environment and development. The discrepancy between the few publications by IEP scholars and the important developments within the world of policy and action should serve to remind us that environmental politics and its study do not always move in tandem. The leading role of the IIED and its collaborator, Earthscan, must be underscored, however, as should the emphasis that the WOMP placed on North–South issues.

While the debates over development and environment clearly heralded the arrival of political economy in IEP research there was very little that directly explored the relations between the internationalization of the economy and the environment (for example, Rubin and Graham, 1982). Finally, the two research areas that received the least attention during the 1970s were those of international societal politics and environmental justice. Both gaps continued to be surprising. By 1972, transnational relations had achieved some prominence in IR while various research and activist organizations were playing an important role in the practice of IEP. Yet there was very little on societal politics during this period (Feraru, 1974; Smith, 1972). Similarly, the North–South divide was central to world politics while radical theories were very prominent. Yet environmental justice and equity were not placed on the agenda, with the possible exception of world order modellers who did so implicitly.

the 1980s: bringing in the south

global scale from above and below

While their origins can be traced back in time, it was the 1980s that provided the two grand narratives that dominate the contemporary study of IEP, that is, sustainable development and global environmental change. This took place as a result of two factors. Pushing from 'below', the South (and its allies in the North) was successful in forcing a synthesis of environment and development. The sustainable development compromise recognized that the environments of the North and South are inexorably tied by natural processes as well as by economic and political processes with their distributive implications (WCED, 1987). From the 'top' there was a major push in large-scale scientific research under the rubric of 'global change', consolidated by the ozone and climate change crises. While the World Climate Research Programme was the first step in this process, the turning point was the International Geosphere-Biosphere Programme (IGBP), formally launched in 1986 (Malone and Roederer, 1985). The social sciences were a secondary component of the IGBP process (Price, 1990). The major social science products out of this collaboration were *Sustainable Development of the Biosphere* (Clark and Munn, 1986) and *The Earth as Transformed by Human Action* (Turner, 1990). These are very interesting and important volumes in their efforts to track the impacts of humanity on the environment but clearly avoided the role of particular historical processes in shaping particular environments. The role of the social sciences was greatly enhanced by the formation of the International Human Dimensions Programme on Global Environmental Change (IHDP) in 1990, which sought to raise the social sciences to the same level as the natural sciences (Jacobson, 1992). There is no doubt that global change and sustainable development are not inimical to each other (Redclift, 1992). Yet as there have been serious ecological doubts raised about the weight that sustainable development places on development, there are serious questions about the emphasis that global change places on management and technology at the expense of equity (Buttel et al., 1990).

The US continued to lead in the study of IEP but other countries in the North were catching up. The World Resources Institute was formed in 1982 and launched its biannual *World Resources* report. The report is now a collaborative effort with the United Nations Environment Programme, the United Nations Development Programme and the World Bank, and has been enormously influential over the years. The Indiana University Political Theory Workshop mentioned earlier played a leading role in forming the Common Property Network (1984) which became the International Association for the Study of Common Property in 1989 and has been programmatically international in its meetings and organization. The ESS was rather stagnant during the 1980s while remaining largely an association of US and Canadian researchers. The first explicitly academic IEP journals were also launched in the US during the late 1980s and early 1990s. These included the *Georgetown International Environmental Law Review* and the now defunct *International Environmental Affairs* (both in 1988), the *Colorado Journal of International Environmental Law and Policy* (1990, US), and the *Journal of Environment and Development* (1991, US).

By the end of this period there is increasing diffusion of IEP beyond the US. The Canadian Global Change Program (1985), the Stockholm Environmental Institute (1989) – which is self-consciously international and plays a central role in the 'global change' research agenda – the Centre for International Climate and Environmental Research (Norway, 1990), and the British Economic and Social Research Council's 'Global Environmental Change Programme' (1990) were set up during this period. What is worth noting here is not only the diffusion of research organizations but, also, the increasing role of the state in setting up and funding these organizations.

This diffusion is also evident in publications. Earthscan was now joined by *Global Environmental Change* (1990, UK), the *Review of European Community and International Environmental Law* (1991, UK) and the *Green Globe Yearbook* (1992, Norway) – now the *Yearbook of International Cooperation on Environment and Development*.

Parallel to this Northern development were the debates and negotiations associated with the UN WCED. The overall process allowed for voices about the South and from the South to be heard, producing increasingly more relevant research along the way (see WCED, 1987, notes and appendices). In addition, a number of research entities in the North adopted development as a central task. Among them were the Pacific Institute for Studies in Development, Environment and Security (US, 1987) and the International Institute for Sustainable Development (Canada, 1990).

At least one journal – the *Journal of Environment and Development* (1991, US) focused explicitly on issues of interest to the South, while other journals, such as the *Journal of Agricultural and Environmental Ethics*, published IEP research dealing with the South. What is interesting is that there was very little on the environment in development and area studies journals until the very late 1980s, when there was an explosion in some of them, such as *World Development* and *Third World Quarterly*. In terms of books there was a growing number, particularly dealing with sustainable development, with Earthscan playing a leading role.

While voices *about* the South were increasing rapidly, there were also more voices from the South during this period, much of it published in the North (for example, Balasubramaniam, 1984; Biswas, 1984; Ghosh, 1984). One important development was the formation of the Centre for Science and the Environment (1982) in New Delhi. Its early work was about India but it increasingly came to be an important Southern voice in IEP (see Jasanoff, 1993, on Indian responses to global environmental change). Towards the end of this period, there were additional research and policy organizations set up in other Southern countries, such as Pakistan and Uruguay – the latter covering the whole of Latin America.

in the shadow of global change and sustainable development

From a disciplinary point of view there was a slowing down of output by IR scholars during this period. This does not mean that the practice of international environmental politics and debates about fundamental issues slowed down, as noted in the previous part. In addition, legal journals continued to publish a great deal in the area of IEP. Thus we are faced with two questions. First, what accounts for the drop in IEP/IR research during the 1980s and its explosion at the end of the period? Second, how is it possible that a period characterized by so little IEP research can be considered pivotal to the study of IEP? I can only offer some plausible answers at this point.

With respect to the limited amount of academic output it is possible that the anti-environmentalism of the Thatcher and Reagan administrations did have a chilling effect on US and UK scholars, who had been the major producers during the previous period. It would be interesting, however, to investigate whether the same happened elsewhere in Europe. According to this view, developments with respect to the regional seas programme of UNEP and the ozone layer were instrumental in bringing forth renewed interest and reversing the decline. A related explanation is that many of the people who eventually published their work in the 1990s were in training during the 1980s. In any event, the discrepancy between the pivotal role of the period and the limited amount of IEP research suggests that the relationship between practice and research depends on various factors.

With respect to worldviews, the most important developments of this period were the emergence of the two grand narratives of 'global environmental change' and 'sustainable development' to replace 'limits to growth' and 'scarcities' – the grand narratives of the 1970s. Global environmental change perspectives have a significant ecological component but have remained less interested in distributive issues, preferring to look at the world in the aggregate (Buttel et al., 1990). Liberal institutionalist analysis is probably the most prominent IEP/IR theoretical force within the global environmental change narrative. The sustainable development narrative, however, did place distributive issues on the agenda of IEP, even though a number of analysts have argued that sustainability has been sacrificed to development. By the end of this period historical materialists also joined the fray, particularly with the launching of the journal *Capitalism, Nature, Socialism*, bringing along their strong attention to redistributive issues.

In terms of research areas there was also a broadening and deepening of the literature as environmental justice finally entered IEP and more analysts addressed issues having to do with the increasing integration of the world political economy.

While the term 'governance' was just beginning to be used, global policy formation received more attention, especially in response to the perceived success of the ozone negotiations (Benedick, 1991; Carroll, 1988; Sand,

1990; Soroos, 1986, 1990). Regime analysis became a prominent research agenda, especially in the US, Norway and Germany, with important projects launched during this period (Wettestad and Andresen, 1991; Young, 1989). The issue of science and knowledge also received increasing attention during this period, propelled by the prominence of large-scale science articulated around global change, climate change and ozone depletion. In addition to more instrumental discussions of the role of scientific evidence (most contributions in Andresen and Østreng, 1989), the epistemic communities approach (Haas, 1989, 1990) extended the sociological and organizational insights evident in earlier work by Ernst Haas (1975) and John Ruggie (1975). Largely, however, both accounts were not addressing questions of social power and the ways it permeates scientific projects (Taylor and Buttel, 1992).

The literature on environmental conflict continued to move beyond its traditional geopolitical foundations and towards a redefinition of human security (Deudney, 1990; Westing, 1977, 1988). In addition to important geopolitical work (Laursen, 1982; Lipschutz, 1989) there was also important work on the impacts of military practices on the environment (Westing, 1988) and on ways of transcending environmental conflicts (Dryzek and Hunter, 1987; Mingst, 1982).

The international political economy of the environment received more attention, as well. At the more structural level, the move from ecodevelopment (Balasubramaniam, 1984; Glaeser, 1984; Riddell, 1981) towards sustainable development involved debates over the relations between North and South, but also an attempt at integrating ecology and society (Biswas, 1984; Ghosh, 1984; Redclift, 1984, 1987). By the end of the 1980s, after the Brundtland Report, sustainable development was also receiving more attention in development and area studies journals. Some of the most interesting work of the 1980s was about the role of multinational corporations (Leonard, 1988; Pearson, 1987), especially with respect to the question of pollution havens.

A major development of the period, in terms of research areas, was the explicit arrival of environmental equity as a result of work by lawyers who, once again, were ahead of conventional IR scholars (D'Amato, 1990; Weiss, 1990).[15] What is interesting to note is that this early work focused on intergenerational justice, a choice that is interesting in light of the intragenerational questions of equity implied in both the North–South debates and the sustainable development compromise. While both intergenerational and intragenerational issues are important, placing the former ahead of the latter obscures existing inequities and avoids questions of current social and geopolitical cleavages by subsuming them into generational cohorts.

from the late 1980s to the present: whose global environmental politics?

towards a global study of iep

With respect to scale the 'global' has achieved a dominant standing, particularly in the North. As we move into the 1990s, moreover, the global framing of the environment found a niche within 'globalization' with certain environmental problems, such as the depletion of the ozone layer and climate change, considered as the examples par excellence of globalization. Despite this discursive move, subglobal issues, such as acid rain, marine pollution or transboundary waters, have not lost their salience and one can argue that they continue to be of more importance to more human beings. The significance of subglobal issues is underscored by the emergence of regional environmental politics to accompany the deepening of regional economic and political integration. The literature on North American and European environmental politics is impressive and growing (Andonova, 2004; Audley, 1997; Johnson and Beaulieu, 1996; Johnson and Corcelle, 1995; Kirton and Maclaren, 2002; Stevis and Mumme, 2000).

Since the early 1990s, there has been a veritable proliferation of the study of IEP into the industrial world while the role of the semi-periphery has also become more prominent. A proliferation in output is evident in both books and journals. While Earthscan, and to some degree W. H. Freeman, Indiana University and Duke University presses, had been loners in terms of publishing environmental books, they are now joined by Ashgate, Edward Elgar, Greenleaf Publishing, Island Press, the MIT Press, Oxford, Rowman and Littlefield, and SUNY, all of which have published IEP series. In addition, other publishers (for example, Cambridge, Columbia University Press, Greenwood, Palgrave, Routledge and the United Nations University Press) publish international environmental literature on a regular basis. IR and development journals routinely publish IEP research, while a number of new IEP journals have been launched, including *International Environmental Agreements* (Netherlands), *Global Environmental Politics* (US) and *Climate Policy*, all in 2001.

The increasing diversity of the study of IEP is also evident in professional associations. In the early 1990s, the ESS was joined by the environmental section of the British International Studies Association (1991), the

Standing Group on Green Politics of the European Consortium for Political Research (1991) and the Open Meetings of the IHDP. Towards the late 1990s, the Research Committee (RC-24): Environment and Society of the International Sociological Association also started paying close attention to IEP, with its members producing a series of articles and books. The IHDP, the International Association for the Study of Common Pool Resources, the RC-24 and the International Association for Ecological Economics tend to have their meetings in various parts of the globe and consciously seek to attract researchers from the South.

While IEP output is still dominated by the North, research about the South or North–South relations has also exploded. Development journals that hardly published anything on the environment in the 1970s and very little in the 1980s now publish as much as IR journals. New journals, such as *Environment, Development and Sustainability* (1999) focus on sustainable development (at all scales), and articles on North–South issues are common in many environmental and all IEP journals.

The number of research organizations in the South is also increasing, whether as a result of local initiatives, government support or support from the North. Older organizations, such as the Centre for Science and the Environment, TERI and the Bariloche Foundation have also continued to produce research on IEP. Worth noting here is the increasing role and the diversity of TERI's projects. Not only has it become a major research organization in IEP, but it is also becoming the key node to international networks throughout the world. In addition to research organizations, IEP education has also been growing. Drummond and Barros (2000) report that there were 19 undergraduate and graduate programmes on environmental politics in Brazil as of 1997, many offering coursework on IEP.

Outlets for Southern voices are also growing, as evident in more articles on IEP in *International Studies* (India), *Foro Internacional* (Mexico) and *Nueva Sociedad* (Venezuela but with German funding). Journals such as the Brazilian *Ambiente e Sociedade* (1998) publish IEP research and offer strong evidence of the growth of IEP in the South. In addition, there are more publications by Southern scholars in Northern journals as well as more participation in global associations. At this point in time there is an identifiable body of IEP research from the South by researchers who work in Southern institutions, albeit mostly from the semi-periphery (for example, Agarwal and Narain, 1991, 1992; Agarwal et al., 1999, 2001;[16] Banuri and Apffel-Marglin, 1993; Banuri and Holmberg, 1992; Dwivedi, 1997, Part II; Guha, 2000; Guha and Martinez-Alier, 1997; Guimarães, 1991; Rajan, 1997; Sachs, 1993; Shiva, 1991, 2002).

pluralist hegemony or hidden pluralism?[17]

The discipline of IR has now arrived completely into the study of IEP. The ESS of the International Studies Association experienced a literal explosion, with membership climbing from about 50 to almost 300, and the number of panels from about 5 to more than 30. Other disciplines, however, such as sociology (for example, Beck, 1999; Mol, 2001; Mol and Spaargaren, 2000; Yearley, 1996) and geography (for example, Dalby, 2002; Low and Gleeson, 1998), are also paying more attention to IEP.

The range of worldviews has broadened significantly since the early 1990s (see Elliott, 1998, 2004; Jacobsen, 1996, 1999; Paterson in this volume). The most important development has been the increasing attention to distributive issues (Low and Gleeson, 1998; Martinez-Alier, 2002; Sachs et al., 1998). Much of it is cast in North–South terms and is often geopolitical or environmentalist in its assumptions (Miller, 1995; see Parks and Roberts in this volume). There is more and more work, however, that breaks away from examining international environmental issues at the level of country or the North–South cleavage, bringing in additional stakeholders (gender, indigenous, localities). This trend reflects the rise of a structural ecopolitics with strong redistributive concerns.[18] At the risk of generalizing, the trend includes work on the risk society and modernity (Beck, 1999; Spaargaren et al., 2000), world-ecosystems (Chew, 2001; Goldfrank et al., 1999; Hornborg, 1998) and 'structural ecologists' (Kütting, 2003; Laferrière and Stoett, 1999; Lipietz, 1997; Lipschutz, 2004; Paterson, 2001; Stevis and Assetto, 2001).

A number of general observations about perspectives in IEP/IR are in order here. First, the prominence of perspectives does vary geographically. While environmentalist views that pay limited attention to distributive issues are more prominent in the United States, Scandinavia and Germany (with notable exceptions, such as Sachs et al., 1998), more ecological and distributive views enjoy greater attention in the rest of Europe as well as Australia and New Zealand. Analyses from the South certainly emphasize distribution. It is not always the case, however, that they move beyond geopolitics or environmentalism or towards redistribution that addresses internal inequities and developmentalism.

The range of research areas and methodological concerns has also broadened and deepened during this period (see Hochstetler and Laituri in this volume). The dominant research area in IR, narrowly defined, is probably that of governance (see Biermann in this volume; Vogler, 2000). Not only is there a journal largely dedicated to environmental governance (*International Environmental Agreements*), but governance is also the largest category of articles published by *Global Environmental Politics* during its first two years of publication (2001–02). In dealing with governance, one must also mention the increasing prominence of common pool resource analysis which has increasingly addressed transboundary issues (for example, Keohane and Ostrom, 1995). While both the regime and the common pool resources approaches are institutional, they vary in the sense that the latter is more open-ended as to the kinds of solutions that are possible – even though both are largely suspicious of the state.[19]

How to build effective international environmental policies has been an issue at least since the early 1970s and relevant questions were raised as far back as the 1949 conferences. In the 1980s, an important project looked at the role of IGOs in IEP. During the early 1990s, however, regime analysts moved in force toward their evaluation (see Wettestad in this volume). Over the last decade a number of collective projects involving mostly American and Scandinavian scholars have sought to evaluate the impacts of regimes. Two important questions have been raised in this regard. First, how can we integrate ecological standards into the provisions and operations of regimes? Second, how can we measure social implications?

While the role of science is often examined in relationship to governance and effectiveness, the debates over knowledge and science have become more interesting in the 1990s as a number of theorists (Boehmer-Christiansen, 1994a, 1994b; Jasanoff and Martello, 2004; Litfin, 1994) have sought to add to the more pluralist institutionalism of epistemic communities.

The global change discourse has clearly moved 'globalization' to the heart of IEP (see Kütting and Rose in this volume; Yearley, 1996). The journal *Global Environmental Change* is explicitly global in discourse, for instance, while world-systemic (Chew, 2001; Goldfrank et al., 1999; see Parks and Roberts in this volume) and world-society (Meyer et al., 1997) approaches are also producing research that is global in perspective. In addition, more historical accounts are also based on globalization assumptions – whether about the economy, governance and global civil society, values and knowledge. International political economy has moved closer to the center of IEP (see Clapp in this volume). The impacts of MNCs, a longer standing issue, have increasingly become the object of systematic research (for example, Levy and Newell, 2005), as has consumption (Princen et al., 2002). The role of economic organizations has received a great deal of attention since the early 1990s (Esty, 1996; Neumeyer, 2000; Williams, 2001) generally as a result of the deepening of global economic integration and its governance.

The issue that has truly exploded during the 1990s is that of societal politics (see Betsill in this volume; Lipschutz, 1992; Princen and Finger, 1994; Wapner, 1995). While societal politics started receiving a great deal of attention in other areas during the late 1980s, IEP scholars were somewhat slower to examine this issue. Yet ever since, there has been a true proliferation with societal politics being the subject of various theoretical perspectives. In general, the major questions are about the instrumental role of NGOs and about their contribution to the emergence of a global civil society, and thus a challenge to the state.

Most of the work in the above research areas has tended to underplay distributive questions, although there are important exceptions and a clear trend towards engaging them. As noted in the beginning to this subsection, it is the attention to redistributive issues that characterizes this period. This is definitely apparent in efforts to redefine of security to include the entitlements of human beings and other species (see Swatuk in this volume; also Deudney and Matthew, 1999). The move towards peace-making offers a promising approach (for example, Conca and Dabelko, 2003; Käkönen, 1992, 1994).

Those who have paid closer attention to sustainability and sustainable development have grappled with redistributive questions as sustainable development involves ecological, social and economic dimensions (see Becker and Jahn, 1999, for the views of most important authors on the subject; also Harrison, 2000). The literature on sustainable development and its measurements (including the footprint of environmental policies) offers an important agenda for research that rectifies the somewhat self-limiting assumptions of the environmental effectiveness literature (see Bruyninckx in this volume). In that sense, the literature on sustainable development can be divided into two categories: that which grapples with the origins and political meaning of the term, and that which focuses on defining and operationalizing sustainable development as a standard for evaluating environmental politics.

As noted earlier, questions of equity have been central to the North–South debates since the late 1960s. However, questions of justice did not enter IEP until the very late 1980s and even then only with respect to intergenerational obligations. Ever since, there has been a growing discussion of international environmental justice (see Parks and Roberts in this volume). Quite conceivably, the issue of international environmental justice can remain isolated. Yet there is a great deal that environmental justice can add to our evaluation of policy effectiveness and the meaning of sustainable development (as well as our understanding of any of the other issues that we have identified). By the late 1990s, intragenerational dimensions of justice were central to IEP (Hampson and Reppy, 1996; Low and Gleeson, 1998; Shue, 1995, 1999).

suggestions for future research

I have anticipated the main findings with respect to the central question of this chapter in the introduction. In general, I have argued that IEP literature has broadened and deepened both substantively and theoretically. A number of developments can obscure the fact that many more flowers are blooming today. One such factor is the predominance, in practice, of liberal environmentalism (Bernstein, 2002). Another factor is that the proliferation of work in IEP leads to a great deal of repetition. A third factor is that the dominance of grand narratives often leads to the use of terms such as 'globalization', 'global governance', 'global change' or 'sustainable development' that do not fit with the actual scale and subject of research. A fourth factor is the tendency of certain research projects with strong institutional support to be self-referential.

I would like to conclude by suggesting some lines of research with respect to the trajectory of the study of IEP that could help us gain a more comprehensive and realistic overview of its diversity and vitality. As we have noted in the introduction, we have chosen a number of topics based on our understanding of the field, our attempt to be as inclusive as we could, and the parameters set by the publisher. We believe that we have covered the study of IEP in the English language more broadly than any of the work that we are familiar with. Yet we can also envision additional chapters to either highlight what now is discussed across chapters, for example, North–South relations or knowledge, or to add additional research areas, such as industrial ecology, population or gender. At the end of the day, we need systematic reviews of more research areas but must also guard against idiographic accounts.

More research on the historical origins on international environmental policy and its study will also help us understand IEP more completely. A number of IR scholars (Boardman, 1981; Caldwell, 1972; McCormick, 1989) have sought to identify the political and intellectual roots of the international environmental problematique over much of the last 100 years. Historians (Grove, 1995; Williams, 2003) have also demonstrated that environmental issues and debates have a longer written history, often not in English (Bramwell, 1989). It stands to reason that such work may not use the same language that we are familiar with as IR scholars of the early twenty-first century. As with other areas of IR scholarship, however, there is value to pushing the historical envelope back in time.

Such an effort is valuable in its own terms but, also, in disrupting the rather ahistorical predilections of much of IR and IEP theorizing. My review of IEP research suggests that the interplay between IR and non-IR scholars is a long-standing and welcome practice, but also one that could benefit from more systematic study. Such studies must pay close attention to the institutional dynamics behind the relations of social and non-social analysts. As with all systematic research enterprises, there are clear intellectual and institutional lineages, the exploration of which will greatly enrich our understanding of contemporary IEP research.

As noted, important early substantive issues, such as resources and population, have given ground to global pollution problems and the whole 'global change' problematique. This, however, may be more apparent in the formal IR literature as various issue areas have carved their own niches with their own professional journals and associations. Systematic research that bridges the gap and identifies how IEP is being treated within issue-centred networks would be very useful. Relatedly, IEP/IR scholars must pay closer attention to how other social science disciplines have addressed international environmental issues. History and archaeology can provide us with longer time horizons, sociology and cultural anthropology can provide important insights into the role of non-state entities, geography and urban and regional studies can enrich our sense of scale, and so on.[20] My research for this chapter has strengthened my belief that there is much that we can learn across social science disciplines and that the disciplinary divides are more often than not easy to bridge.

Finally, it is clearly time for a systematic study of how the study of IEP has developed in different countries and parts of the world. The collective volume suggests that important traditions have developed in various industrial countries. Australians and New Zealanders, for instance, have played a central role in the study of environmental politics in general and IEP in particular (for example, Eckersley, 2004; Low and Gleeson, 1998; for an inclusive account see Elliott, 2004). I have also noted that neither Northern nor Southern literatures are internally homogeneous. Increasingly, in fact, there are commonalities that cross these broad geopolitical divides. Research that traces similarities and differences, as well as the intellectual and institutional lineages, would help us move beyond the North–South reification. More research on the study of IEP in the South would certainly help in that direction, while also helping IR to become less centred on particular views from the North.

notes

1. I would like to thank Steinar Andresen, Dennis Pirages, Marvin Soroos and John Vogler for answering a number of questions on the study of international environmental politics. Dennis, Marvin, John and Lorraine Elliott were kind enough to read and comment on a draft of the chapter, as were my co-editors who read multiple versions. I thank Steven Bernstein, Fred Buttel, Elisabeth Corell, Radoslav Dimitrov, Gabriela Kütting, Ron Mitchell and Detlef Sprinz for providing me with important information in response to a question placed on gep-ed.
2. The editors are aware of the debates over the differences between the 'global' and 'international'. The latter term is used in a broad heuristic sense here to cover both global and international issues. I will note whenever the terms are used in more precise fashion.
3. While the chapter focuses primarily on the English-language literature, I cannot claim that I have done justice to all of it. In addition to the English-language literature, however, I have also engaged in a preliminary review of French- (for example, Le Prestre, 1997) and Spanish-language research on IEP and, with the help of my colleague Kathryn Hochstetler, Portuguese (Brazilian) literature. A number of the contributors also reference works in other European languages. As I indicate in the conclusion to this chapter the time is ripe for in-depth reviews of the study of IEP in languages other than English.
4. These numbers are drawn from an advanced search of WorldCat. They certainly include a broad variety of volumes on international environmental issues, well beyond what may be disciplinary IR. Every effort was made, however, to narrow the search using a number of permutations of terms. Thus while not an accurate measure of IEP publications, these numbers are strongly indicative of changes that were taking place in the study of IEP.
5. By broadening I am referring to the addition of issues, theoretical approaches, broad research areas, and specific research agendas. Quite possibly, however, there can be a proliferation which remains within a specific worldview or perspective and does not address questions of relevance to important stakeholders. Thus deepening refers to the addition of theoretical approaches, issues, and so on, that bring in hitherto excluded stakeholders as well as alternative worldviews. Quite possibly, a stakeholder can be included without contributing to the deepening of theoretical discourse.
6. The scale and specific issues addressed reflect the priorities of those countries, and groups within countries, best able to shape the agenda during a particular historical period. Thus they are not the product of some apolitical process.
7. These are broad descriptions of the study of international environmental politics over the last 60 years.
8. The North–South divide has often led to thinking of Northern and Southern voices as homogeneous. For the purposes of this aspect of political geography I have limited myself to the geographic diffusion of IEP. In dealing with perspectives and research areas I address non-geographic diffusion.
9. While I will refer to identifiable theories in the course of the discussion I am not equating perspectives with theories. In seeking to understand the characteristics of each period I have focused on perspectives or worldviews

which can accommodate various theories (see Paterson in this volume for a similar approach).
10. One could conceivably employ more or different dimensions, for example structural and non-structural. However, I think that over its course, environmental politics, in general, has had to deal with the tensions between ecological imperatives and human equity. In my view, the two are not irreconcilable (Stevis, 2000) and the heuristic employed here can serve the purpose for which it was chosen. Geopolitical refers to perspectives that focus on the strategic uses of the environment; environmental refers to perspectives that focus on ameliorating the environment; ecopolitical refers to perspectives that attribute some autonomous value to the environment (for a similar view see Alker and Haas, 1993). Some approaches pay no attention to distributive issues; others are distributive in the technical sense, in their emphasis on allocation within the existing order; finally, others focus on redistribution, which implies some change in the order of things.
11. The works of Murray Bookchin (under a pseudonym) and Rachel Carson put the use of pesticides and other chemicals on the agenda of environmental politics, but there was little IEP research on the subject during this period. Work on the Green Revolution became more prominent during the subsequent period.
12. To my knowledge this is the first explicitly theoretical review of the study of IEP and reflects the predominance of legal and non-IR literature. Both strands have continued strong during the subsequent period.
13. I use the term 'populationist' to refer to arguments that assign causal weight to population. Most of these arguments have been Malthusian, both in their mechanistic geometric growth assumptions and their disregard for equity issues. Many analysts, however, may diverge from this Malthusian model in one or both of its basic assumptions. While Brown (1954), Hardin (1968) and Ophuls (1977) may be considered strong Malthusians, the Ehrlichs can be considered weaker Malthusians because their more recent work considers equity. In short, not all population-based theories are Malthusian.
14. Emphasis on the impacts of growth does not have to be associated with absolute scarcities. Growth, and other practices, can have adverse environmental impacts, sometimes on particular groups more than others, without pushing ecological limits in an absolute sense.
15. As noted, the World Order Models Project was deeply normative and IR scholars influenced by it were very aware of questions of equity (see Orr and Soroos, 1979). Surprisingly, however, the issue of environmental justice is not addressed explicitly.
16. The Agarwal et al. (1999, 2001) volumes are very useful as sources of additional references of IEP research from or about the South.
17. The contributions to this volume provide in-depth accounts of the theoretical developments of the last 15 years. As a result I will keep my point and references to a minimum in relation to the ground that has to be covered.
18. Many of the most prominent works of the 1970s were also structural and ecopolitical but paid little attention to redistributive issues, particularly in any concrete historical sense.

19. The role of the state under conditions of global integration and governance has also received increasing attention (for example, Eckersley, 2004; Litfin, 1998; Spaargaren et al., 2006).
20. The debates within archaeology over the impacts of natural processes on civilizations can help us understand some of the issue associated with connecting broad natural processes to actual social outcomes (see Van Buren, 2002).

bibliography

Afriat, S. N. (1965) 'People and Population', *World Politics*, 17, 3, pp. 431–39.

Agarwal, Anil, and Sunita Narain (1991) *Global Warming in an Unequal World: A Case of Environmental Colonialism*, New Delhi: Centre for Science and Environment.

Agarwal, Anil, and Sunita Narain (1992) *Towards a Green World: Should Global Environmental Management be Built on Legal Conventions or Human Rights?*, New Delhi: Centre for Science and Environment.

Agarwal, Anil, Sunita Narain and Anju Sharma (eds) (1999) *Green Politics*, New Delhi: Centre for Science and Environment.

Agarwal, Anil, Sunita Narain, Anju Sharma and Achila Imchen (eds) (2001) *Poles Apart*, New Delhi: Centre for Science and Environment.

Alker, Hayward Jr, and Peter Haas (1993) 'The Rise of Global Ecopolitics', in Nazli Choucri (ed.), *Global Accord*, Cambridge, Mass.: MIT Press, pp. 133–71.

Andonova, Liliana B. (2004) *Transnational Politics of the Environment: The European Union and Environmental Policy in Central and Eastern Europe*, Cambridge, Mass: MIT Press.

Andresen, Steinar, and Willy Østreng (eds) (1989) *International Resource Management: The Role of Science and Politics*, London: Belhaven Press.

Audley, John J. (1997) *Green Politics and Global Trade: NAFTA and the Future of Environmental Politics*, Washington, DC: Georgetown University Press.

Bain, H. Foster (1930) 'Mineral Resources and Their Effect on International Relations', *Journal of the Royal Institute of International Affairs* 9, 5, pp. 664–79.

Balasubramaniam, Arun (1984) *Ecodevelopment: Towards a Philosophy of Environmental Education*, Singapore: Konigswinter Friedrich Naumann Stiftung.

Banuri, Tariq, and Frédérique Apffel-Marglin (1993) *Who Will Save the Forests?: Knowledge, Power, and Environmental Destruction*, London: Zed Books.

Banuri, Tariq, and Johan Holmberg (1992) *Governance for Sustainable Development: A Southern Perspective*, London: International Institute for Environment and Development.

Beck, Ulrich (1999) *World Risk Society*, Cambridge: Polity Press.

Becker, Egon, and Thomas Jahn (eds) (1999) *Sustainability and the Social Sciences*, Paris: UNESCO.

Beckerman, Wilfred (1974) *In Defense of Economic Growth*, London: J. Cape.

Benedick, Richard Elliot (1991) *Ozone Diplomacy: New Directions in Safeguarding the Planet*, Cambridge, Mass.: Harvard University Press.

Beres, Louis René, and Harry R. Targ (1975) *Planning Alternative World Futures: Values, Methods, and Models*, New York: Praeger.

Bernstein, Steven (2002) *The Compromise of Liberal Environmentalism*, New York: Columbia University Press.
Biswas, Asit K. (1984) *Climate and Development*, Dublin: Tycooly International Publishers.
Boardman, Robert (1981) *International Organization and the Conservation of Nature*, Bloomington: Indiana University Press.
Boehmer-Christiansen, Sonja (1994a) 'Global Climate Protection Policy: The Limits of Scientific Advice: Part I', *Global Environmental Change* 4, 2, pp. 140–59.
Boehmer-Christiansen, Sonja (1994b) 'Global Climate Protection Policy: The Limits of Scientific Advice: Part 2', *Global Environmental Change*, 4, 3, pp. 185–200.
Bookchin, Murray (1974[1962]) *Our Synthetic Environment*, New York: Harper Colophon Books.
Boulding, Kenneth E. (1966) *Environmental Quality in a Growing Economy*, Baltimore: Johns Hopkins University Press and Resources for the Future.
Bramwell, Anna (1989) *Ecology in the 29th Century: A History*, New Haven: Yale University Press.
Brenton, Tony (1994) *The Greening of Machiavelli: The Evolution of International Environmental Politics*, London: Royal Institute of International Affairs.
Brown, Harrison (1954) *The Challenge of Man's Future; an Inquiry Concerning the Condition of Man During the Years That Lie Ahead*, New York: Viking Press.
Buttel, Frederick H., Ann P. Hawkins and Alison G. Power (1990) 'From Limits to Growth to Global Change: Constraints and Contradictions in the Evolution of Environmental Science and Ideology', *Global Environmental Change*, 1, 1, pp. 57–66.
Caldwell, Lynton Keith (1972) *In Defense of Earth: International Protection of the Biosphere*, Bloomington: Indiana University Press.
Caldwell, Lynton Keith, and Paul Stanley Weiland (1996) *International Environmental Policy: From the Twentieth to the Twenty-First Century*, Durham, NC: Duke University Press.
Carroll, John E. (1988) *International Environmental Diplomacy: The Management and Resolution of Transfrontier Environmental Problems*, Cambridge: Cambridge University Press.
Charney, Jonathan I. (1982) *The New Nationalism and the Use of Common Spaces: Issues in Marine Pollution and the Exploitation of Antarctica*, Totowa, NJ: Allanheld Osmun.
Chasek, Pamela S. (ed.) (2000) *The Global Environment in the Twenty-First Century: Prospects for International Cooperation*, Tokyo: United Nations University Press.
Chew, Sing C. (2001) *World Ecological Degradation: Accumulation, Urbanization, and Deforestation, 3000 B.C.–A.D. 2000*, Walnut Creek: AltaMira Press.
Choucri, Nazli (1993) *Global Accord: Environmental Challenges and International Responses*, Cambridge, Mass.: MIT Press.
Choucri, Nazli, and James P. Bennett (1972) 'Population, Resources, and Technology: Political Implications of the Environmental Crisis', *International Organization* 26, 2 (International Institutions and the Environmental Crisis), pp. 175–212.
Choucri, Nazli, and Robert Carver North (1975) *Nations in Conflict: National Growth and International Violence*, San Francisco: W. H. Freeman.
Choucri, Nazli, and Thomas W. Robinson (1978) *Forecasting in International Relations: Theory, Methods, Problems, Prospects*, San Francisco: W. H. Freeman.

Christoff, Peter (1996) 'Ecological Modernisation, Ecological Modernities', *Environmental Politics* 5, 3, pp. 476–500.
Clark, William C., and R. E. Munn (1986) *Sustainable Development of the Biosphere*, Cambridge: Cambridge University Press.
Cole, H. S. D. (ed.) (1973) *Models of Doom: A Critique of the Limits to Growth*, New York: Universe Books.
Conca, Ken, and Geoffrey D. Dabelko (eds) (2003) *Environmental Peacemaking*, Washington, DC: Woodrow Wilson Center Press.
Conca, Ken, and Geoffrey Dabelko (eds) (2004) *Green Planet Blues: Environmental Politics from Stockholm to Johannesburg*, Boulder: Westview Press.
Dahlberg, Kenneth A. (1979) *Beyond the Green Revolution: The Ecology and Politics of Global Agricultural Development*, New York: Plenum Press.
Dahlberg, Kenneth A. (1973) 'The Technological Ethic and the Spirit of International Relations', *International Studies Quarterly* 17, 1, pp. 55–88.
Dalby, Simon (2002) *Environmental Security*, Minneapolis: University of Minnesota Press.
Daly, Herman E. (ed.) (1973) *Toward a Steady-State Economy*, San Francisco: W. H. Freeman.
Daly, Herman E., and Kenneth N. Townsend (eds) (1993) *Valuing the Earth: Economics, Ecology, Ethics*, Cambridge, Mass.: MIT Press.
D'Amato, Anthony (1990) 'Do We Owe a Duty to Future Generations to Preserve the Global Environment?', *American Journal of International Law* 84, 1, pp. 190–8.
de Araujo Castro, Joao Augusto (1972) 'Environment and Development: The Case of the Developing Countries', *International Organization* 26, 2 (International Institutions and the Environmental Crisis), pp. 401–16.
Deudney, Daniel (1990) 'The Case against Linking Environmental Degradation and National Security', *Millennium* 19, 3, pp. 461–76.
Deudney, Daniel, and Richard Anthony Matthew (eds) (1999) *Contested Grounds: Security and Conflict in the New Environmental Politics*, Albany: State University of New York Press.
Deutch, Karl (ed.) (1977) *Ecosocial Systems and Ecopolitics: A Reader on Human and Social Implications of Environmental Management in Developing Countries*, Paris: UNESCO.
di Castri, Francesco (1985) 'Twenty Years of International Programmes on Ecosystems and the Biosphere: An Overview of Achievements, Shortcomings and Possible New Perspectives', in Thomas F. Malone and Juan G. Roederer (eds) *Global Change*, Cambridge: Cambridge University Press.
Drummond, Jose Augusto, and Luzimar Ramos Barros (2000) 'O Ensino De Temas Socio-Ambientais Nas Universidades Brasileiras – Uma Amonstra Comentada De Programas De Disciplinas', *Ambiente & Sociedade* 3, 6/7, pp. 185–269.
Dryzek, John S., and Susan Hunter (1987) 'Environmental Mediation for International Problems', *International Studies Quarterly* 31, 1, pp. 87–102.
Dwivedi, O. P. (1997) *India's Environmental Policies, Programmes and Stewardship*, Basingstoke: Macmillan.
Eckersley, Robyn (2004) *The Green State: Rethinking Democracy and Sovereignty*, Cambridge, Mass.: MIT Press.
Ehrlich, Paul R. (1968) *The Population Bomb*, New York: Ballantine Books.

Ehrlich, Paul, and Anne Ehrlich (eds) (1972) *Population, Resources, Environment: Issues in Human Ecology*, 2nd edition, San Francisco: W. H. Freeman.
Elliott, Lorraine (1998) *The Global Politics of the Environment*, New York: New York University Press.
Elliott, Lorraine M. (2004) *The Global Politics of the Environment*, 2nd edn, New York: New York University Press.
Falk, Richard A. (1971) *This Endangered Planet: Prospects and Proposals for Human Survival*, New York: Random House.
Falk, Richard A. (1975) *A Study of Future Worlds*, New York: Free Press.
Farvar, M. Taghi (1974) *International Development and the Human Environment: An Annotated Bibliography*, New York: Macmillan Information.
Farvar, M. Taghi, and John P. Milton (eds) (1972) *The Careless Technology: Ecology and International Development*, Garden City, NY: Natural History Press.
Feraru, Anne Thompson (1974) 'Transnational Political Interests and the Global Environment', *International Organization* 28, 1, pp. 31–60.
Founex Report (1972) *Development and Environment* (Report and working papers of a panel of experts convened by the Secretary-General of the United Nations Conference on the Human Environment, Founex, Switzerland, 4–12 June 1971), Paris: Mouton.
Fox, William T. R. (1968) 'Science, Technology and International Politics', *International Studies Quarterly* 12, 1, pp. 1–15.
Frankel, Francine R. (1971) *India's Green Revolution; Economic Gains and Political Costs*, Princeton: Princeton University Press.
Gardner, Richard N. (1972) 'The Role of the UN in Environmental Problems', *International Organization* 26, 2, pp. 237–54.
Gardner, Richard, and Max Millikan (eds) (1968) *The Global Partnership: International Agencies and Economic Development*. Special issue of *International Organization*, 22, 1.
Ghosh, Pradip K. (1984) *Population, Environment and Resources, and Third World Development*, Westport: Greenwood Press.
Gjessing, Gutorm (1967) 'Ecology and Peace Research', *Journal of Peace Research* 4, 2, pp. 125–39.
Glaeser, Bernhard (1984) *Ecodevelopment: Concepts, Projects, Strategies*, Oxford: Oxford University Press.
Goldfrank, Walter L., David Goodman and Andrew Szasz (eds) (1999) *Ecology and the World-System*, Westport: Greenwood Press.
Goldsmith, Edward (1972) *Blueprint for Survival*, Boston: Houghton Mifflin.
Golley, Frank B. (1993) *A History of the Ecosystem Concept in Ecology: More Than the Sum of the Parts*, New Haven: Yale University Press.
Goodrich, Carter (1951) 'The United Nations Conference on Resources', *International Organization* 5, 1, pp. 48–60.
Grove, Richard (1995) *Green Imperialism: Colonial Expansion, Tropical Island Edens, and the Origins of Environmentalism, 1600–1860*, Cambridge: Cambridge University Press.
Guha, Ramachandra (2000) *Environmentalism: A Global History*, New York: Longman.
Guha, Ramachandra, and Joan Martinez-Alier (1997) *Varieties of Environmentalism: Essays North and South*, London: Earthscan.

Guimarães, Roberto Pereira (1991) *The Ecopolitics of Development in the Third World: Politics & Environment in Brazil*, Boulder: Lynne Rienner Publishers.

Haas, Ernst B. (1975) 'Is There a Hole in the Whole? Knowledge, Technology, Interdependence, and the Construction of International Regimes', *International Organization* 29, 3, pp. 827–76.

Haas, Peter M. (1989) 'Do Regimes Matter? Epistemic Communities and Mediterranean Pollution Control', *International Organization* 43, 3, pp. 377–403.

Haas, Peter M. (1990) *Saving the Mediterranean: The Politics of International Environmental Cooperation*, New York: Columbia University Press.

Hampson, Fen Osler, and Judith Reppy (eds) (1996) *Earthly Goods: Environmental Change and Social Justice*, Ithaca, NY: Cornell University Press.

Hardin, Garrett James (1968) 'The Tragedy of the Commons', *Science* 162, pp. 1243–48.

Hardin, Garrett James, and John Baden (1977) *Managing the Commons*, San Francisco: W. H. Freeman.

Hargrove, John Lawrence (ed.) (1972) *Law, Institutions, and the Global Environment; Papers and Analysis of the Proceedings*, Dobbs Ferry, NY: Oceana Publications.

Harrison, Neil (2000) *Constructing Sustainable Development*, Albany, NY: SUNY Press.

Herrera, Amílcar Oscar (1976) *Catastrophe or New Society?: A Latin American World Model*, Ottawa: International Development Research Centre.

Homer-Dixon, Thomas (1999) *Environment, Scarcity, and Violence*, Princeton: Princeton University Press.

Hornborg, Alf (1998) 'Towards an Ecological Theory of Unequal Exchange: Articulating World-Systems Theory and Ecological Economics', *Ecological Economics*, 25, 1, pp. 127–36.

Hughes, Barry (1980) *World-Modeling: The Mesarovic-Pestel World Model in the Context of its Contemporaries*, Lexington: Lexington Books.

Hurrell, Andrew, and Benedict Kingsbury (1992) *The International Politics of the Environment: Actors, Interests, and Institutions*, Oxford: Clarendon Press.

Hveem, Helge (1979) 'Militarization of Nature: Conflict and Control over Strategic Resources and Some Implications for Peace Policies', *Journal of Peace Research* 16, 1, pp. 1–26.

IUCN, UNEP, WWF, FAO and UNESCO (1980) *World Conservation Strategy*, Gland, Switzerland: IUCN.

Jacobsen, Susanne (1996) *North–South Relations and Global Environmental Issues: A Review of the Literature*, Copenhagen: Centre for Development Research.

Jacobsen, Susanne (1999) 'International Relations and Global Environmental Change: Review of the Burgeoning Literature on the Environment', *Cooperation and Conflict* 34, 2, pp. 205–36.

Jacobson, Harold K. (1992) Institutions – Human Dimensions of Global Environmental Change Program, *Environment*, 34, 5, pp. 44–5.

Jacobson, Harold Karan, and Eric Stein (1966) *Diplomats, Scientists, and Politicians: the United States and the Nuclear Test Ban Negotiations*, Ann Arbor: University of Michigan Press.

James, Jeffrey (1978) 'Growth, Technology and the Environment in Less Developed Countries: A Survey', *World Development* 6, pp. 937–65.

Jancar, Barbara (1991/92) 'Environmental Studies: State of the Discipline', *International Studies Notes* 16/17, 1/3, pp. 25–31.

Jasanoff, Sheila (1993) India at the Crossroads in Global Environmental Policy. *Global Environmental Change* 3, 1, pp. 32–52.

Jasanoff, Sheila, and Marybeth Long Martello (eds) (2004) *Earthly Politics: Local and Global in Environmental Governance*, Cambridge, Mass.: MIT Press.

Jenks, C. Wilfred (1956) 'An International Regime for Antarctica?', *International Affairs* 32, 4, pp. 414–26.

Johnson, Pierre-Marc, and André Beaulieu (1996) *The Environment and NAFTA: Understanding and Implementing the New Continental Law*, Washington, DC: Island Press.

Johnson, Stanley, and Guy Corcelle (1995) *The Environmental Policy of the European Communities*, London: Kluwer Law International.

Juda, Lawrence (1978) 'Negotiating a Treaty on Environmental Modification Warfare: The Convention on Environmental Warfare and its Impact Upon Arms Control Negotiations', *International Organization* 32, 4, pp. 975–91.

Käkönen, Jyrki (1992) *Perspectives on Environmental Conflict and International Relations*, London: Pinter Publishers.

Käkönen, Jyrki (1994) *Green Security or Militarized Environment?*, Brookfield, Vermont: Dartmouth.

Kay, David, and Harold Jacobson (eds) (1983) *Environmental Protection: The International Dimension*, Totowa, NJ: Allanheld, Osmun and Co.

Kay, David, and Eugene Skolnikoff (eds) (1972) *World Eco-crisis: International Organizations in Response*, Madison: University of Wisconsin Press.

Kelley, Donald R. (1977) *The Energy Crisis and the Environment: An International Perspective*, New York: Praeger.

Kelley, Donald, Kenneth Stunkel and Richard Wescott (1976) *The Economic Superpowers and the Environment: the United States, the Soviet Union, and Japan*, San Francisco: W. H. Freeman.

Kennan, George (1970) 'To Prevent a World Wasteland: A Proposal', *Foreign Affairs* 48, 3, pp. 401–13.

Keohane, Robert O., and Elinor Ostrom (1995) *Local Commons and Global Interdependence: Heterogeneity and Cooperation in Two Domains*, London: Sage.

Kirton, John J., and Virginia White Maclaren (2002) *Linking Trade, Environment, and Social Cohesion: Nafta Experiences, Global Challenges*, Burlington, Vermont: Ashgate.

Konigsberg, Charles (1960) 'Climate and Society: A Review of the Literature', *Journal of Conflict Resolution* 4, 1 (The Geography of Conflict), pp. 67–82.

Kuczynski, R. R. (1944) 'World Population Problems', *International Affairs* 20, 4, pp. 449–57.

Kütting, Gabriela (2003) Globalization, Poverty and the Environment in West Africa: Too Poor to Pollute? *Global Environmental Politics* 3, 4, pp. 42–60.

Laferrière, Eric, and Peter J. Stoett (1999) *International Relations Theory and Ecological Thought: Towards a Synthesis*, London: Routledge.

Laursen, Finn (1982) 'Security Versus Access to Resources: Explaining a Decade of U.S. Ocean Policy', *World Politics* 34, 2, pp. 197–229.

Le Prestre, Philippe G. (1997) *Écopolitique Internationale*, Montréal: Guérin universitaire.

Leith, C. K. (1926) 'Mineral Resources in Certain International Relations', *Journal of the Royal Institute of International Affairs* 5, 3, pp. 155–9.

Leonard, H. Jeffrey (1988) *Pollution and the Struggle for the World Product: Multinational Corporations, Environment, and International Comparative Advantage*, New York: Cambridge University Press.

Levy, David, and Peter Newell (eds) (2005) *The Business of Global Environmental Governance*, Cambridge, Mass.: MIT Press.

Lipietz, Alain (1997) 'The Post-Fordist World: Labour Relations, International Hierarchy and Global Ecology', *Review of International Political Economy* 4, 1, pp. 1–41.

Lipschutz, Ronnie D. (1989) *When Nations Clash: Raw Materials, Ideology, and Foreign Policy*, New York: Ballinger.

Lipschutz, Ronnie (1992) 'Reconstructing World Politics: The Emergence of Global Civil Society', *Millennium: Journal of International Studies* 21, 3, pp. 389–420.

Lipschutz, Ronnie D. (2004) *Global Environmental Politics: Power, Perspectives, and Practice*, Washington, DC: CQ Press.

Litfin, Karen (ed.) (1994) *Ozone Discourses: Science and Politics in Global Environmental Cooperation*, New York: Columbia University Press.

Litfin, Karen (1998) *The Greening of Sovereignty in World Politics*, Cambridge, Mass.: MIT Press.

Low, Nicholas, and Brendan Gleeson (1998) *Justice, Society and Nature: An Exploration in Political Ecology*, London: Routledge.

Malone, Thomas F., and Juan G. Roederer (eds) (1985) *Global Change*, Cambridge: Cambridge University Press.

Martinez-Alier, Joan (2002) *The Environmentalism of the Poor: A Study of Ecological Conflicts and Valuation*, Cheltenham: Edward Elgar.

McCormick, John (1989) *Reclaiming Paradise: The Global Environmental Movement*, Bloomington: Indiana University Press.

McCormick, John (1995) *The Global Environmental Movement*, New York: Wiley.

Meadows, Donella H. et al. (1972) *The Limits to Growth: a Report for the Club of Rome's Project on the Predicament of Mankind*, New York: Universe Books.

Meadows, Donella H., John M. Richardson and Gerhart Bruckmann (1982) *Groping in the Dark: The First Decade of Global Modelling*, New York: Wiley.

Meyer, John, David Frank, Ann Hironaka, Evan Schofer and Nancy Tuma (1997) The Structuring of a World Political Regime, 1870–1990', *International Organization*, 51, 4, pp. 623–51.

Miller, Marian (1995) *The Third World in Global Environmental Politics*, Boulder: Lynne Rienner, Publishers.

Mingst, Karen (1982) 'Evaluating Public and Private Approaches to International Disputes: Statist and Transnational Solutions to Acid Rain', *Natural Resources Journal* 22, 1, pp. 5–20.

Mitchell, Ronald (2001) 'International Environment', in Walter Carlsnaes, Thomas Risse and Beth A. Simmons (eds) *Handbook of International Relations*, London: Sage, pp. 500–16.

Mol, Arthur (2001) *Globalization and Environmental Reform: The Ecological Modernization of the Global Economy*, Cambridge, Mass.: MIT Press.

Mol, Arthur, and Gert Spaargaren (2000) 'Ecological Modernization Theory in Debate', *Environmental Politics* 9, 1, pp. 17–49.

Mooney, Harold (1999) 'On the Road to Global Ecology', *Annual Review of Energy and the Environment*, 24, 1, pp. 1–31.

Neumayer, Eric (2000) 'Trade and the Environment: A Critical Assessment and Some Suggestions for Reconciliation', *Journal of Environment & Development* 9, 2, pp. 138–59.

Nicholson, Max (1970) *The Environmental Revolution: A Guide for the New Masters of the World*, New York: McGraw-Hill.

North, Robert C. (1977) 'Toward a Framework for the Analysis of Scarcity and Conflict', *International Studies Quarterly* 21, 4 (Special Issue on International Politics of Scarcity) pp. 569–91.

Onuf, Nicholas (1983) 'Reports to the Club of Rome', *World Politics* 36, 1, pp. 121–46.

Ophuls, William (1977) *Ecology and the Politics of Scarcity: Prologue to a Political Theory of the Steady State*, San Francisco: W. H. Freeman.

Orr, David W., and Marvin S. Soroos (eds) (1979) *The Global Predicament: Ecological Perspectives on World Order*, Chapel Hill: University of North Carolina Press.

Osborn, Fairfield (1948) *Our Plundered Planet*, Boston: Little, Brown.

Osborn, Fairfield (1953) *The Limits of the Earth*, Boston: Little, Brown.

Paterson, Matthew (2001) *Understanding Global Environmental Politics: Domination, Accumulation, Resistance*, Basingstoke: Palgrave.

Pearce, David (2002) 'An Intellectual History of Environmental Economics', *Annual Review of Energy and the Environment* 27, pp. 57–81.

Pearson, Charles S. (1987) *Multinational Corporations, Environment, and the Third World: Business Matters*, Durham, NC: Duke University Press.

Peccei, Aurelio (1969) *The Chasm Ahead*, New York Macmillan.

Pirages, Dennis (1978) *Global Ecopolitics: The New Context for International Relations*, North Scituate, Mass.: Duxbury Press.

Pirages, Dennis (1983) 'The Ecological Perspective and the Social Sciences', *International Studies Quarterly* 27, 3, pp. 243–55.

Poleszynski, Dag (1977) 'Waste Production and Overdevelopment: An Approach to Ecological Indicators', *Journal of Peace Research* 14, 4, pp. 285–98.

Porter, Gareth, Janet Welsh Brown and Pamela S. Chasek (2000) *Global Environmental Politics*, Boulder: Westview Press.

Price, Martin F. (1990) 'Humankind in the Biosphere: The Evolution of International Interdisciplinary Research', *Global Environmental Change* 1, 1, pp. 3–13.

Princen, Thomas, and Matthias Finger (eds) (1994) *Environmental NGOs in World Politics: Linking the Local and the Global*, London: Routledge.

Princen, Thomas, Michael Maniates and Ken Conca (eds) (2002) *Confronting Consumption*, Cambridge, Mass.: MIT Press.

Rajan, Mukund Govind (1997) *Global Environmental Politics: India and the North–South Politics of Global Environmental Issues*, New Delhi: Oxford University Press.

Redclift, Michael R. (1984) *Development and the Environmental Crisis: Red or Green Alternatives?*, London: Methuen.

Redclift, Michael R. (1987) *Sustainable Development: Exploring the Contradictions*, London: Methuen.

Redclift, Michael (1992) 'Sustainable Development and Global Environmental Change: Implications of a Changing Agenda', *Global Environmental Change: Human and Policy Dimensions*, 2, 1, pp. 32–42.

Riddell, Robert (1981) *Ecodevelopment: Economics, Ecology, and Development: An Alternative to Growth Imperative Models*, New York: St Martin's Press.

Rubin, Seymour J., and Thomas R. Graham (1982) *Environment and Trade: The Relation of International Trade and Environmental Policy*, Totowa, NJ: Allanheld Osmun.

Ruggie, John Gerard (1975) 'International Responses to Technology: Concepts and Trends', *International Organization* 29, 3, pp. 557–83.

Sachs, Ignacy (1974) 'Environnment et palnification: quelques pistes de recherches et d'action', *Information sur les Sciences Sociales* 13, 6, pp. 17–29.

Sachs, Ignacy (1980) *Stratégies de l'écodéveloppement*, Paris: Éditions Économie et humanisme Éditions ouvrières.

Sachs, Wolfgang (ed.) (1993) *Global Ecology: A New Arena of Political Conflict*, London: Zed Books.

Sachs, Wolfgang, Reinhard Loske, Manfred Linz, et al. (1998) *Greening the North: A Post-Industrial Blueprint for Ecology and Equity*, London: Zed Books.

Sand, Peter H. (1990) *Lessons Learned in Environmental Governance*, Washington, DC: World Resources Institute.

Shiva, Vandana (1991) *The Violence of the Green Revolution: Third World Agriculture, Ecology, and Politics*, London: Zed Books.

Shiva, Vandana (2002) *Water Wars: Privatization, Pollution and Profit*, Cambridge, Mass.: South End Press.

Shue, Henry (1995) 'Ethics, the Environment and the Changing International Order', *International Affairs* 71, 3 (Ethics, the Environment and the Changing International Order), pp. 453–61.

Shue, Henry A. (1999) 'Global Environment and International Inequality', *International Affairs* 75, 3, pp. 531–45.

Simon, Julian Lincoln (1981) *The Ultimate Resource*, Princeton: Princeton University Press.

Skolnikoff, Eugene B. (1971) 'Science and Technology: The Implications for International Institutions', *International Organization* 25, 4, pp. 759–75.

Smith, H. A. (1949) 'The Waters of the Jordan: A Problem of International Water Control', *International Affairs* 25, 4, pp. 415–25.

Smith, J. Eric (1972) 'The Role of Special Purpose and Nongovernmental Organizations in the Environmental Crisis', *International Organization* 26, 2, pp. 302–26.

Soroos, Marvin S. (1977) 'The Commons and Lifeboat as Guides for International Ecological Policy', *International Studies Quarterly* 21, 4, pp. 647–74.

Soroos, Marvin S. (1986) *Beyond Sovereignty: The Challenge of Global Policy*, Columbia: University of South Carolina Press.

Soroos, Marvin (1990) 'Global Policy Studies and Peace Research', *Journal of Peace Research*, 27, 2, pp. 117–25.

Soroos, Marvin (ed.) (1991) 'Special Issue: The Human Dimensions of Global Environmental Change', *International Studies Notes* 16, 1.

Spaargaren, Gert, Arthur P. J. Mol and Frederick H. Buttel (eds) (2000) *Environment and Global Modernity*, London: Sage.

Spaargaren, Gert, Arthur P.J. Mol and Fred Buttel (eds) (2006) *Governing Environmental Flows: Global Challenges to Social Theory*, Cambridge, Mass.: MIT Press.

Sprout, Harold, and Margaret Sprout (1957) 'Environmental Factors in the Study of International Politics', *Journal of Conflict Resolution* 1, 4, pp. 309–28.

Sprout, Harold Hance, and Margaret Tuttle Sprout (1965) *The Ecological Perspective on Human Affairs, with Special Reference to International Politics*, Princeton: Princeton University Press.
Sprout, Harold Hance, and Margaret Tuttle Sprout (1971) *Toward a Politics of the Planet Earth*, New York: Van Nostrand Reinhold Co.
Stevis, Dimitris (2000) 'Whose Ecological Justice?', *Strategies* 13, 1, pp. 63–76.
Stevis, Dimitris, and Valerie J. Assetto (eds) (2001) *The International Political Economy of the Environment: Critical Perspectives*, Boulder: Lynne Rienner Publishers.
Stevis, Dimitris, Valerie Assetto and Stephen P. Mumme (1989) 'International Environmental Politics: A Theoretical Review of the Literature', in James P. Lester (ed.) *Environmental Politics and Policy*, Durham: Duke University Press, pp. 289–313.
Stevis, Dimitris, and Stephen P. Mumme (2000) 'Rules and Politics in International Integration: Environmental Regulation in NAFTA and the EU', *Environmental Politics* 9, 4, pp. 20–42.
Taylor, Peter, and Frederick H. Buttel (1992) 'How do we know we have Global Environmental Problems? Science and the Globalization of Environmental Discourse', *Geoforum* 23, 3, pp. 405–16.
Thomas, William Leroy (ed.) (1956) *Man's Role in Changing the Face of the Earth*, Chicago: University of Chicago Press.
Turner, B. L. (ed.) (1990) *The Earth as Transformed by Human Action: Global and Regional Changes in the Biosphere over the Past 300 Years*, Cambridge: Cambridge University Press.
World Commission on Environment and Development (WCED) (1987) *Our Common Future*, Oxford: Oxford University Press.
Van Buren, Mary (2002) 'The Archaeology of El Niño Events and Other "Natural" Disasters' *Journal of Archaeological Method and Theory* 8, 2, pp. 129–49.
Vogler, John (2000) *The Global Commons: Environmental and Technological Governance*, New York: Wiley.
Vogler, John, and Mark Imber (eds) (1996) *The Environment and International Relations*, London: Routledge.
Vogt, William (1948) *Road to Survival*, New York: William Sloane Associates, Inc.
Wapner, Paul (1995) 'Politics Beyond the State: Environmental Activism and World Civic Politics', *World Politics* 47, 3, pp. 311–40.
Ward, Barbara (1966) *Spaceship Earth*, New York: Columbia University Press.
Ward, Barbara, and René J. Dubos (1972) *Only One Earth: The Care and Maintenance of a Small Planet*, New York: Norton.
Weiss, Edith Brown (1990) 'Our Rights and Obligations to Future Generations for the Environment', *American Journal of International Law* 84, 1, pp. 198–207.
Westing, Arthur H. (1977) *Weapons of Mass Destruction and the Environment*, London: Taylor & Francis.
Westing, Arthur H. (1988) *Cultural Norms, War and the Environment*, Oxford: Oxford University Press.
Wettestad, Jørgen, and Steinar Andresen (1991) *The Effectiveness of International Resource Cooperation: Some Preliminary Findings*, Lysaker, Norway: Fridtjof Nansen Institute.
Wijkman, Magnus (1982) 'Managing the Global Commons', *International Organization* 36, 3, pp. 511–36.

Williams, Marc (2001) 'In Search of Global Standards: The Political Economy of Trade and the Environment', in Dimitris Stevis and Valerie Assetto (eds) *The International Political Economy of the Environment: Critical Perspectives*, Boulder: Lynne Rienner Publishers, pp. 39–61.

Williams, Michael (2003) *Deforesting the Earth: From Prehistory to Global Crisis*, Chicago: University of Chicago Press.

Wilson, Thomas, Jr (1971) *International Environmental Action: A Global Survey*, Cambridge, Mass.: Dunellen.

Woodhouse, Edward J. (1972) 'Re-Visioning the Future of the Third World: An Ecological Perspective on Development', *World Politics* 25, 1, pp. 1–33.

Yearley, Steven (1996) *Sociology, Environmentalism, Globalization: Reinventing the Globe*, London and Thousand Oaks, Calif.: Sage.

Young, Oran R. (1977) *Resource Management at the International Level: The Case of the North Pacific*, London F. Pinter: New York.

Young, Oran R. (1989) *International Cooperation: Building Regimes for Natural Resources and the Environment*, Ithaca: Cornell University Press.

annotated bibliography

Elliot, Lorraine (2004) *The Global Politics of the Environment*, 2nd edn, Basingstoke: Palgrave. This volume tracks the major issue areas in global environmental politics since the early 1970s while it is also sensitive to the relevant theoretical literature. This is also a very good source of IEP literature from Australia and New Zealand.

Jacobsen, Susanne (1996) *North–South Relations and Global Environmental Issues: A Review of the Literature*, Copenhagen: Centre for Development Research, and (1999) 'International Relations and Global Environmental Change: Review of the Burgeoning Literature on the Environment', *Cooperation and Conflict* 34, 2, pp. 205–36. Combined, these two items offer one of the most complete accounts of the study of IEP. The author draws upon the English language literature but is also well grounded in European work on the subject.

Laferrière, Eric and Peter J. Stoett (1999) *International Relations Theory and Ecological Thought: Towards a Synthesis*, London: Routledge. This is a very incisive discussion of the challenges that ecological thought presents for conventional IR. The authors offer an account that is well grounded in both ecological theory and IR theory. Also, it is a good source of Canadian literature on IEP.

Vogler, John, and Mark Imber (eds) (1996) *The Environment and International Relations*, London: Routledge. This edited volume offers a broad coverage of the major theoretical issues at the interface of the environment and international relations as seen from the UK. All of the theoretical issues raised by the authors remain timely.

3
theoretical perspectives on international environmental politics[1]
matthew paterson

Reviews of the theories of international environmental politics, like those of international relations (IR) more generally, tend to be organized around different perspectives, commonly realism, liberalism/institutionalism/ pluralism, structuralism/Marxism and 'critical theories' (variously Frankfurt school critical theory, poststructuralism, feminism, green thought) (see, for example, Hovden, 1998; Laferrière and Stoett, 1999; Mantle, 1999; Paterson, 2000, chs 2–3). Such ways of organizing tend to create the sense of homogeneous, internally consistent and uncontested perspectives, and perhaps more importantly fail to investigate the specifically *theoretical* aspects of the ideas – that is, they describe the arguments offered by differing perspectives, but do not get to the heart of the assumptions underpinning them or ask questions about the internal logic and how one gets to the perspective from these assumptions.

I therefore propose to organize this journey by the fundamental starting points of each group of perspectives. I will investigate the nature of these starting points and the ways that various perspectives arise out of them. These starting points are not all of the same character or nature: some for example are ontologies – basic assumptions about what the world is like – while others are normative commitments. They are not all therefore commensurate with each other, making comparison between them complicated. The perspectives which arise out of these starting points also exceed their limits, and many specific theorists or perspectives could fall into more than one group. Some perspectives we are used to treating as distinct fall of course into more than one category, and I will try to keep to the question of what this particular starting point for theorizing enables the perspective to analyse or interpret. Starting with these basic assumptions enables us to show how a perspective has been built on them, what different perspectives have been built on the same assumption, and how persuasive each case might then be taken to be. It seems to me that there are six principal starting points for enquiry that guide most analyses of IEP. In roughly the order that they have appeared in contemporary debates (see also the historical account offered by Stevis in this volume), they are:

- international anarchy,
- knowledge processes,
- pluralism,
- structural inequalities,
- capital accumulation, and
- sustainability.

international anarchy

The proposition that the central starting point for analysing IEP is the anarchic structure of international politics is one which unites three traditions which many would suggest are the dominant traditions in the subject: realism, liberal institutionalism and ecoauthoritarianism. It also in my view, although this is a more contentious claim, remains the underpinning of constructivism, increasingly influential within the study of IEP as well as more broadly in IR.

Anarchy remains the central guiding assumption underpinning the work of most scholars in IR. It is not necessarily taken to refer to continual chaos, but simply to the absence of a central authority in world politics. Thus world politics is taken to be composed principally of states – political institutions defined in the conventional Weberian sense in terms of their 'legitimate monopoly of violence within a given territory' – more commonly in IR referred to as the condition of sovereignty. States thus recognize no authority over them which can legitimately impose its will on them, and also recognize each other as sovereign in this sense.

But by contrast, environmental degradation is typically transnational. 'The earth is one, the world is not', runs a standard assumption behind much environmental analysis. In this, the environment is similar to many other aspects of world politics which cannot be neatly bounded to territorial states – the economy, telecommunications, religious movements, for example, are also similarly deterritorial in principle. As a consequence, IEP is defined in terms of the collective action problem – how do sovereign actors interact when faced with problems which they cannot individually resolve but need to deal with each other to address them?

Three principal responses to this anarchy problematique have been articulated in IEP. For ecoauthoritarians (Ophuls, 1977), amongst the earliest writers in IEP (early 1970s), this is a fundamental contradiction. International anarchy means that states pursue their own interests, and for Ophuls this inevitably means that common resources will be overused. The principal theoretical logic is that of the 'tragedy of the commons', articulated famously by Garrett Hardin (1968). This involves an account of collective action problems which emphasizes the uncooperative nature of common (or more precisely open access) property regimes.[2] In an open access resource, the resource itself has overall limits on sustainable use while individual users pursue their own interests. In this situation, each user has an incentive to use the resource more than their 'fair share' to pursue immediate interests in increasing wealth. The result is overuse and degradation of the resource overall. For Hardin, Ophuls and others, this is an inevitable result of the way that ownership of many resources is structured. The consequence for Ophuls is that structural overhaul of global politics becomes necessary, involving the establishment of a world government. If the 'problem' is the lack of central authority and thus the pursuit of individual interests which collide with the collective good of sustainability, then the solution is to shift authority to the global level where there is no longer any contradiction between the collective interests of the world and the individual interests of the principal global political institution. Connected to this is the general premise that the problem of environmental degradation is one of an 'excess of freedom' – individuals, corporations and states need to have their liberal 'freedoms' restrained in the service of the common good – hence the term 'ecoauthoritarian'.

Realism, usually referred to as the dominant approach in IR, has many similarities to ecoauthoritarianism. Resources are overused because of the contradiction between individual state interests and those of the common good, and the same contradiction prevents sufficient cooperation to alleviate the resulting problems in any significant manner. The notion of the 'tragedy of the commons' is very similar structurally to realist interpretations of the Prisoner's Dilemma (PD) in game theory, which similarly focuses on the disjuncture between individual and collective interests to explain non-cooperation and suboptimal outcomes.

The difference is that realists tend to eschew the normative conclusion promoted by ecoauthoritarians. They tend to argue that the structural overhaul of world politics is simply impossible to achieve because states will not cede authority to any such global institutions (Bull, 1977, pp. 293–5; Shields and Ott, 1974). Therefore the world is for realists 'doomed' to unsustainability and crisis. On the one hand, on specific environmental problems, sufficient cooperation will never be achieved, as individual state interests[3] prevent such cooperation from emerging.[4] On the other hand, environmental degradation gets nationalized, resulting in the emergence of environmental security discourse.[5] While this has a range of meanings, one origin of this discourse is the realist assumption that states are the primary actors and their principal motivation for action is individual state interests, frequently understood in terms of security. In this sense, environmental security refers to the attempt to overcome the anarchy/environment contradiction by rendering environment in nationalist terms.

Liberal institutionalism has been the mainstay of much analysis in IEP since the late 1980s. Institutionalists agree with realists that an overarching change in the organizing principles of world politics is impossible to achieve, but disagree with both realists and ecoauthoritarians concerning the implications of international anarchy. The starting point is a critique of realist accounts of collective action problems, especially developed theoretically in terms of game theory. The principal theoretical development, most famously by Robert Axelrod (1984), was to show that the standard model of PD was entirely consistent with substantial cooperation between actors, if one assumes both that the game was iterated allowing strategic interaction ('tit for tat', or conditional cooperation strategies) to elicit cooperation from other actors, and that communication between actors may generate trust. Both are absent from the classical PD situation articulated by realists, but both are present in most situations of international cooperation, including regarding environmental questions. Liberal institutionalists then strengthen this argument by arguing that states act as absolute-gains maximizers rather than relative-gains maximizers (especially Keohane, 1989), and thus the potential 'zone of agreement' is much larger than the single point assumed by realists focusing on relative gains. If 'cooperation under anarchy' (Oye, 1986) is possible, a further consequence is that the role of institutions is significantly greater than realists accept.[6] For realists international institutions are fundamentally epiphenomena, while for liberals they can play significant roles in forging cooperation between states, acting as entrepreneurial leaders, helping to find points of agreement, reducing transaction costs, facilitating information flows, building trust, and so on. None of these alter the fundamentally state-centric nature of international politics for institutionalists, but they are significant in promoting interstate cooperation. This is the case even when some analysts (for example, Haas et al., 1993; Ostrom, 1990; Young, 1997a) appear to make the institutions themselves the focus of the analysis. These institutions are for them fundamentally inter*state* institutions which arise because of the logic of anarchy.

Institutionalists refer to the structured patterns of interstate cooperation which result as international regimes. IEP has been one of the most significant sites of regime research, contributing greatly to empirical research and theoretical refinement. Research elaborating different 'regime stages' – formation, development, implementation – has been developed through research in IEP (for example, Haas et al., 1993; Rowlands, 1995; Young, 1994). General explanations of regimes (a triad of emphases on power, interests and cognitive factors is common, for example, Rowlands, 1995) have emerged as have more 'meso' level analyses of specific factors favouring regime success (Bernauer, 1995; Hahn and Richards, 1989). The concept of regime effectiveness has been elaborated principally in relation to environmental research, in part because of the specific conceptual problems in evaluating effectiveness of environmental regimes (Hovi et al., 2003; Miles et al., 2002; Mitchell, 2002; Victor et al., 1998; Wettestad in this volume; Young, 1999, 2001).

For all three of these perspectives, anarchy means that the actors in world politics behave in a manner to be understood in terms of rational choice. They are actors who pursue their individual interests, who do not (except for tactical or strategic reasons) take the interests of other actors into account in deciding what to do or how they should act. They differ in what they think this means in terms of concrete actions; for realists and ecoauthoritarians this means sufficient cooperation to alleviate environmental problems is impossible to achieve, while for liberal institutionalists, interdependence (in environmental matters as elsewhere in world politics) makes it rational in many instances to cooperate.

There is a fourth perspective in IR which also arises principally out of the anarchy problematique, but has rather different assumptions about how actors behave than the rational choice assumptions underpinning the three previous perspectives. Constructivists argue that agents need to be understood not as engaged in the pursuit of clearly defined goals, but rather as acting on the basis of their interpretation of the meanings of the actions of others, and on a reflexivity regarding their own and others' identities. In IR, constructivists nevertheless rely on a move which asserts the centrality of states to their analysis (for example, Wendt, 1999). What are thus socially constructed (or at least the site at which the social construction of identities interests them) are the meanings of *state* identities (McSweeney, 1999), sovereignty (Litfin, 1998; Weber and Biersteker, 1996), and so on. So while in their social constructivism regarding the meanings of sovereignty and anarchy they differ from realists and institutionalists, they share with those perspectives an assumption that it is with analyses of sovereignty and that we need to start. Wendt (1999) thus develops his argument to explain the possibility of different sorts of international orders (he calls these Hobbesian, Lockean and Kantian) all consistent with the principle of anarchy.

In the environmental field Steven Bernstein (2001) has elaborated the constructivist position best, focusing on shifts in dominant norms regarding the environment. His point of departure is to argue that regime research based in liberal institutionalism is unable to account for the specific content of the regimes themselves, and for broad shifts in the norms underpinning environmental governance. He describes a shift from norms about environmental governance from limits to growth in the 1970s, to sustainable development in the 1980s, to liberal environmentalism in the 1990s, and shows how these underpin the various specific environmental regimes that emerge in the respective periods. Theoretically, the point is Wendt's, that anarchy has no necessary consequences. The point is to theorize how different sorts of international orders, different forms of interstate interaction, can emerge. It is the norms of international environmental regimes which can show us this in concrete form, hence Bernstein's attempt to chart such shifts. Bernstein shows (persuasively, in my view) the shifts in the norms underpinning environmental regimes, and as a consequence shifts in the environmental identities of the states participating in them. But it remains states to which such identities and norms are attached.

knowledge processes

The centrality of knowledge is clear in environmental policy debates, where ideas concerning uncertainty, risk, claims about expertise, and the importance of 'sound science' predominate in considerations about how policy decisions can be best made. Others have also made a different, and often contrasting argument, that underlying environmental degradation are precisely the universalizing, reductionist, sort of knowledge often regarded as integral to science. This centrality of knowledge claims to environmental politics has meant that a range of perspectives in IEP have emerged taking knowledge processes as the starting point. I will analyse three principal perspectives here, which I will label (recognizing the oversimplification involved) rationalist, constructivist and critical.

The rationalist project takes the importance of science to 'good' environmental policy as self-evident and seeks to identify the conditions under which scientific advice is taken seriously and acted on by policy-makers. Mainstream IR views focus on how scientific knowledge and 'rational management' are essential for successful responses to global environmental problems. Good scientific knowledge is necessary both to be able to identify environmental problems and to provide the tools to respond effectively to them. Thus a prevalent argument is that international cooperation on environmental problems depends on sufficient availability of scientific information to be able to assess the rationality and effectiveness of various strategies as well as the existence of an epistemic consensus among the relevant scientific experts (Andresen and Østreng 1989; Andresen et al., 2000; Mitchell et al., forthcoming).

Peter Haas (1989, 1990) first articulated the most popular way of analysing IEP through such a lens in his epistemic communities model which he developed as part of his analysis of the Mediterranean Action Plan. Haas takes as his point of departure the inadequacy of established models in IR, and their inability to identify the agents who promote cooperation. For Haas and others, epistemic communities are the agents pursuing the solutions to collective action problems as in the liberal institutionalist accounts of cooperation. Haas and colleagues define epistemic communities as 'a network of individuals or groups with an authoritative claim to policy-relevant knowledge in their domain of expertise' (Adler, 1992, p. 101). They are regarded as central to IEP because they are the actors who policy-makers turn to under the conditions of uncertainty and the requirements for expert knowledge pervading environmental politics, especially for 'new' issues that established state bureaucracies have little experience of dealing with. They thus get treated as the authoritative experts which then enables states to pursue 'solutions'. Despite Haas' (1999) more recent constructivist tendencies, there is a propensity to be fairly rationalist about knowledge claims themselves and the production of scientific knowledge as a process. The assumption is that scientists are neutral politically, at least as concerned their production of scientific knowledge, and that scientific 'truth' is easily translated into policy processes, which is rather problematic.

The relationship between science and policy is conceptualised differently by other works which nevertheless take knowledge to be central to IEP. Using Foucault, Karen Litfin (1994) subjects the claims made by scientists regarding their neutrality to critique, and finds the relations between scientists and politicians are more complex than in the rationalist model. The rationalist perspective tends to assume that improvements in scientific knowledge lead to improvements in environmental policy, whereas Litfin shows that they are just as likely to harden existing policy positions and be used to political advantage.[7] Others take this scepticism further, and are highly critical of scientists involved in environmental policy-making, verging on conspiracy theory. Boehmer-Christiansen's (1993, 1995a, 1995b; Boehmer-Christiansen and Kellow, 2002) work on the Intergovernmental Panel on Climate Change (IPCC) and climate change science-policy processes exemplifies such tendencies. Boehmer-Christiansen argues that IPCC scientists have been involved in a set of strategies of exaggerating climate change threats to get it on the political agenda and secure increased funding for high profile and high prestige research activities, and then of emphasizing remaining scientific uncertainties for the same purpose. However, one doesn't need to hold to the conspiratorial tone of her arguments to recognize that scientists have interests of their own in addition to their cognitive/professional commitments, and that there is a complex interplay between this and the use of scientists by political elites to legitimize policy outcomes.

A third perspective takes a critical point of view with regard to the role of science in producing environmental degradation and the marginalization of alternative forms of knowledge about the world involved in the hegemony of (Western) science (for example, Broadhead, 2002). Many critics interpret modern scientific rationality and scientific institutions as underlying structural causes of environmental problems. There are two principal aspects to this argument. Firstly, modern science was founded on the dualistic assumption of human separation from and domination over the rest of the natural world, and for many scientists its purpose has been precisely to further this separation and domination. Many writers suggest that this has led to anti-ecological attitudes and practices because the rest of the natural world has been reduced to an object for human instrumental use, whereas conceiving it as an end in itself would produce less ecologically damaging behaviour. It is also because of the way in which science (or at least, dominant traditions within science) has adopted a reductionist methodology, where phenomena are reduced to their constituent parts, and analysed as individuals. Science has therefore been less well focused (perhaps at least until recently) on the interactions between things, yet it is primarily in these interactions that environmental problems emerge (Merchant, 1980; Plumwood, 1993).[8]

The second aspect to this critical perspective is that the emergence of science has transferred legitimacy concerning knowledge about environmental problems to particular elites. Science (especially in conjunction with other power structures) has become a way in which control over environments has been transferred from individuals or communities to experts, who increasingly live away from the environments which they are charged with managing, and thus have no personal interest in whether the management of those environments is sustainable, or whether it meets the needs of those who do depend on it. But if successful responses to environmental problems rely on those who depend on resources being able to control how they are used, then at the very least the particular organization of modern science (that is, elitist rather than democratic) is problematic from an environmental point of view (Banuri and Apffel-Marglin, 1993; Beck, 1995 ch. 7; *Ecologist*, 1993, pp. 67–9, 183–6; Gorz, 1994). One common argument is that this concentration of power amongst those adopting a Western scientific episteme involves the marginalization of other forms of knowledge claims, such as indigenous knowledge claims, based more on direct experience and a more 'embedded' account of human interactions with ecosystems (Banuri and Apffel-Marglin, 1993; Martello, 2001). There is a corresponding argument that appropriate forms of environmental action are rooted in such embedded forms of knowledge, with of course a counter-argument that this frequently romanticizes indigenous peoples and their knowledge systems.

There is a range of specific theoretical questions regarding approaches focusing on the politics of knowledge. One concerns assumptions regarding the nature of scientific knowledge – in particular the politics of the production of knowledge. Is it to be thought of as a 'purely' cognitive, rational process, or are the politics of funding, and thus the importance of science/technology to broader patterns of social reproduction and political power important in influencing the production of particular knowledge claims? A second concerns the politics of translating scientific knowledge into policy processes. Do we think of this as a 'rational' process, where 'better' truth claims win out over 'worse' ones? Or do those claims that succeed do so because of their fit with dominant ideologies or the interests of other political elites, and their ability to legitimize policies those elites want to pursue, often for other, 'non-environmental' or 'non-scientific' reasons? Behind both of these is a broader question of whether focusing on scientific knowledge is sufficient as a basis for analysing outcomes in IEP. If, for example, science is simply something which political elites use as part of a legitimating strategy, then it is important also (perhaps more important) to analyse what it is they are trying to legitimize.

pluralism

For some, IEP exemplifies the shift in the structure of international politics away from the 'anarchy' as emphasized by realists and others to the emergence of new actors in the international system. Underlying this approach is the fundamental assumption of international interdependence, as understood by 1970s pluralists such as Keohane and Nye (1977) and Rosenau (1990). Interdependence signifies not only the mutual dependence of states with each other, but also a multiplicity of intersocietal contacts which breakdown the exclusively state-centred nature of international politics. 'New' actors come onto the international scene, notably multilateral organizations, transnational corporations (TNCs), and non-governmental organizations (NGOs).

In IEP, most attention from writers in this vein has been on NGOs, often branching out into more conceptual discussions about 'transnational (or global) civil society' (see Betsill in this volume). Wapner (1996) and Lipschutz (with Mayer, 1996) are the paradigm cases here.[9] The central theoretical claim is not only that NGOs have become more important in affecting interstate regimes,[10] but that they are important both in producing new forms of governance in IEP and in providing new models of politics. Wapner's (1996) threefold image of World Wide Fund for Nature (WWF)/Friends of the Earth (FoE)/Greenpeace as different political models is illustrative here – WWF representing traditional interstate models of governance, Greenpeace being associated with notions of world government, and FoE advocating a form of environmental politics focused on localism with transnational networking between local communities. Another commonly mentioned example here (for example, Lipschutz, 1999) is the Forest Stewardship Council (FSC), representing a model of governance that is fundamentally deterritorialized. The Council, established by WWF, attempts (for most commentators, extremely successfully) to govern the practices of forestry companies through a labelling scheme with which it then works with major timber retailers and the construction industry to create consumer demand for FSC-approved wood.

The other important component in this approach is the claim that governance practices in IEP (as in global politics more generally) are 'bifurcating', 'fragmegrating', 'glocalizing', or undergoing some other such transformation (the phrases are Rosenau's). Rosenau (1993) has made this claim specifically with regard to IEP, and Hempel's *Environmental Governance* (1996) is centred on this idea. More recently, the notion of 'multilevel governance' (for example, Vogler, 2003) also conveys a notion that the patterns of authority in global politics, driven at least in part by IEP processes, are moving upwards to regional and global levels, and at the same time downwards to subnational and local levels. Governance from this perspective then ceases to follow a 'sovereign' model with final authority residing at a specific point, instead being distributed at various levels, operating across a range of levels, and organized through networks rather than hierarchies.[11]

Empirically, the central debate within the pluralist perspective is whether or not such phenomena can legitimately be held to constitute breaks with patterns of IEP (and IR more widely) centred around sovereign states. Young (1997b), in direct response to Wapner (1997), argues that the importance of NGOs is still solely in his view to do with the supportive role they play in developing interstate regimes, providing information, lobbying and shaping public support for regimes, and there is nothing fundamentally transformative of major global political structures in what they do. Similarly, for many the sovereign state remains the central site of governance in IEP, and global and regional organizations, or local institutions, still play a secondary role.

structural inequalities

While agreeing that there are many other actors in IEP than simply states, others are unconvinced that this plurality of actors is usefully thought of in terms of pluralism. Rather, the relations among these groups are structured and relational. Marxists, dependency theorists, feminists and most Greens all, in differing ways, start with the assumption that the world is politically organized in terms of structural inequalities (of class, gender, race, core/periphery, principally) which are both at the root of the generation of environmental degradation and of the conflicts which pervade attempts to resolve them.

The most immediate and obvious of these inequalities in IEP concerns global economic inequalities broadly along 'North–South' lines.[12] The starting point here is the assumption that the dynamics of IEP are driven by conflict between states along a fault line between North and South[13] which both structures the possible bargains between states on specific issues, and is itself normally regarded as structural in nature. The common, if often implicit, theoretical background here is dependency theory, which along with its related framework, world-systems analysis suggests that the world economy has developed historically in such a way that inequalities are integral to its operation and tend to be self-reproducing. This offers an explanation of IEP in terms of the structuring of the global political economy into core and periphery, with resulting geopolitical conflict over IEP as well as the structuring of certain types of environmental degradation (deforestation and biodiversity/biotechnology comprise the paradigm cases) by such global inequalities. We can see the former of these, for example, in the debates about the toxic-waste trade (Clapp, 1994), ozone depletion (Miller, 1996), climate change (Agarwal and Narain, 1990; Paterson, 1996a, ch. 3), biodiversity (Guha and Martinez-Alier, 1997; Shiva, 1993), or generally over 'sustainable development' (Redclift, 1987) which are fundamentally structured by North–South inequalities.

Normatively, such analyses are often closely connected to a concern with justice as a central ethical question for students and practitioners alike in IEP. The systemic inequalities involved in both the production of environmental degradation and its impacts, as well as the way that the global economy constrains the actions of developing countries in particular with regard to environmental problems, has helped produce the dominance of the discourse of distributive justice in arguments about how responses to global environmental problems can be legitimized. This is perhaps most widely discussed in relation to climate change (for example, Athanasiou and Baer, 2002; Grubb, 1995; Harris, 2003; Paterson, 1996c; Shue, 1992; Wiegandt, 2001), but is also more generally discussed in relation to IEP (Hampson and Reppy, 1996; Low and Gleeson, 1998).

In addition to these inequalities that largely follow national boundaries,[14] other inequalities endemic within many societies also structure IEP in important ways. These inequalities, broadly along the lines of gender, race and class, are taken as starting points by feminists, environmental justice advocates and Marxists. Feminists locate the origins of environmental degradation in power relations in patriarchal societies, emphasizing the gendered nature of such inequalities. For many, the origins of this are in the dualistic philosophy of Western society (Merchant, 1980; Plumwood, 1993) with mind/matter, nature/culture, male/female as principal dualisms, and the transformation of social relations between men and women set in train in early modernity associated with such philosophical shifts. This both structures everyday relations, meaning that men can often insulate themselves from the environmental impacts of their activities, and acts as a conceptual block, producing the reductionist forms of knowledge which prevent attempts to look holistically at ecosystems. Concretely, this starting point has produced analyses of the gendering of specific environmental debates such as for desertification and deforestation, with gendered divisions of labour both meaning the impacts of environmental change are significantly gendered, but also with male power helping to produce such change in the first place (Dankelman and Davidson, 1988; Sontheimer, 1991). It has also produced much work on the gender politics of environmental movements (for example, Bretherton, 1998, 2003).

Environmental justice movements, starting in the US but spreading elsewhere, have tended to focus on the way that racial inequalities have been used as well as intensified by environmental degradation (see also Parks and Roberts in this volume). The most prominent trigger for such movements has been over the location of toxic waste dumps, which have disproportionately been placed in ethnic minority communities (Bullard, 1990; Szasz, 1994). At the same time, mainstream environmental NGOs have widely been regarded to have failed to develop campaigns to deal with this sort of environmental injustice. These inequalities have thus structured both how environmental degradation is organized and legitimized, and how movements to campaign against such degradation have emerged.

These racial inequalities often also intersect with class inequalities. Both in a loose use of the term 'class', regarding the extreme income inequalities which are prevalent in many countries, as well as regarding a more precise usage concerning the relation to the means of production, environmental politics are conditioned for many by class relations. Early developments of such arguments (Enzensberger, 1974; more recently, Harvey, 1993) have tended to regard environmentalism with some suspicion, suggesting that it is a middle-class movement one of whose effects (even if unintentionally) is to 'pull up the ladder' behind them and prevent working-class people from enjoying the benefits of wealth that they themselves enjoy. This is similar in structure to arguments about 'ecocolonialism' in relation to North–South inequalities (Agarwal and Narain, 1990).

More generally, this focus on structural inequalities emphasises that attempting to respond to a range of environmental problems without thinking carefully about those inequalities and how they structure the possible bargains between different social groups, the burdens they may disproportionately impose on some over others, and so on, will doom such responses to failure.

capital accumulation[15]

For Marxists, the central starting point is the dynamics of capitalist society, which is understood as fundamentally constituted in terms of class relations – the division of society into differing classes according to their access to the means of production. This is the theoretical background behind the focus on class and environmental politics as above. But the other dimension of such a theoretical position is the focus on accumulation. The specific class relation to capitalism is the wage labour form – that workers contract a portion of their time, or the labour needed for a specific piece of work, to an employer. This generates a way of extracting surplus value from labour which is particularly efficient, and thus creates the enormous technological dynamism and unprecedentedly rapid economic growth of capitalist societies.

At the same time, accumulation (or crudely, economic growth, although these are not precisely the same thing) is widely regarded to lie at the origins of many of the environmental problems the world faces. Whether or not one subscribes strictly to a 'limits to growth' thesis, specific patterns of growth in material production and consumption clearly underlie particular patterns of environmental degradation, from CO_2 emissions and climate change to toxic-waste generation. Taking these two aspects of growth as a starting point generates what are sometimes called eco-Marxist approaches to environmental politics (for example, O'Connor, 1994).

The political point here is that the dynamism of capitalist societies is unstable – it is highly uneven and unequal in the distribution of its product, but class relations also create political conflict and contribute to the instability of accumulation because of a lack of effective demand for the products of capitalist industry. Thus one of the central functions of states in capitalist societies is to create the political conditions for promoting capital accumulation. As a consequence, promoting growth has become the political imperative for elites throughout the world, even in countries that are not ideologically inclined to capitalism.[16] Conversely, those who organize growth (capital, as a class; businesses, as individual enterprises) gain structural power with respect to policy-making, thus structuring environmental policy-making in particular directions. For some, environmental degradation and its politics thus appear as a 'second contradiction of capital' (O'Connor, 1991), between the mode of production and the (ecological) conditions of production. Increasingly, capital reaches limits in terms of how environmental degradation itself creates costs for capital and society as a whole, and as environmental movements articulate growth itself as a problem. This then creates contradictions for policy-makers, as they simultaneously face the need to intervene to promote growth and to legitimize themselves to environmentally aware electorates to secure their rule, which frequently involves intervention to limit growth (Hay, 1994).

These features of capitalist societies provide explanations for a range of processes in IEP. They enable explanations of the origins of environmental degradation, with the principal drivers identified as economic growth and the externalization of environmental costs. In addition to providing an explanation for the social conflicts and inequalities which dominate many environmental policy arenas as alluded above, it enables an explanation of the way that particular environmental policy projects are structured by these class conflicts, the dominant ideologies through which class dominance is legitimized, and large-scale transformations in the global economy.

This generates a number of research foci (although not all of the authors mentioned below would strictly subscribe to the perspective just outlined). One such focus is the power of and importance of business in IEP (for example, Clapp, 1998, and in this volume; Falkner, 2001; Finger and Kilcoyne, 1997; Levy, 1997; Levy and Newell, 2005; Newell, 2000; Paterson, 2001b). A second focus is on the interrelation between IEP and broader patterns of global politics, especially global political economy. Debates over 'global governance', both in terms of the governance of the global economy for globalizing elites, and in terms of resistance against such capitalist globalization, involve discussions about how IEP is at the forefront of producing new forms of governance for capitalism and at the same time closely connected in patterns of resistance to such 'globalization', where attempts to move towards sustainable societies intermingle closely with resistance to corporate power and projects (for example, Paterson et al., 2003). We can see the former, for example, in the International Organization for Standardization (ISO) and its relationship to the World Trade Organization (WTO) (Clapp, 1998; Finger and Tamiotti, 1999), in WTO trade-environment debates (Eckersley, 2004a; Williams, 2001), in the use of biotechnology (often legitimized in relation to biodiversity, especially in the Convention on Biological Diversity) as a way to embed Western intellectual property regimes globally, in the formation of the World Business Council for Sustainable Development as a key organization of globalizing capital (Sklair, 2000; van der Pijl, 1998). We can see the latter in the plethora of resistance movements – Northern environmentalists, Southern subsistence farmers and fair trade movements, many of whose agenda comes together in a concern to resist neoliberal forms of economic management in part because of their socioecological impacts, and to build alternative, sustainable, forms of economy and policy (Klein, 2002; O'Brien et al., 2000; Starr, 2000). Such movements simultaneously arise out of the injustices in IEP discussed in the previous section, but also more immediately out of the way that accumulation disrupts the daily lives and subsistence potential of many around the world.

sustainability

Finally, there are groups of authors who, rather than coming out of some more general perspective in IR, arise out of environmentalist concerns and aim to build a theory of global politics from this perspective. Green politics is now the name most often given to such an enterprise, but ecoauthoritarians such as Ophuls also started from this point of view, and others (notably Sprout and Sprout, 1965; Pirages, 1978) also engaged global politics in this manner.

This approach should be distinguished from literatures on sustainable development (see Bruyninckx in this volume). Sustainable development (and its sister discourse, ecological modernization[17]) tends to work with a 'weak' notion of sustainability. The key distinction between weak and strong notions of sustainability is that the former assumes that substitution between 'natural' and 'human' capital is by and large possible, while a strong notion rejects this and insists that sustainability requires that societies work within limits to the use of a range of natural resources (see, for example, Ekins, 2000). Weak notions of sustainability do not, in my view, generate a distinct perspective on IEP, being containable within institutionalist or constructivist frames as the norms that underpin international environmental governance, as in Bernstein (2001) or Harrison (2000).

It is only with 'strong' notions of sustainability that we get distinctively different perspectives arising, as this generates fundamental challenges to existing political institutions. Strong sustainability insists on the non-substitutability of human and natural capital, the importance of 'critical natural capital' of particular sorts of ecological disruptions that are irreversible and threaten whole ecosystems. As such, it generates a focus on the scale and character of human use of ecological resources and services, and an argument that such use needs to be radically cut back.

Here the intention is to attempt to think ecologically, starting with a question such as 'What does sustainability require politically?' (It is perhaps a surprise to many that most of the literatures on IEP are fundamentally not concerned with such a question.) We have a range of analyses consistent with this basic question, but they are by no means homogeneous in how they address it. Laferrière and Stoett (1999) employ ecological critiques of existing theories of IR. Kütting (2000) approaches the subject through a critique of regime theory focusing on notions of time and complexity. Princen (2003, 2005) focuses on the value of sufficiency, which he suggests is required to underpin a politics of sustainability. Paterson (2000) attempts to develop an approach that starts from what it is about global power structures that engenders environmental degradation and thus where a politics of sustainability might start.

For most who do address this question directly, an attempt to think through the political implications of limits to economic growth, reducing the throughput of non-renewable resources, reducing global inequalities, and the complexities of many ecosystems involves radical challenges to most contemporary political institutions. This generates, for example, empirical analyses and concepts such as the 'ecological footprint', a measure which tries to account for the total ecological impact of a particular country or region (Wackernagel and Rees, 1996). For some (Eckersley, 1992; Naess, 1989; Plumwood, 1993) it also involves a deep philosophical shift, a rejection of anthropocentric ethics which puts priority on meeting human wants and needs over those of either other organisms or ecosystems, in favour of ecocentric ethics which prioritize the needs of non-humans and emphasize the (potential) compatibility of these with human interests.

While in the 1970s, this combination of views tended to result in ecoauthoritarian arguments as outlined above, the most common image now is of substantial decentralization of authority and social organization, with political institutions embedded in a pattern of global relations in a 'post-sovereign' manner (Dalby, 2002, ch. 7; *Ecologist*, 1993; Helleiner, 1996; Paterson, 1996b; Sachs, 1993). The general argument is that the dynamics of unsustainability are characterized by a disjuncture not so much between the territorial state and a global environmental crisis, as between an overcentralized polity and the inevitably local character of environmental problems. The argument for decentralization of power is expressed particularly well by Dryzek (1987). He shows that small-scale political institutions have short feedback loops, meaning that the distance and time between problems appearing and responses being developed is much shorter than with larger-scale institutions. This is in addition to the advantages of small-scale institutions in fostering direct democracy (or at least much less heavily mediated representative democracies), meaning that those making decisions regarding sustainability and those affected by them can be understood to share many more institutions than in large representative systems.

There is of course a connection back to institutionalist arguments here, similar to those developed by Keohane discussed above. Ostrom (1990),

Berkes (1989) and McCay and Acheson (1987) all make similar arguments about the importance of scale in determining patterns of successful cooperation over resource use. But such analyses tend to be organized around rational choice assumptions. Political analyses of sustainability emphasize not only that the scale of political and social institutions needs to be radically reorganized, but that such reorganization also involves changes in the character of political institutions (direct instead of representative democratic forms), of property relations (communal rather than private) and of social norms (sufficiency rather than accumulation oriented) (*Ecologist*, 1993).

conclusions

The strengths and weaknesses of these analyses rest to a large extent on the nature of the questions we want to ask of IEP. In many contexts, where we may want to attempt to influence particular patterns of political behaviour, the limited, cautious, careful, sets of propositions as developed by regime theory can clearly be useful in working out the dynamics of regime building, and thus how and where actors may try to intervene to affect outcomes. But even in this pragmatic mode, there remain questions about how the fundamental assumptions of regime theory create presumptions both about 'how' intervention might take place and 'where' it might take place. From more structuralist perspectives, the basic point that political coalitions to support environmental measures are required which can legitimize policy projects in terms of their capacity to pursue accumulation is a crucial point missed in institutionalist lenses. Also, for structuralists (as for pluralists) the sites of intervention may in many instances be different, as governance of the environment occurs (increasingly?) by corporations directly, or by different sorts of international organisations than those with which we are familiar (the ISO instead of the United Nations Environment Programme, for example). So even in a pragmatic, policy-oriented mode, institutionalism does not have a monopoly on utility.

But of course, there is no reason why social scientists should be tied to such a pragmatic project in any case. To interrogate critically the basic assumptions of theoretical positions, such as the principles of sovereignty, anarchy and interdependence, is an important component of social scientific enquiry. We can and should thus ask basic questions of the adequacy of taking these six assumptions as starting points. Does it really make sense to characterize world politics in terms of interstate anarchy, whichever analysis one then generates about the implications of anarchy? Is scientific knowledge so critical to environmental politics that it makes sense to develop whole theories out of a focus on it? Are the inequalities in power and wealth structural in nature as Marxists and other argue, and if so, do they determine outcomes in IEP so strongly as they suggest? Such critical interrogations intersect, of course, with the point above, which is that what fundamentally matters is the question we ask of IEP.

The range of perspectives available to scholars working in IEP has broadened significantly in the last decade, with the development of perspectives and debates from a situation of dominance by institutionalist perspectives with occasional dialogue between this perspective and realists and ecoauthoritarians. Closely related to this development is that the IR dimensions of environmental politics are now much more closely connected to debates in other fields in politics (social movements, political theory, in particular) and to debates beyond the discipline of politics. For the former of these, Robyn Eckersley's recent *The Green State* (2004b) is a shining example of how the recent debates in IR and IEP (particularly some of the analyses by structuralists and constructivists as discussed above) are having an important impact outside IR, but also of how ecological political theory can be used to inform the development of research in IEP. Particularly useful to how our attention might be directed in the future is the attempt to integrate the insights of critical perspectives into more 'practically' oriented research. So, for example, research which retains a critical edge on the possibilities of sustainability in a continually growing world but which at the same time focuses on how ecological modernization processes are starting to transform global capitalism, and how such processes might be advanced more fully, is of considerable importance in the coming years.

notes

1. Thanks are due to the participants at the ISA workshop for this book in Portland, Oregon, in March 2003, for useful feedback, and especially to the editors of the volume for very useful comments on earlier drafts which forced me to think more carefully about some of the arguments here and the overall structure of the chapter.
2. This is not the place necessarily to go into great detail on the distinction here. Hardin referred to a mythical English commons as a property regime where users were under no restrictions on use of property. He thus failed to distinguish between regimes where there were community-based restrictions on use (commons), and those where no property was asserted in a resource (open access). His account of the commons is now usually referred to as an open access resource. For various on this conceptual debate, see Berkes (1989), Ostrom (1990) and Vogler (2000). 'Open access resources' is thus the

more precise conceptual term. For a fuller account along the lines here, see Paterson (2000, ch. 2).
3. I will leave aside the various conceptual problems to do with the ascription of interests to collective bodies like states, or the other complication introduced by realists' highly problematic conflation of nation and state in their usage of the term 'national interests'.
4. Soroos' recent analysis of the climate change regime (2001) is a good example of such realist logic, as well as a good example of the obvious limits of the approach. See my response in Paterson (2001a).
5. See also Swatuk (in this volume) on environmental security. Of course not all authors using the 'environmental security' label start from realist premises. Some, for example, focus explicitly on 'security for the biosphere' rather than 'the nation' (Dalby, 2002). But the dominant discourse of environmental security remains realist in orientation, and even critical writers on the topic tend to take as their starting point precisely a critique of the realist frame.
6. It is worth noting and emphasizing that institutions and organizations are not the same thing in institutionalist theory. Young sums up the distinction well. For him, institutions are 'social practices consisting of easily recognized roles coupled with clusters of rules or conventions governing relations between occupants of these roles', while organizations are 'material entities possessing physical locations (or seats), offices, personnel, equipment and budgets' (1989, p. 32).
7. For analyses similar to Litfin's, see Dimitrov (2003), Harrison and Bryner (2003), Jasanoff and Wynne (1998), Miller and Edwards (2001), Parson (2003) or Shackley and Wynne (1996).
8. On the emergence of more ecological, holistic, approaches within science, see in particular Worster (1994).
9. Lipschutz's (2004) work has more recently become less pluralist and more structuralist in orientation. It is worth noting also that some within this perspective do focus on TNCs (Garcia-Johnson, 2000), but it would be fair to say that most attention is on environmental NGOs.
10. A claim that is already commonplace in IEP and IR more widely (see, for example, Princen and Finger (1994), Clark et al. (1998) and Friedman et al. (2005)) and has been discussed even before regime theory became widely adopted in IEP (Boardman, 1981; Kay and Jacobson, 1984). This work is more appropriately in my view seen as a secondary literature to institutionalist debates about regimes, as NGOs in this view play a role in supporting and furthering interstate regimes. The central theoretical point about pluralism is that interstate forms of environmental governance are no longer the only form of governance available.
11. A good example of this is Bulkeley and Betsill's (2003) account of 'translocal' cooperation amongst cities over climate change.
12. For general accounts of IEP focusing on this dimension, see Bhaskar and Glyn (1995), Miller (1996), Thomas (1992), or various chapters in Stevis and Assetto (2001).
13. This terminology of course has its own politics, which I will have to avoid dealing with explicitly here.
14. Or at least this is what Southern negotiators and many commentators maintain rhetorically in environmental regimes.

15. Arguably, these two themes of structural inequality and capital accumulation could be treated together, and clearly there are close connections between them. But I keep them distinct since it is clearly possible, and many in practice do this, to analyse IEP as a set of structural conflicts without making the connection back to the fundamental structure and dynamic of capitalist society. Furthermore, the analysis of accumulation creates its own focuses and at least in principle, can be divorced from some of its Marxist background to examine the way that growth and its political imperatives produce environmental degradation and structure international responses to such degradation.
16. I do not mean to get into a largely fruitless debate about whether socialist/communist states are 'really' capitalist or not. My point is that such states did (and the remaining ones still do) organize their politics around accumulation, at least in part due to military competition with Western states. The principal contemporary alternative ideological orientation that nevertheless supports the point made here is the notion of 'development'. Southern states, which might ideologically position themselves as non- or anti-capitalist, nevertheless articulate their political programmes towards a project that at its core promotes accumulation as understood here.
17. On ecological modernization, see, for example, Dryzek et al. (2003), Hajer (1995), Mol (2001) or Weale (1992).

bibliography

Adler, E. (1992) 'The Emergence of Cooperation: National Epistemic Communities and the International Evolution of the Idea of Nuclear Arms Control', *International Organization* 46, 1, pp. 101–46.

Agarwal, A., and Narain, S. (1990) *Global Warming in an Unequal World: A Case of Environmental Colonialism*, New Delhi: Centre for Science and Environment.

Andresen, D., and W. Østreng (eds) (1989) *International Resource Management: The Role of Science and Politics*, London: Belhaven Press.

Andresen, Steinar, Tora Skodvin, Arild Underdal and Jorgen Wettestad (2000) *Science and Politics in International Environmental Regimes: Between Integrity and Involvement*, Manchester: Manchester University Press.

Athanasiou, Tom, and Paul Baer (2002) *Dead Heat: Global Justice and Global Warming*, New York: Seven Stories Press.

Axelrod, Robert (1984) *The Evolution of Cooperation*, New York: Basic Books.

Banuri, T., and Apffel-Marglin, F. (eds) (1993) *Who Will Save the Forests? Political Resistance, Systems of Knowledge, and the Environmental Crisis*, London: Zed Books.

Beck, Ulrich (1995) *Ecological Politics in an Age of Risk*, Cambridge: Polity Press.

Berkes, Fikrit (ed.) (1989) *Common Property Resources: Ecology and Community-Based Sustainable Development*, London: Belhaven Press.

Bernauer, Thomas (1995) 'The Effect of International Environmental Institutions: How We Might Learn More', *International Organization* 49, 2, pp. 351–77.

Bernstein, Steven (2001) *The Compromise of Liberal Environmentalism*, New York: Columbia University Press.

Bhaskar, V., and Glyn, A. (1995) *The North, the South and the Environment: Ecological Constraints and the Global Economy*, London: Earthscan.

Boardman, Robert (1981) *International Organization and the Conservation of Nature*, Bloomington: Indiana University Press.

Boehmer-Christiansen, Sonja (1993) 'Science Policy, the IPCC and the Climate Convention: The Codification of a Global Research Agenda', *Energy and Environment* 4, 4, pp. 362–408.

Boehmer-Christiansen, Sonja (1995a) 'Britain and the International Panel on Climate Change: The Impacts of Scientific Advice on Global Warming Part I: Integrated Policy Analysis and the Global Dimension', *Environmental Politics* 4, 1, pp. 1–18.

Boehmer-Christiansen, Sonja (1995b) 'Britain and the International Panel on Climate Change: The Impacts of Scientific Advice on Global Warming Part II: The Domestic Story of the British Response to Climate Change', *Environmental Politics* 4, 2, pp. 175–96.

Boehmer-Christiansen, Sonja, and Aynsley Kellow (2002) *International Environmental Policy: Interests and the Failure of the Kyoto Process*, London: Edward Elgar.

Bretherton, Charlotte (1998) 'Global Environmental Politics: Putting Gender on the Agenda?', *Review of International Studies* 24, 1, pp. 85–100.

Bretherton, Charlotte (2003) 'Movements, Networks, Hierarchies: A Gender Perspective on Global Environmental Governance', *Global Environmental Politics* 3, 2, pp. 103–19.

Broadhead, Lee-Anne (2002) *International Environmental Politics: The Limits of Green Diplomacy*, Boulder: Lynne Rienner Publishers.

Bulkeley, Harriet, and Michele Betsill (2003) *Cities and Climate Change: Urban sustainability and Global Environmental Governance*, London: Routledge.

Bull, Hedley (1977) *The Anarchical Society: A Study of Order in World Politics*, London: Macmillan.

Bullard, Robert (1990) *Dumping in Dixie*, Boulder: Westview Press.

Clapp, Jennifer (1994) 'Africa, NGOs and the International Toxic Waste Trade', *Journal of Environment and Development* 3, 2, pp. 17–46.

Clapp, Jennifer (1998) 'The Privatization of Global Environmental Governance: ISO 14000 and the Developing World', *Global Governance* 4, 3, pp. 295–316.

Clark, Ann Marie, Elisabeth Jay Friedman and Kathryn Hochstetler (1998) 'The Sovereign Limits of Global Civil Society: A Comparison of NGO Participation in UN World Conferences on the Environment, Women and Human Rights', *World Politics* 51, 1, pp. 1–35.

Dalby, Simon (2002) *Environmental Security*, Minneapolis: University of Minnesota Press.

Dankelman, Irene, and Davidson, Joan (1988) *Women and Environment in the Third World*, London: Earthscan.

Dimitrov, Radoslav S. (2003) 'Knowledge, Power, and Interests in Environmental Regime Formation', *International Studies Quarterly* 47, pp. 123–50.

Dryzek, John (1987) *Rational Ecology: Environment and Political Economy*, Oxford: Blackwell.

Dryzek, J., D. Downes, C. Hunold and D. Schlosberg, with H. K. Hernes (2003) *Green States and Social Movements*, Oxford: Oxford University Press.

Eckersley, Robyn (1992) *Environmentalism and Political Theory: Towards an Ecocentric Approach*, London: UCL Press.

Eckersley, Robyn (2004a) 'The Big Chill: The WTO and Multilateral Environmental Agreements', *Global Environmental Politics* 4, 2, pp. 25–50.

Eckersley, Robyn (2004b) *The Green State: Rethinking Democracy and Sovereignty*, Cambridge, Mass.: MIT Press.

Ecologist (1993) *Whose Common Future? Reclaiming the Commons*, London: Earthscan.

Ekins, Paul (2000) *Economic Growth and Environmental Sustainability: The Prospects for Green Growth*, London: Routledge.

Enzensberger, Hans Magnus (1974) 'A Critique of Political Ecology', *New Left Review* 84, pp. 3–31.

Falkner, Robert (2001) 'Business Conflict and U.S. International Environmental Policy: Ozone, Climate, and Biodiversity', in Paul G. Harris (ed.) *The Environment, International Relations, and U.S. Foreign Policy*, Washington, DC: Georgetown University Press, pp. 157–77.

Finger, Matthias, and James Kilcoyne (1997) 'Why Transnational Corporations are Organizing to Save the Global Environment', *Ecologist* 27, 4, pp. 138–42.

Finger, Matthias, and Ludivine Tamiotti (1999) 'New Global Regulatory Mechanisms and the Environment: The Emerging Linkage Between the WTO and the ISO', *IDS Bulletin* 30, 3, pp. 8–16.

Friedman, Elisabeth Jay, Kathryn Hochstetler and Ann Marie Clark (2005) *Sovereignty, Democracy and Global Civil Society: State–Society Relations at UN World Conferences*, Albany: SUNY Press.

Garcia-Johnson, Ronie (2000) *Exporting Environmentalism*, Cambridge, Mass.: MIT Press.

Gorz, André (1994) 'Political Ecology: Expertocracy versus Self-Limitation', *New Left Review*, 202, pp. 55–67.

Grubb, Michael (1995), 'Seeking Fair Weather: Ethics and the International Debate on Climate Change', *International Affairs* 71, 3, pp. 463–96.

Guha, Ramachandra, and Juan Martinez-Alier (1997) 'The Merchandising of Biodiversity', in their *Varieties of Environmentalism: Essays North and South*, London: Earthscan.

Haas, Peter M. (1989) 'Do regimes matter? Epistemic Communities and Mediterranean Pollution Control', *International Organization* 43, 3, pp. 377–403.

Haas, Peter M. (1990) *Saving the Mediterranean: The Politics of International Environmental Cooperation*, New York: Columbia University Press.

Haas, Peter (1999) 'Social Constructivism and the Evolution of Multilateral Environmental Governance', in Aseem Prakash and Jeffrey Hart (eds) *Globalization and Governance*, London: Routledge, pp. 103–33.

Haas, P. M., R. O. Keohane and M. A. Levy (1993) *Institutions for the Earth: Sources of Effective Environmental Protection*, Cambridge, Mass: MIT Press.

Hahn, Robert, and Kenneth Richards (1989) 'The Internationalization of Environmental Regulation', *Harvard International Law Journal* 30, pp. 421–46.

Hajer, M. (1995) *The Politics of Environmental Discourse: Ecological Modernisation and the Policy Process*, Oxford: Clarendon.

Hampson, Fen Osler, and Judith Reppy (eds) (1996) *Earthly Goods: Environmental Change and Social Justice*, Ithaca: Cornell University Press.

Hardin, Garrett (1968) 'The Tragedy of the Commons', *Science* 162, pp. 1243–48.

Harris, Paul (2003) 'Fairness, Responsibility, and Climate Change', *Ethics and International Affairs* 17, 1, pp. 149–56.

Harrison, Neil E. (2000) *Constructing Sustainable Development* Albany: SUNY Press.

Harrison, Neil E., and Gary Bryner (eds) (2003) *Science and Politics in the International Environment*, Lanham, MD: Rowman & Littlefield.

Harvey, David (1993) 'The Nature of the Environment: The Dialectics of Social and Environmental Change', in Leo Panitch and Ralph Miliband (eds) *The Socialist Register* pp. 1–51.
Hay, C. (1994) 'Environmental Security and State Legitimacy', in M. O'Connor (ed.) *Is Capitalism Sustainable? Political Economy and the Politics of Ecology*, New York: Guilford Press.
Helleiner, Eric (1996) 'International Political Economy and the Greens', *New Political Economy* 1, 1, pp. 59–78.
Hempel, Lamont (1996) *Environmental Governance: The Global Challenge*, Washington, DC: Island Press.
Hovden, Eivind (1998) 'The Problem of Anthropocentrism: A Critique of Institutionalist, Marxist, and Reflective International Relations Theoretical Approaches to Environment and Development'. Unpublished PhD thesis, University of London.
Hovi, Jon, Detlef F. Sprinz and Arild Underdal (2003) 'The Oslo-Potsdam Solution to Measuring Regime Effectiveness: Critique, Response, and the Road Ahead', *Global Environmental Politics* 3, 3, pp. 74–96.
Jasanoff, Sheila, and Brian Wynne (1998) 'Science and Decision-Making', in S. Rayner and E. Malone (eds) *Human Choice and Climate Change: The Societal Framework*, Columbus, Ohio: Batelle Press.
Kay, David, and Jacobson, Harold (eds) (1984) *Environmental Protection: The International Dimension*, Totowa, NJ: Allanheld, Osmun.
Keohane, Robert O. (1989) *International Institutions and State Power: Essays in International Relations Theory*, Boulder: Westview Press.
Keohane, R., and J. Nye (1977) *Power and Interdependence: World Politics in Transition*, Boston: Little Brown and Company.
Klein, Naomi (2002) *Fences and Windows: Dispatches from the Frontlines of the Globalization Debate*, London: Flamingo.
Kütting, Gabriela (2000) *Environment, Society and International Relations: Towards More Effective International Agreements*, London: Routledge.
Laferrière, Eric, and Peter Stoett (1999) *Ecological Thought and International Relations Theory*, London: Routledge.
Levy, David (1997) 'Business and International Environmental Treaties: Ozone Depletion and Climate Change', *California Management Review*, 39, 3, pp. 54–71.
Levy, David, and Peter Newell (eds) (2005) *Business in International Environmental Governance*, Cambridge, Mass.: MIT Press.
Lipschutz, Ronnie (1999) 'Why is there no International Forestry Law? An Examination of International Forestry Regulation, both Public and Private', available at: <www.seweb.uci.edu/users/dimento/Lipschutz.htm>.
Lipschutz, Ronnie (2004) *Global Environmental Politics: Power, Perspectives, and Practice*, Washington, DC: Congressional Quarterly Press.
Lipschutz, Ronnie D., with Judith Mayer (1996) *Global Civil Society and Global Environmental Governance: The Politics of Nature from Place to Planet*, Albany: SUNY Press.
Litfin, Karen (1994) *Ozone Discourses: Science and Politics in Global Environmental Cooperation*, New York: Columbia University Press.
Litfin, Karen (ed.) (1998) *The Greening of Sovereignty in World Politics*, Cambridge, Mass.: MIT Press.
Low, Nicholas, and Brendan Gleeson (1998) *Justice, Society and Nature: An Exploration of Political Ecology*, London: Routledge.

Mantle, Deborah (1999) 'Critical Green Political Theory and International Relations Theory – Compatibility or Conflict'. Unpublished PhD thesis, Keele University.
Martello, Marybeth Long (2001) 'A Paradox of Virtue?: "Other" Knowledges and Environment-Development Politics', *Global Environmental Politics* 1, 3, 114–41.
McCay, Bonnie, and James Acheson (eds) (1987) *The Question of the Commons: The Culture and Ecology of Communal Resources*, Tuscon: University of Arizona Press.
McSweeney, Bill (1999) *Security, Identity and Interests*, Cambridge: Cambridge University Press.
Merchant, Carolyn (1980) *The Death of Nature: Women, Ecology and the Scientific Revolution*, San Francisco: Harper & Row.
Miles, Edward L., Arild Underdal, Steinar Andresen, Jørgen Wettestad, Jon Birger Skjærseth and Elaine M. Carlin (2002) *Environmental Regime Effectiveness: Confronting Theory with Evidence*, Cambridge, Mass.: MIT Press.
Miller, Clark, and Paul N. Edwards (eds) (2001) *Changing the Atmosphere: Expert Knowledge and Environmental Governance*, Cambridge, Mass.: MIT Press.
Miller, Marian (1996) *The Third World in Global Environmental Politics*, Buckingham: Open University Press.
Mitchell, Ronald B. (2002) 'A Quantitative Approach to Evaluating International Environmental Regimes', *Global Environmental Politics* 2, 4, pp. 58–83.
Mitchell, Ronald B., William C. Clark, David W. Cash and Frank Alcock (eds) (forthcoming) *Global Environmental Assessments: Information, Institutions, and Influence*, Cambridge, Mass.: MIT Press.
Mol, A. (2001) *Globalization and Environmental Reform: The Ecological Modernisation of the Global Economy*, Cambridge, Mass.: MIT Press.
Naess, Arne (1989) *Ecology, Community and Lifestyle*, Cambridge: Cambridge University Press.
Newell, Peter (2000) 'Environmental NGOs, TNCs, and the Question of Governance', in Dimitris Stevis and Valerie J. Assetto (eds) *The International Political Economy of the Environment*, Boulder: Lynne Rienner, pp. 85–107.
O'Brien, Robert, Anne Marie Goetz, Jan Aart Scholte and Marc Williams (2000) *Contesting Global Governance*, Cambridge: Cambridge University Press.
O'Connor, James (1991) 'On the Two Contradictions of Capitalism', *Capitalism, Nature, Socialism* 2, 3, pp. 107–9.
O'Connor, M. (ed.) (1994) *Is Capitalism Sustainable? Political Economy and the Politics of Ecology*, New York: Guilford Press.
Ophuls, William (1977) *Ecology and the Politics of Scarcity*, San Francisco: W. H. Freeman & Co.
Ostrom, Elinor (1990) *Governing the Commons: The Evolution of Institutions for Collective Action*, Cambridge: Cambridge University Press.
Oye, Kenneth (ed.) (1986) *Cooperation under Anarchy*, Princeton: Princeton University Press.
Parson, Edward (2003) *Protecting the Ozone Layer: Science and Strategy*, Oxford: Oxford University Press.
Paterson, Matthew (1996a) *Global Warming and Global Politics*, London: Routledge.
Paterson, Matthew (1996b) 'Green Politics', in Scott Burchill (ed.) *Theories of International Relations*, London: Macmillan.

Paterson, Matthew (1996c), 'International Justice and Global Warming', in B. Holden (ed.) *The Ethical Dimensions of Global Change*, London: Macmillan, pp. 181–204.
Paterson, Matthew (2000) *Understanding Global Environmental Politics: Domination, Accumulation, Resistance*, London: Macmillan.
Paterson, Matthew (2001a) 'Climate Policy as Accumulation Strategy: The Failure of COP6 and Emerging Trends in Climate Politics', *Global Environmental Politics* 1, 2, pp. 10–17.
Paterson, Matthew (2001b) 'Risky Business: Insurance Companies in Global Warming Politics', *Global Environmental Politics* 1, 4, pp. 8–42.
Paterson, Matthew, David Humphreys and Lloyd Pettiford (2003) 'Conceptualizing Global Environmental Governance: From Interstate Regimes To Counter-Hegemonic Struggles', *Global Environmental Politics* 3, 2, pp. 1–10.
Plumwood, Val (1993) *Feminism and the Mastery of Nature*, London: Routledge.
Pirages, Dennis (1978) *Global Ecopolitics: The New Context for International Relations*, North Scituate, Mass: Duxbury Press.
Princen, Thomas (2003) 'Principles for Sustainability: From Cooperation and Efficiency to Sufficiency', *Global Environmental Politics* 3, 1, pp. 33–50.
Princen, Thomas and Matthias Finger (eds) (1994) *Environmental NGOs in World Politics*, London: Routledge.
Princen, Tom (2005) *The Logic of Sufficiency*, Cambridge, Mass.: MIT Press.
Redclift, Michael (1987) *Sustainable Development: Exploring the Contradictions*, London: Routledge.
Rosenau, James N. (1990) *Turbulence in World Politics: A Theory of Change and Continuity*, Princeton: Princeton University Press.
Rosenau, James N. (1993) 'Environmental Challenges in a Turbulent World', in Ken Conca and Ronnie Lipschutz (eds) *The State and Social Power in Global Environmental Politics*, New York: Columbia University Press.
Rowlands, I. H. (1995) *The Politics of Global Atmospheric Change*, Manchester: Manchester University Press.
Sachs, Wolfgang (ed.) (1993) *Global Ecology*, London: Zed Books.
Shackley, S. and B. Wynne (1996) 'Representing Uncertainty in Global Climate-Change Science and Policy – Boundary-Ordering Devices and Authority', *Science, Technology & Human Values*, 21, 3, pp. 275–302.
Shields, Linda and Marvin Ott (1974) 'The Environmental Crisis: International and Supranational Approaches', *International Relations* 4, 6.
Shiva, Vandana (1993) *Monocultures of the Mind: Perspectives on Biodiversity and Biotechnology*, London: Zed Books.
Shue, H. (1992) 'The Unavoidability of Justice', in A. Hurrell and B. Kingsbury (eds) *The International Politics of the Environment*, Oxford: Clarendon.
Sklair, Leslie (2000) *The Transnational Capitalist Class*, Oxford: Blackwell.
Sontheimer, Sally (ed.) (1991) *Women and the Environment: A Reader*, London: Earthscan.
Soroos, Marvin (2001) 'Global Climate Change and the Futility of the Kyoto Process', *Global Environmental Politics* 1, 2, pp. 1–9.
Sprout, H., and M. Sprout (1965) *The Ecological Perspective on Human Affairs*, Princeton: Princeton University Press.
Starr, Amory (2000) *Naming the Enemy: Anti-Corporate Movements Confront Globalization*, London: Zed Books.
Stevis, Dimitris, and Valerie Assetto (eds) (2001) *The International Political Economy of the Environment*, Boulder: Lynne Rienner.

Szasz, Andrew (1994) *Ecopopulism: Toxic Waste and the Movement for Environmental Justice*, Minneapolis: Minnesota University Press.
Thomas, Caroline (1992) *The Environment in International Relations*, London: Royal Institute of International Affairs.
van der Pijl, Kees (1998) *Transnational Classes and International Relations*, London: Routledge.
Victor, David, Kal Raustiala and Eugene Skolnikoff (eds) (1998) *The Implementation and Effectiveness of International Environmental Commitments*, Cambridge, Mass.: MIT Press.
Vogler, John (2000) *The Global Commons: Environmental and Technological Governance*, London: John Wiley.
Vogler, John (2003) 'Taking Institutions Seriously: How Regime Analysis can be Relevant to Multilevel Environmental Governance', *Global Environmental Politics* 3, 2, pp. 25–39.
Wackernagel, Mathias, and William Rees (1996) *Our Ecological Footprint: Reducing Human Impact on the Earth*, Gabriola Island, British Columbia: New Society Publishers.
Wapner, Paul (1996) *Environmental Activism and World Civic Politics*, Albany: SUNY Press.
Wapner, Paul (1997) 'Governance in Global Civil Society', in Oran R. Young (ed.) *Global Governance: Drawing Insights from the Environmental Experience*, Ithaca: Cornell University Press, pp. 65–84.
Weale, A. (1992) *The New Politics of Pollution*, Manchester: Manchester University Press.
Weber, Cynthia, and Thomas Biersteker (eds) (1996) *State Sovereignty as Social Construct*, Cambridge: Cambridge University Press.
Wendt, Alex (1999) *Social Theory of International Politics*, Cambridge: Cambridge University Press.
Wiegandt, E. (2001) 'Climate Change, Equity, and International Negotiations', in U. Luterbacher and D. Sprinz (eds), *International Relations and Global Climate Change*, Cambridge, Mass.: MIT Press, pp. 127–50.
Williams, Marc (2001) 'Trade and Environment in the World Trading System: A Decade of Stalemate?', *Global Environmental Politics* 1, 4, pp. 1–9.
World Commission on Environment and Development (1987) *Our Common Future*, Oxford: Oxford University Press.
Worster, Donald (1994) *Nature's Economy: A History of Ecological Ideas*, 2nd edn, Cambridge: Cambridge University Press.
Young, Oran R. (1989) *International Cooperation: Building Regimes for Natural Resources and the Environment*, Ithaca: Cornell University Press.
Young, Oran R. (1994) *International Governance: Protecting the Environment in a Stateless Society*, Ithaca: Cornell University Press.
Young, Oran R. (ed.) (1997a) *Global Governance: Drawing Insights from the Environmental Experience*, Ithaca: Cornell University Press.
Young, Oran R. (1997b) 'Global Governance: Towards a Theory of Decentralized World Order', in Oran R. Young (ed.) *Global Governance: Drawing Insights from the Environmental Experience*, Ithaca: Cornell University Press, pp. 272–99.
Young, Oran R. (ed.) (1999) *The Effectiveness of International Environmental Regimes: Causal Connections and Behavioral Mechanisms*, Cambridge, Mass.: MIT Press.
Young, Oran R. (2001) 'Inferences and Indices: Evaluating the Effectiveness of International Environmental Regimes', *Global Environmental Politics* 1, 1, pp. 99–121.

annotated bibliography

The Ecologist (1993) *Whose Common Future? Reclaiming the Commons*, London: Earthscan. This book both critiques the UNCED process and provides the framework for understanding on what political principle sustainable societies might be organized. Their central argument, a direct inversion of Hardin, is that commons are in fact this principle.

Haas, Peter M. (1990) *Saving the Mediterranean: The Politics of International Environmental Cooperation*, New York: Columbia University Press. This is the original 'epistemic communities' argument. While many disagree with Haas' conclusions and framework, his way of focusing on knowledge has been a point of departure for debates in this field.

Hardin, Garrett (1968) 'The Tragedy of the Commons', *Science*, 162, pp. 1243–48. The classic statement, from which much later theorizing in IR (and elsewhere) has developed. Hardin uses a metaphor from a mythical medieval English village to provide an explanation for environmental degradation that it arises primarily out of the incentives to overuse resources created by open access (he misnamed them commons) property regimes.

Hay, Colin (1994) 'Environmental Security and State Legitimacy', in Martin O'Connor (ed.) *Is Capitalism Sustainable? Political Economy and the Politics of Ecology*, New York: Guilford Press. Hay shows that states are faced with the twin, contradictory pressures of promoting accumulation/growth to secure the continuation of capitalist society and legitimizing themselves in the face of mobilization by environmental movements. This provides a framework for understanding the emergence of sustainable development and ecological modernization as policy frameworks.

Laferrière, Eric, and Peter Stoett (1999) *Ecological Thought and International Relations Theory*, London: Routledge. This is the fullest survey of the relationship between IR theory and ecological thought in print. It provides a useful overview of the complexities of green thought, and then runs through chapters organized around a series of perspectives in International Relations theory (realism, pluralism, structuralism, critical theories).

Wapner, Paul (1996) *Environmental Activism and World Civic Politics*, Albany: SUNY Press. This succinct, clearly argued book outlines a pluralist case that global environmental governance is undergoing important shifts away from being centred on states. Wapner analyses the activities of three international environmental NGOs and shows that they embody different principles of world political organization.

Young, Oran R. (1989) *International Cooperation: Building Regimes for Natural Resources and the Environment*, Ithaca: Cornell University Press. This is the standard text (with a series of updates since then) for liberal institutionalist 'regime theory' in IEP. Young provides the overview of the core concepts in regime theory – the explanations for why states enter into regimes, how regimes affect state behaviour, how regimes enable cooperation, in particular – and applies them to cases to illustrate his argument.

4
methods in international environmental politics
kathryn hochstetler and melinda laituri

> The various methods available to us make up a diverse set of arrows in the quiver of the social scientist, and we should choose the arrow most likely to hit our target.
>
> (Schwartz et al., 2000, p. 89)

Researchers in international environmental politics (IEP) have devoted little extended attention to the methods that they use in their field. With a few exceptions that are discussed below, they have simply carried out their research without exploring which methods are best-suited to the field as a whole. This is a laudable approach to an area of research whose data can range from the cultural discourses in global negotiations about climate change to a time-series data set of measurements of CO_2 in the atmosphere. The absence of a hegemonic methodological discourse in the field fits its diversity well, and this chapter does not aim to establish any such hegemony. On the other hand, the lack of extended reflection about the methodologies appropriate to the field may prevent IEP researchers from thinking more creatively about their research designs and approaches. Greater attention to research design and methodology would help them avoid unnecessary and unintended weaknesses in their studies. To that end, this chapter outlines a number of different approaches and specifies how they are used and for which kinds of analytical projects, focusing on issues of research design. It also identifies characteristic pitfalls and critiques of the different methods.

The IEP field as a whole needs a full methodological toolbox. While individual researchers may specialize in particular methods – and few can master all methods – different kinds of research questions within the study of IEP demand different kinds of methodological approaches. Traditional qualitative methods and newer discursive analyses are especially useful for studying the *processes* of IEP, including various kinds of formal and informal negotiations. Interviews, observation and documentary research provide the data for this set of methods. These methods are also especially useful for identifying the *purposes* and worldviews of different actors and policies. Rational choice approaches focus on similar questions, while using quite different analytical models. For identifying empirical *patterns* or tendencies across larger populations, various kinds of quantitative methods are more appropriate. Thus survey methods help map public values and statistical methods trace causal relationships over larger numbers of cases. Geospatial and statistical data are especially important for viewing the impact of human choices on the physical environment itself.

Much of the existing literature on methods in IEP focuses on the question of whether qualitative or quantitative studies are more appropriate for its research questions. The predominant mode of analysis of IEP so far has been largely qualitative, with most researchers selecting a small number of case studies for close study. A few researchers have laid out strategies for successful qualitative study of IEP (Homer-Dixon, 1996; Mitchell and Bernauer, 1998), but there needs to be more attention to these methodologies that so many in the field use. Perhaps because of this dominance of qualitative methods in IEP, many of the more explicit discussions of methods in IEP are arguing for the use of other kinds of methods. Detlef Sprinz has argued in a number of pieces for more use of quantitative and modelling research approaches (Sprinz, 1999a, 1999b, 2004), a call recently joined by others (Kilgour and Wolinsky-Nahmias, 2004; Mitchell, 2002).

The quantitative/qualitative divide is an important one that will be addressed here, but it is cross-cut by a more fundamental division that maps better onto the different theoretical projects of IEP scholars. This more fundamental division is the epistemological divide between approaches that are aiming for positivist causal explanations of the phenomena they study and those (here called critical theory or postpositivism) that reject at least part of the possibility or desirability of achieving such explanations. Because this methodological review sees two dimensions differentiating current IEP research, its organization departs somewhat from other summaries. An initial set of sections introduces different methodologies as they relate to the positivist project of causal explanation. The first section introduces the standards and procedures that are relevant for all positivist approaches and shows how they appear in their qualitative form. It is then followed by sections focused on the other primarily positivist methodologies – quantitative/statistical, rational choice and geospatial technologies – which are divided by the kinds of data they require and the ways they manipulate that data. The second major part of the chapter examines the alternative standards for research design, evidence, and argument of critical approaches. Again, the standards are introduced in their qualitative form, which is how they are usually used in IEP. A final section takes up the small subset of critical scholars who use non-qualitative methods and discusses how they differ from their counterparts.

The international relations field as a whole is similarly divided between positivist and critical approaches. While there are elective affinities between standard international relations theories and the various methodological approaches, there is no one to one correspondence among them. Constructivists, for example, often use qualitative methods, as these textual and discursive methods are compatible with the constructivists' focus on meanings. Nonetheless, constructivists are divided between positivist and postpositivist approaches, and collectively use most of the methodologies discussed (Finnemore and Sikkink, 2001). Realists are disproportionately present among users of quantitative and formal methodologies, but they use a full range of the positivist methodologies, as do liberal institutionalists. Critical theorists of various kinds are often drawn to qualitative methodologies, but they also sometimes use quantitative and formal methods (R. Morrow, 1994).

How are methodological issues different when studying international *environmental* politics as compared to international relations more generally? The clearest point of distinction is that environmental issues organically link to the natural world and its associated physical and natural sciences in ways that, for example, human rights and security issues do not. In the course of an IEP project, the scholar may well need to grapple with the complexities of climate change models, debates about how to measure forest biodiversity or the inner workings of a two-stroke engine. IEP scholars often find themselves in dialogue or even in institutions with natural scientists, who bring their disciplines' standards and approaches with them. The nature of that dialogue depends quite a bit on the methodological approach taken by the IEP scholar. The tendency-finding approaches of statistical and geospatial data analysis offer the IEP scholar the most straightforward opportunities for dialogue with natural and physical scientists, as their disciplines tend to place a high value on these kinds of methods. The other approaches, however, intersect in more complicated ways with non-social scientists. A minority of natural scientists uses positivist qualitative methods.[1] In the section on such methods below, authors of various disciplines argue that qualitative methods are especially useful for disentangling the complex and interactive relationships which are uncovered in studies that look seriously at both natural and social processes. The arguments in this section might help provide the basis for a conversation with natural scientists about the value of qualitative methods in a positivist research agenda. Postpositivist scholars have the most tendentious relations with natural scientists, since one of the postpositivists' major concerns is to undercut the claimed special expertise of science and its practitioners. Finally, the human purposiveness at the root of rational choice modelling gives it few analogues in the natural and physical sciences, although formal rational choice approaches obviously draw on the mathematics discipline. So far, rational choice scholars have not reflected directly on whether environmental topics call on characteristic kinds of rational decision-making.

positivist approaches to methodology

positivist standards for research and positivist qualitative methods

The most common empirical approach to IEP so far has been what are called positivist – or causal or rationalist – qualitative methods. As the label suggests, this approach shares a general orientation toward research standards and aims with a number of non-qualitative methodologies, while collecting and analysing its data in qualitative (non-numerical) ways.[2] In particular, it embraces the overall project of developing causal explanations that can be used for understanding general patterns of IEP. To this end, positivist qualitative scholars often adopt the language of statistical analysis and define independent, dependent, and control variables, developing hypotheses about the relationships among these. One of the key dilemmas for this approach is that an individual qualitative study does not provide the systematic assessment of proposed causal relationships for which positivism aims because qualitative methods tend to produce a great deal of information about a relatively small number of cases. In the debates between quantitative and qualitative scholars, qualitative scholars have delineated what their methods can add to the positivist explanatory project instead, usually focusing on qualitative research's role in generating hypotheses and in tracing causal processes. These points are developed in this section.

Much of the important analytical work of positivist qualitative methods is done in the research design phase of the project, where a small number of cases are selected on the basis of proposed causal variables (George, 1979; Peters, 1998). After the cases are chosen, the researcher systematically collects information about the variables and the relationships among them, using documents and other archival records, interviews and observation (see Fenno, 1990; Hill, 1993; Taylor and Bogdan, 1998). During the data collection phase, further analysis is carried out as the researcher codes and classifies the case-specific information as examples of more general categories. In the final analytical stage, the researcher relates the observed and collected data to the hypothesized relationship. An initial step is the basic one of assessing whether the independent and dependent variables do take on the expected values. In assessing this fit, the depth of case knowledge available in qualitative analysis can provide interesting qualifiers to the frequently more superficial categories used in quantitative analysis. For example, quantitative researchers have been unable to find systematic causal relationships between crucial variables like democracy (defined and operationalized in narrow, formal institutional terms) and environmental protection despite numerous theories suggesting that the relationship exists (Midlarsky, 1998). Midlarsky himself concludes that case study research of a few of the most important cases such as Brazil or Russia is needed to refine the hypothesis (Midlarsky, 1998, p. 359).

While assessing correlational fit is the most important analytical step in quantitative methods, it is less important for a qualitative project since there are not enough cases to establish causal variation directly. Qualitative researchers have commonly argued that their most important analytical contribution is process-tracing, an operation that statistical analysis cannot itself do (George, 1979). Process tracing requires the analyst to break 'down an overarching causal relationship into a set of smaller causal links in a larger causal chain' (Mitchell and Bernauer, 1998, p. 22). The analysis is often presented as a chronological narrative or schematic map of the causal relationship. Moving to this level can help identify scope conditions for the causal relationship and locate intermediate processes and variables (Homer-Dixon, 1996, p. 144; Schwartz et al., 2000, p. 85). 'Temporal succession' and 'contiguity' are the causal forces at work here according to Hume's categories, with evidence of causal relationships not limited to the 'constant conjunction' of statistics (Schwartz et al., 2000, p. 84, citing Andrew Bennett). Process tracing can also help to argue against alternative hypotheses, as the information-dense format allows considering these alternatives and finding them less consistent (Schwartz et al., 2000, p. 86).

As noted, positivist qualitative methods have been the most common empirical approach to IEP. In the field currently there are two basic uses of qualitative studies, reflecting different views on the primary purposes of qualitative research. The first use sees qualitative methods as most important early in research programmes, for generating hypotheses that can later be evaluated more systematically using many more cases. This pre-statistical conception of qualitative research is well-illustrated by research on the effectiveness of international environmental regimes (see Wettestad in this volume). After qualitative case studies of individual regimes accumulated over a decade, researchers then began to use that qualitative data to create systematic data bases on regimes. These have now been used for both quantitative/statistical analysis (Breitmeier et al., 1996; Miles et al., 2002) and comparative analysis that explores the usefulness of rational choice-type assumptions of utility-maximizing behaviour (Young, 1999). Similarly, Ostrom drew on more than 5,000 qualitative case studies of communities who managed common pool resources to develop her rational choice models of their effects (Ostrom, 1990, p. xv). Some researchers in the area of environmental security have urged this step there (Gleditsch, 1998), but others strongly argue for continuing qualitative (Schwartz et al., 2000) or critical (Peluso and Watts, 2001) approaches.

The other approach to qualitative methods sees them as the likely final methodological stage of many IEP research programmes. One area of IEP that largely concurs with this point of view is the extensive body of work on non-state actors in IEP (see Betsill in this volume). In a recent article that proposes a framework for analysing the influence of non-governmental organizations (NGOs), Betsill and Corell explicitly argue for qualitative approaches: 'We argue that precise quantification of influence is futile and would only create a false impression of measurability for a phenomenon that is highly complex and intangible' (Betsill and Corell, 2001, p. 80). Other writers agree that qualitative methods are likely to have enduring value for certain kinds of IEP research problems. Mitchell and Bernauer's list (1998, pp. 6–7) includes the following instances: '(a) important but difficult-to-quantify variables (such as power, interests, or leadership); (b) theoretically important, empirically rare, or previously ignored cases; (c) innovative (but, by their nature, rare) international environmental policy strategies; and (d) causal rather than merely correlational relationships'. Homer-Dixon argues that complex ecological-political systems involve interactive, non-linear, and reciprocal causal relations that may always require qualitative methods because the case variation cannot be systematically controlled (Homer-Dixon, 1996, pp. 132–4). Schrader-Frechette and McCoy agree that ecology – a base science of IEP – is also prone to problems of uniqueness and interaction that make all research cases partially unique and its concepts imprecise (Schrader-Frechette and McCoy, 1993, pp. 114–16).

Critics of positivist qualitative methods argue that such imprecision and inability to systematically control variation are attributes of the approach rather than the empirical material (for example, Gleditsch, 1998, pp. 391–2). One project, which attempts to systematize qualitative information to use it for quantitative studies of environmental regime effectiveness, complains that too many qualitative studies fail to select their cases on theoretical grounds, involve idiosyncratic choices of variables and operationalizations, and are consequently not generalizable (Breitmeier et al., 1996, p. 1). Even proponents of qualitative methods admit that, in practice, qualitative scholars often fail to pay sufficient attention to various selection or practitioner biases or are inadequately precise and critical in their analyses (McKeown, 1999, p. 178; Schrader-Frechette and McCoy, 1993, pp. 139–47). Such flaws can make their work difficult to generalize and not replicable, or can raise questions about how representative their sample of cases is. Beyond these errors of execution, positivist qualitative methods are inherently limited in several ways with respect to the aims such scholars themselves claim. As indicated a number of times already, qualitative case studies in and of themselves cannot provide the large-number statistical correlations which are the 'gold standard' of positivist causal analysis. Thus they are limited in the crucial aim of generalization even while they might provide other causal pieces that statistical analyses cannot. The tracing of causal mechanisms also cannot formally weight causal variables, even as it shows their interaction (Schwartz et al., 2000, pp. 87–8).

quantitative methodologies

Quantitative causal analysis shares its basic purposes with positivist qualitative approaches, but uses larger numbers of cases. This allows the researcher to speak with greater confidence about patterns across the phenomenon of interest, and about the probability of any particular outcome or cause. A second difference between quantitative and qualitative positivist research is that quantitative researchers use statistical models of the hypothesized relationship among their variables. This is often a regression equation that must include the major possible explanations of changes in the dependent variable, with its variables defined in 'relatively generic and generalizable' terms (Mitchell, 2002, p. 74). Thus while a qualitative researcher often defines variables such as regime effectiveness in terms that are quite specific to the case at hand, for example, the regime governing trade in endangered species, the quantitative researcher will need to define regime effectiveness in a way that can be measured comparably across a broad set of regimes.

A primary empirical task of the quantitative researcher is to create (or locate existing) databases, coded with detailed data protocols, which collect data on the variables of interest in the cases of interest. This is partially a conceptual issue, as different definitions of a concept may exist in the literature, and the decision to operationalize and collect data on one as opposed to another can produce quite different quantitative results (for example, Midlarsky, 1998, on differing definitions of democracy). There is also a secondary issue more unique to quantitative research, which is its requirement that all conceptual variables be operationalized in ways that can be represented with data with numerical values, a process that can raise issues of construct validity – 'measuring what we think we are measuring' (King et al., 1994, p. 25). Some data such as trade statistics, gross national product or levels of pollutants are widely available in numerical form, although even these may require some manipulation such as calculating annual percentage change in absolute levels to make the data comparable enough for analysis (Mitchell, 2002, pp. 70–1).

Quantitative researchers use a variety of statistical procedures, which are beyond the scope of this chapter. They should match the underlying assumptions of the causal model; thus a simple linear regression can be usefully performed only when there are good reasons to think that its assumptions of linearity and independence of the variables are met. Other kinds of procedures such as time-series analysis or multivariate regression are needed for more complicated relationships. What all these forms of analysis share is the use of correlational analysis 'using sizable numbers of cases to frame and test generalizations about relationships between and among variables thought of as dependent variables and independent variables' (Breitmeier et al., 1996, p. 13). Statistical correlations are then the foundation for general causal claims about the relationships among the variables in the analysis.

Quantitative analysis is still comparatively rare in IEP. Perhaps not surprisingly, these studies first appeared in some of the subject areas of international relations where quantitative data is most readily available, especially in the intersection with economic issues. Thus there are studies that link gross domestic product (GDP) (as an operationalization of levels of development) to environmental protection as well as others that examine the relationship of trade and the environment (see Sprinz, 2004). Several efforts are underway to create databases that will speak to the relationship of international conflict and the environment (Gleditsch, 1998) and to the effectiveness of international environmental regimes (Breitmeier et al., 1996; Mitchell, 2002; Wettestad in this volume). These emerging efforts display substantial debate to this point on exactly how to operationalize and model key concepts, variables and relationships. Political variables are especially difficult to include in quantitative analyses, but several studies have taken up the challenge. Midlarsky (1998) found few solid conclusions in his study of the relationship between democracy and environmental protection, but Frank et al. (2000) found strong correlations between major international political events like the Stockholm conference and the creation of domestic environmental protections. In addition, as governments, intergovernmental organizations (IGOs) and NGOs collect and report data more systematically on the physical environment in countries across the world, quantitative IEP scholars can take advantage of that data for more direct measures of environmental outcomes.

Both advocates and critics of a quantitative approach to IEP agree that systematic data collection is a crucial requirement for this approach. Gleditsch even calls for 'a Correlates of War project for the environment' (Gleditsch, 1998, p. 396). Without good data behind them, sophisticated statistical techniques are simply misleading. Quantitative IEP analysts also need better data on control groups and cases to make their analyses less biased toward positive findings, for example, inclusion of cases of non-regimes (Dimitrov, 2003). For proponents of quantitative methods, this is a temporary stage in quantitative analysis, which can be overcome with collaborative efforts and some substantial research grants. For critics, the quality of quantitative data is a more enduring problem. The construct validity issue is a perennial one, with non-quantitative scholars arguing that crucial variables are very difficult to operationalize in numerical terms, for example, ingenuity in response to environmental scarcity (Schwartz et al., 2000, p. 88). Mitchell and Bernauer (1998, p. 5) voice this critique as the charge that quantitative scholars only study what can be quantified, use data that are too simple to be valid measures of complex constructs, and produce 'precise but unreliable or irrelevant results with sophisticated statistical techniques but data of poor quality'. More fundamentally, the generalized and probabilistic nature of the conclusions based on quantitative research may not help to explain outcomes in single cases of particular interest. Thus quantitative and qualitative approaches illustrate a clear tradeoff between the ability to reach general conclusions and the ability to explain one case well (Mitchell, 2002, p. 59).

rational choice approaches

The label 'rational choice approaches' is used here to refer to a family of related literatures that assume that political phenomena can be explained with reference to the choices of an individual 'whose behavior springs from individual self-interest and conscious choice. He or she is credited with extensive and clear knowledge of the environment, a well-organized and stable system of preferences, and computational skills that allow the actor to calculate the best choice (given individual preferences) of the alternatives available' (Monroe, 1991, p. 4). In an important sense, this assertion of the explanatory force of the microfoundations of individual choice is the underlying hypothesis of every rational choice analysis, although the research design will not necessarily test that hypothesis directly. At the highest level of abstraction, choice situations may either be parametric, where an individual faces given external constraints from the structure of the situation, or strategic, where the decisions of two or more individuals are interdependent (Elster, 1986, pp. 7–8). In strategic situations, game theory is the methodological tool used to find the equilibrium choices of the set of interdependent actors (Kilgour and Wolinsky-Nahmias, 2004; Tsebelis, 1990).

To begin their analyses, rational choice scholars develop logical models of the incentives and constraints facing an individual in the situation of interest and make hypotheses about the choices the individual(s) will make in that situation. In doing so, they aim for a parsimonious statement of the main stimuli to which the decision-maker responds, not a complete description of the particular situation. These statements may be taken deductively from existing research, for example, fundamental assumptions like the common one that elected officials prefer to be re-elected, or could be developed inductively from a particular situation. Rational choice scholars often pay special attention to institutional rules and procedures, such as those governing treaty ratification or legislative committee processes. Especially when analysing strategic interactions through game theory, this initial stage may involve identifying a specific situation as an example of one of the standard problems of strategic interaction, such as the tragedy of the commons and its free-rider problem, the collective action problem, the 'Chicken' problem, and so on (J. Morrow, 1994). The nature of the bargaining process, that is, whether it is iterative or one-off, and the amount of information decision-makers are assumed to have are also critical for determining the interactive strategy. The model of the situation may be presented through a narrative description or in formal terms, such as a matrix or tree form. For formal models, significant mathematical modelling skills may be needed. Rational choice scholars argue that one of the advantages of their approach is this requirement to clearly and precisely present the relevant actors/players, their alternative choices and preferences, and the various outcomes (Kilgour and Wolinsky-Nahmias, 2004).

The next stage in a rational choice analysis is to move on to compare the model's 'predictions with actual outcomes in a situation thought to be relevant to assessing the performance of the formal model' (McKeown, 1999, p. 177). At this stage, rational choice modellers can and do turn to any of the positivist methodologies. Assessing the model's fit statistically to historical quantitative data is the most-preferred method, but carefully selected qualitative case studies also provide initial tests, as can counterfactual historical analyses. Toke (2002) illustrates a qualitative version of rational choice analysis in IEP, using contrasting institutional arrangements to explain successful collective action on wind power in Denmark versus collective action failure in the UK. While a model is often developed out of a description of a particular empirical situation, as Toke does, a full test that follows positivist standards needs to evaluate it in an additional setting or in a whole array of them.

One of the most famous articles in the study of the environment, Hardin's story of the tragedy of the commons, is a non-formal analysis of a strategic choice situation (Hardin, 1968). His 'tragedy' of collective failure based on individual best choices has become a motivating metaphor for an entire sub-area of rational choice analysis that addresses such social dilemmas and free-riding problems. More recently, Ostrom has refined Hardin's argument to develop a new area of research on common pool resource management. Drawing on more than 5,000 such cases, she concluded that Hardin's tragedy was just one possible version of the game, noting that users of natural resources are able to change their constraints in ways that prisoners in PD game theory cannot (Ostrom, 1990, p. 7). She argues that they can overcome the dilemma with self-financed contract enforcement rules and, with others, has since proceeded to study the workings of those rules in numerous contexts (Gibson et al., 2000; Ostrom et al., 1994). Recent work on regime effectiveness extends and evaluates these insights in an international context (Young, 1999).

To this point, the contributions of rational choice theory to IEP in most other areas of study are more potential than actualized. A number of IEP puzzles would be susceptible to rational choice analysis, especially those involving bargaining and strategic interaction. Kilgour and Wolinsky-Nahmias (2004)[3] suggest one area of application would be using non-cooperative game theory models to study bargaining over shared natural systems, both globally, as in the climate change issue (Grundig et al., 2001), as well as across state boundaries. Non-cooperative game theory has also been used to model governments' simultaneous weighing of domestic and international incentives (two-level games) in international environmental negotiations (Wolinsky, 1997). Cooperative game theory could be used to extend Ostrom's largely domestic model of cooperation in the commons to study the development of international environmental regimes, modelling how regime participants allocate the surplus of their cooperation to overcome collective action problems. Much international cooperation, or lack thereof, to find environmental solutions can be modelled as a collective action problem. In general, Kilgour and Wolinsky-Nahmias (2004, p. 318) suggest that the field of international environmental politics presents a full range of classic rational choice problems, including the 'management of common resources, environmental negotiation, enforcement of environmental agreements, and the balance of domestic and international incentives'. We propose that rational choice scholars should also take up questions about whether the general approach might need to be modified to work with IEP, as environmental issues' link to the natural world might affect assumptions about information, risk and uncertainty that are central to rational choice modelling.

One reason for the scarcity of empirical rational choice studies in IEP may be the difficulty in gathering the information required to develop and test rational choice models. Empirical rational choice analysis requires data about the components of its explanations, such as institutional arrangements, preference structures, and so on. Because other approaches are not as interested in decisional microfoundations, rational choice scholars often need to develop their own datasets and the lack of adequate data is a continuing problem for this approach. In addition, preference structures and expected utilities are central concepts that are notoriously difficult to measure in any non-post hoc way. Changes in preferences or in the structure of the game are even more complicated to trace (McKeown, 1999, p. 179). In their broad critique of the empirical applications of rational choice theory, Green and Shapiro (1994, p. 41) conclude bluntly that 'the formal precision of rational choice models greatly outstrips political scientists' capacity to measure'. As with other methodologies and approaches, many actual uses of rational choice suffer from regular problems: 'Hypotheses are formulated in empirically intractable ways; evidence is selected and tested in a biased fashion; conclusions are drawn without serious attention to competing explanations; empirical anomalies and discordant facts are often either ignored or circumvented by way of post hoc alterations to deductive arguments' (Green and Shapiro, 1994, p. 6).

An enduring critique specific to the rational choice approach is that there are profound tensions between the many simplifications necessary to produce models that are mathematically tractable and what non-rational choice scholars consider adequate representations of 'empirically encountered situations' (McKeown, 1999, pp. 177–8). This is a debate among scholars who are all committed to positivism, as well as a critique levied by non-positivists. In a related point, critics question how many political situations are actually comprehensible through the simplifying assumptions required for rational choice analysis, arguing that the domain of rational choice applications may be quite limited (Weyland, 2002). Finally, non-rational choice scholars often criticize the excessive weight to theoretically driven model development, including models which will not be empirically testable in any foreseeable future.

geospatial information technologies/methods

Geospatial information technologies (GIT) refers to various tools that inventory, assess and analyse geographic information in a computerized environment. A geographic information system (GIS) allows the computer to 'think' it is a map – a map with the ability to analyse geographic information and tell its users about any part of the world. A GIS specifies three characteristics of a location: (1) where it is geographically (for example, latitude and longitude coordinates); (2) what it is (road, lake, well, city, and so on), and (3) its relationship to other locations. Remotely sensed data (RS, that is, satellite imagery, aerial photography) are an example of the types of information that a GIS may use for classification of vegetative types or land cover. Global positioning systems (GPS) allow the user to know where he/she is on the earth (for example, latitude and longitude coordinates) by consulting a radio receiver and using the constellation of 24 satellites orbiting the earth.

GIT approaches research questions from a spatial perspective allowing for the explicit mapping of the landscape relationships of multiple spatial arenas (cities, nation-states, watersheds, ecosystems). Spatial relationships are examined in a myriad of ways: locational and inventory queries, pattern analysis, complex modelling of socioeconomic and environmental processes, trend analysis and what-if scenarios. Additionally, integration of diverse datasets effectively lends GIT to interdisciplinary and multidisciplinary perspectives. Physical or environmental data can be analysed in conjunction with socioeconomic or political conditions of specific locations. Despite these advantages, geospatial technologies have made very few inroads into the politics and international relations disciplines, perhaps because of a lag in methodological training. The use of GIT requires special expertise in spatial literacy (Goodchild, 1995; Veregin, 1995). Spatial literacy refers to a suite of skills in using computers for geographic analysis (downloading digital data, model development), an understanding of digital geographic data and their underlying concepts (scale, projection, datums, data collection that is properly geocoded for use in a geographic database) and experience with the principles of map design and visual representation. With the increasing availability and lower cost of desktop computers, digital data, GIS software and the plethora of GIT certificates and online courses, this expertise is well within the grasp of many. Alternatively, individuals with GIT skills can be hired or included in research projects to create appropriate outputs.

GIT have proved to be powerful tools for the examination and analysis of physical environmental data. Socioeconomic data, such as census data, have also been used effectively in GIS applications. A key output of such data compilation has been the development of numerous atlases with CD-ROMs that are compendiums of data integrated using GIS to create innovative products such as the World Atlas of Biodiversity (2002); the World Atlas of Coral Reefs (2002); The Penguin Atlas of War and Peace (2003), the Atlas of the New West (1997) and the Atlas of International Freshwater Agreements (2002). These products identify the outcomes of environmental policy through visual representation and tabular descriptions.

Other types of data, such as values, cultural perspectives and indigenous knowledge, require innovative methods for accessing, capturing and using information. An emerging literature in public participation GIS (PPGIS) identifies methods for data collection, data sharing and negotiated research that respects host communities and sensitive information (Craig et al., 2002; Aboriginal Mapping Network, n.d.). Participatory mapping methodology has been developed that capture values and identify areas of potential conflict and integrate community perspectives. Often, these methods are used in land use planning charetes (Craig et al., 2002).

An important characteristic of GIT is that it can be used in both quantitative and qualitative studies. Once the data is collected, it is examined for associations or correlations of factors that are specifically geographic in nature. This examination is conducted through the development of algorithms or equations that define relationships or rules for how data is analysed. Proximity analysis yields descriptive maps that identify how far, how near, or what is the distance to some object or objects of interest. Qualitative analysis is well developed within the GIT literature for reclassification of data into categories (vegetative communities, political affiliations) and ranking of information (income levels, education, soil erosion classes). Quantitative methodologies include integration of data (for example, map algebra), interpolation of new information from existing datasets (kriging and inverse distance weighting, for instance), and modelling future scenarios (Lo and Yeung, 2002).

GIT is particularly salient to international environmental politics and can enhance interdisciplinary research through collaboration on and development of projects that include GIT. Transboundary analysis has been conducted that addresses water flows, animal movements, air pollution, and acid rain. These analyses have focused on the physical parameters of the phenomenon (Reed et al., 1996). The logical next step would be the inclusion of the political landscape and its relationship to environmental outcomes. This would involve integration of physical and political data to create a visual representation of enviro-political issues. One possible topic where these could be integrated is in the study of climate change and climate policy, as climate change models are already heavily dependent on GIT in ways that international relations scholars have not addressed.

Environmental security is another critical area for research and examination. Of particular interest are international relief work and humanitarian aid after and during warfare as well as in response to natural disasters. In 1998, Hurricane Mitch had a devastating effect upon Central America and precipitated the Digital Atlas of Central America that assisted in locating locations for humanitarian aid (Greene, 2000). The tragedy of September 11, 2001, resulted in a heightened awareness of national security. The newly created Homeland Security Department in the United States has identified GIT as critical for emergency response management and for modelling the effects of terrorism on the environment in terms of biological, chemical and nuclear weapons (ESRI, 2001). The unilateral approach to terrorism by the United States provides a rich arena in which to assess and understand alternative geographic perspectives with regard to terrorism and its impact on the environment. Treaty negotiations are another area for exploring GIT and international environmental politics. For example, GIS was an important tool in crafting the partition boundaries and calculating territory apportionments of Bosnia for the Ohio Accord (Greene, 2000).

The drawbacks to this approach are numerous. Data and access to data is a key issue. Data may be sporadically collected and result in inconsistent spatial coverage within a country as well as between countries. There is no standardization between countries for data collection, creation and maintenance, although there are efforts to address this (that is, the Federal Geographic Data Committee, <www.fgdc.gov/>). Changing political boundaries result in obsolete datasets. Different nations have different policies with regard to public access to data that may be considered sensitive to national security interests.

The digital divide is a real phenomenon and is manifested in several ways. There is inconsistent digital data in many developing countries. This is further compounded by a lack of access to hardware and software for both developing countries and marginalized groups such as indigenous peoples.

Conceptually, GIT enforces a particular approach to environmental analysis. In fact, several global studies cite the need to identify key datasets to describe environmental conditions and trends (University Consortium for Geographic Information Science, 1996). Complex environmental issues are necessarily simplified within a computerized environment. Pattern identification is an important output of GIT analysis; yet how well does pattern infer and identify process? What happens to unique data about particular places? Common methodologies in analysis and data construction embed points of views and assumptions that are not transparent and may affect outcomes. Issues related to culture, values and language represent another arena in which GIT is as yet poorly developed.

These drawbacks should not prevent researchers from using GIT in conjunction with international environmental politics. If anything, they represent an exciting new area for research and collaboration across boundaries of various kinds, including disciplinary. Information is increasingly viewed as an important currency of the future. GIT provides a medium for analysis that is dynamic, complicated and difficult. The global political environment presents a challenging but rewarding landscape for these new analytical techniques.

postpositivist approaches to methodology

critical qualitative methods

Critical – or postpositivist – qualitative methods share just one crucial area of intersection with positivist ones: they too find their data in the form of documents and other archival records, interviews and observation. In their research gathering practices, all qualitative methods users will find themselves engaged in activities like open-ended interviews, in observation of meetings and events, or in reading and analysing texts of various kinds. Sharp differences in how positivist and critical scholars design research programmes and analyse their data during and after collecting it distinguish the two approaches and often emerge in the information collection stage as well. Most importantly, most critical scholars question the value and possibility of the entire positivist enterprise of cumulatively building a body of replicable, unbiased, causal explanations of generalized phenomena that constitute an objective reality independent of language or theory. All of them agree that positivist research itself fails to achieve these stated aims. As a group, they work instead to show the power relations inherent in both academic and policy constructions of phenomena, identifying dominant constructions and undermining them. Critical theorists are especially interested in how theoretical and policy problems are defined as problems, and also in how solutions reach the status of solutions (Stevis and Assetto, 2001a). The positivists' claims about the kind of knowledge they produce are seen as themselves an assertion of power, a managerial power that moves from explaining the world to manipulating it (Fay, 1975).

A number of different approaches challenge positivism, including interpretive, poststructuralist, structuralist and feminist approaches. This section emphasizes their common threads, with somewhat more attention to the variants that stress the role of ideas and identity, as virtually all scholars who follow those variants use qualitative methods. The next section separates out the approaches that are more structuralist and materialist as most likely to use non-qualitative methods. These divisions are preliminary and possibly controversial, and deserve much more attention from scholars working in these areas. There are comparatively few explicit discussions of methodology in the critical traditions (with the exception of interpretivism – which is not necessarily critical – and feminism) and none written by IEP scholars. In fact, some critical scholars have argued in the past against methodological discussions for fear of replacing the orthodoxies of positivism with new methodological orthodoxies. More recently others have counter-argued that this is a position that too-easily grants positivists the right to define the standards of good research (see Milliken, 2001, on critical discourse analysis).

Critical research projects often begin by identifying a particular received understanding for further analysis. The careful strictures on representativity and variation control of positivist analysis are not relevant here, as any specific context to be studied will embody the power dynamics of meaning construction, reproducing social orders in some ways while remaining open to the possibility of change through struggle in others. This is a logic of discovery or methodology based on the argument that 'because actions and practices are dependent on rules, embedded in the context of a game [broadly defined], we can discover the structure of these meanings in the context itself' (Fierke, 2001, p. 127) – where context is a referent to what other methods would call a case. Rather than seeing cases as representative examples of generalized patterns in an objective reality, the critical theorist sees specific contexts as providing glimpses of underlying systemic relations. Thus 'case selection' relies on locating particularly telling contexts that illuminate these processes or even in virtuoso analyses of quite common phenomena, as in Fiske's presentation of the eating of a hamburger as 'a practice of a system' (Fiske, 1998, p. 371).

Description is central in critical analysis even though many postpositivists are uneasy about the word 'empirical' and its implicit claim that there are data independent of theory. For the postpositivist, the purpose of description is related to the critical retelling of received constructions:

> Description, while not inherently critical, becomes so if it makes us look again, in a fresh way, at that which we assume about the world because it has become overly familiar … . In this way, new spaces are opened for thinking about the past and the present and, therefore, how we construct the world. (Fierke, 2001, p. 122)

Thus while a positivist might consider two or more alternate readings of a particular event or decision with the intent of figuring out which is correct, critical theorists will often simultaneously engage multiple readings, with the intent of showing the purposes and interests that each reading serves. Some postpositivists, such as poststructuralists, argue that there can never be a final reading and that all readings are indeterminate, while critical and feminist theorists often make politically- and practice-grounded choices about the most useful interpretations of a situation (Fanow and Cook, 1991).

Critical theorists usually use qualitative techniques for recovering alternate descriptions of the world. One common strategy is to use qualitative interviews to recreate the voices of those left out in dominant accounts of policy or economic decisions – to interview women, indigenous peoples or peasants to hear their experiences of large dams, free trade agreements, agricultural policies, and so on (for example, Peet and Watts, 1996). Such interviews can provide direct and vivid narratives that stress considerations beyond national and international power dynamics or global market forces. Direct observation of meetings and negotiations, another classic qualitative method, can also deepen understanding of the different positions and identities expressed by participants as they interact. The very immediacy of these kinds of methods can raise dilemmas for postpositivist scholars, however. Some fieldwork can be dangerous for either the researcher or the research subjects (Peritore, 1990), and the researcher may face both political and ethical dilemmas (Punch, 1986) about how to analyse and publish the research.

Qualitative documentary research is even more common for critical scholars. Documentary research carries less of the political risks of the more direct methods, but the comparative distance of documentary research raises a different issue: most documents are created and stored by dominant actors and so tend to reflect dominant constructions. As a result, postpositivists disagree with the classic use of archives that sees them as a 'repository of "facts"' (Spivak, 1985, p. 248), and question the related project of getting the historical story 'right' in any case. Archives and documents are still important for the postpositivist, however, who uses them while remaining alert to these issues. One analytical strategy is to focus more on unprocessed rather than processed or final documents, as the former are more likely to contain the traces of alternate points of view or evidence that undercuts official positions that interest the critical scholar. Another strategy is to use the official documents and their presentations of reality as analytical subjects in their own right, turning attention to how discourses create systems of meaning, produce actors and other phenomena related to the discourse, and are played out in practice (Milliken, 2001, pp. 138–9). Increasingly, documents of the less-powerful are also available, and can be used to create alternative accounts. For example, international negotiations have been challenged in recent years by the ability of non-governmental actors to disseminate alternate accounts immediately through electronic means, and the internet generally is a powerful tool for those who have access to it – as well as a rich source of data for researchers.

Any issue in IEP could be studied from a critical perspective. Because of other interests and commitments, topics related to international political economy are prominent for critical theorists (Stevis and Assetto, 2001b). The related theme of globalization is also central (see Kütting in this volume), with its companion topic of resistance to neoliberal globalization (Dryzek, 2001; Paehlke, 2001). Critical theorists have researched the relationship between the environment and numerous kinds of social inequity, especially in the guise of studies of environmental justice (see Parks and Roberts in this volume) and a gender perspective on the international environment (Bretherton, 2003). Another line of focus is on the discourses and identities created in environmental politics, with Karen Litfin's study of the discourses around ozone standing out as an early example (Litfin, 1994). Many of these studies look especially closely at the role of science and scientific discourses.

Critical IEP scholars need to go on developing more explicit methodological discussions that would give greater guidance to new scholars and offer clearer standards for evaluating their research. Without such discussions, postpositivist scholars have a hard time responding to the critiques of mainstream scholars and risk leaving the appearance of there being only one – positivist – set of standards for research. Jettisoning the positivist aims of the mainstream means that critical theorists are often working outside what is recognized as good research, especially as indicated by adjectives like 'systematic' or 'scientific'. Thus common criticisms include postpositivists' failure to produce precise, neutral and generalizable findings. As this chapter has shown, positivists and critical theorists have genuinely different approaches to research and it is unlikely that they will ever evaluate research in the same ways. More explicit methodological discussion would at least allow positivists to try to evaluate postpositivist research by its own standards if they were so inclined and would encourage postpositivists to reflect on what kinds of research would move their theoretical agendas forward.

structuralist and non-qualitative critical approaches

A final approach to IEP research joins some of postpositivism's critiques of positivism while accepting more of positivism's research aims and tools, especially the non-qualitative ones. As it is a hybrid approach with little explicit methodological discussion, the boundaries of this kind of approach are not very clear and this section is necessarily more preliminary than the others. Most scholars in this category are a variant of critical theorists, who agree with the argument outlined above that positivists' refusal to take power into account makes their empirical conclusions biased. In their view, positivists implicitly accept the existing international order and do not consider alternatives to it, but seek to manage it. On the other hand, this category of critical scholars is more inclined to both structuralist and materialist views of the world, which limit their attention to language and interpretation compared to other postpositivists (Buttel, 2002, p. 39). Instead, they posit the existence of objective underlying processes that produce empirical outcomes that can be described and measured. This means they are likely to accept the project of cumulative building of causal explanations, so long as the research does consider those underlying relations of power. It also means that they are more willing to use non-qualitative methods, since they believe that there should be visible and systematic material manifestations of the power dynamics they identify.

In empirical research, structural analysis bears a family resemblance to rational choice modelling, although the differences are equally marked. Structurally oriented theorists similarly try to work out abstract models of underlying processes that they see as the ultimate foundation of specific events or occurrences. Critical theorists will often look to the structural dynamics of capitalism as the foundation of their models. In one example of this kind of approach, Marian Miller traces the process and effects of the 'enclosure of knowledge', developing her arguments from a fundamental logic of capitalism: 'Integral to the logic of capitalism is the need for repetitive expansion, involving a continuous reinvestment of profits to create more profits' (Miller, 2001, p. 113). From this starting point, she develops a model of capitalist expansion based on the enclosure of land and knowledge, which she then further develops empirically with studies of these linked processes in Kenya and India.

As with rational choice theorists, some structuralist scholars choose to focus on the logical working out of their models while others are more interested in investigating the ways that abstract processes appear in actual practices – including by using quantitative methods. The approach that has gone furthest in joining such rationalist and structuralist presumptions with non-qualitative research on the environment is world-systems theory. World-systems theory focuses on the structural forces produced by 'the historical legacy of a country's "incorporation" into the global economy' (Roberts and Grimes, 2002, p. 167), as well as other materialist and historical concerns. Researchers have begun to try to link this causal argument to environmental outcomes using a variety of quantitative and qualitative methods in addition to modeling likely outcomes of structural processes (for example, Goldfrank et al., 1999). Roberts (1996), for example, directly tested world-systems theory's central argument about the importance of the legacy of incorporation for political outcomes. He concluded on the basis of aggregate cross-national data analysis that world-system position was a better predictor than gross national product (GNP)/capita of which countries would sign international environmental treaties. Similarly, Bergesen and Parisi (1999) used quantitative data to conclude that economic dependency – a measure of world-system position – was a useful predictor of rates of toxic emissions.

This work lacks the extended methodological discussions and body of actual research that would allow full evaluation of the approach.

In the abstract, efforts to join critical epistemologies and ontologies with methods more often associated with positivism might fall prey to the problems of both, display the virtues of both, or fall somewhere in between.

conclusion

As this chapter should make clear, IEP scholars have a wide variety of methodological tools they can use to study their diverse topics of interest. While these tools are divided between positivist and critical epistemologies and between qualitative and quantitative methods, all of them make distinct contributions to the field as a whole. All also raise some serious potential problems that researchers should try to avoid, where possible. Since these drawbacks often appear as reversed tradeoffs when one method is compared to another, the field – and individual research projects – would benefit from the use of multiple methods to approach any given topic. Beyond its general call for pluralism, this chapter has identified several directions for further methodological development that seem especially likely to be fruitful.

The largest gap calls on critical and postpositivist IEP scholars to engage in more explicit methodological discussions. Such approaches are quite common in the study of international environmental politics, but the methodological discussions have not kept pace. The overarching purpose of these discussions should be to provide guidance for graduate students and other scholars looking to begin critical projects, even if scholars do not aim for a single authoritative approach or set of standards. Among the crucial questions that should be addressed are the purposes and use of empirical materials in such approaches, including directives for the kinds and nature of evidence to be presented. In addition to allowing scholars who use these approaches to proceed with clearer self-understanding, such discussions will allow positivists to evaluate postpositivist research on its own terms and to see the internal consistency of critical theory (or, more likely, variants of critical theory) as an approach to research.

One of the more promising developments in recent methodological discussions has been the effort to consider the relationships of different positivist methodologies to each other. This has been done in two ways. The first way is exemplified in the volume edited by Sprinz and Wolinsky-Nahmias (2004), which consists of explicit methodological discussion of the varying contributions of qualitative, quantitative, and rational choice methodologies to causal explanation and the possible pitfalls of each. This chapter has similar aims. The second approach has taken place within particular substantive areas of research and consists of collective discussion of their progress to date in methodological terms. While this is useful for all approaches, it is especially critical for researchers who aim to cumulate knowledge and causal conclusions. Such discussions have taken place among scholars studying regime effectiveness and are currently unfolding among scholars of environmental security and, to a lesser extent, non-state actors. They should be undertaken in other substantive issue areas as well. These discussions should be ongoing and open, with the rancorous methodological debates among scholars of environmental security providing an example to be avoided. Even when large systematizing projects are providing a period of assessment of the field as a whole, there must be room for alternative projects.

Finally, this chapter has identified several potentially relevant methodologies that have not been used very often in IEP research to date, notably geospatial information technologies and rational choice methodologies. Methodological choices should not be made on the basis of novelty, of course, and these approaches do have the drawback of requiring both methodological training and political data in forms that may not yet exist. Nonetheless, the existence of these tools should be considered by IEP scholars formulating new projects or considering new kinds of methodological training.

notes

1. In our university, students in the College of Liberal Arts must cross the college divide to take GIS courses in the College of Natural Resources. Correspondingly, the graduate course in qualitative methods in Political Science often fills a third of its spaces with Natural Resource College students, who have no such course in their college curriculum. Students in the College of Natural Science have no qualitative methods option of their own, nor do they take the Liberal Arts College courses. All the colleges teach statistical methods, however.
2. Mitchell and Bernauer (1998) provide an explicit enumeration of the steps involved in carrying out a positivist qualitative project in IEP. King et al. (1994) do the same for political science more generally.
3. The chapter is an excellent starting point for relating game theory to IEP, and this paragraph relies heavily on its summary of the potential uses of game theory in this area. Kilgour and Wolinsky-Nahmias also illustrate the specific steps game theoretic modellers take in a hypothetical study of river water sharing.

bibliography

Aboriginal Mapping Network (n.d.) 'Aboriginal Mapping Network – Networking the Aboriginal Mapping Community', <www.nativemaps.org/news/GlobalizationMap.html>.

Bergesen, Albert J., and Laura Parisi (1999) 'Ecosociology and Toxic Emissions', in Walter L. Goldfrank, David Goodman and Andrew Szasz (eds) *Ecology and the World-System*, Westport, Conn., and London: Greenwood Press.

Betsill, Michele, and Elisabeth Corell (2001) 'NGO Influence in International Environmental Negotiations: A Framework for Analysis', *Global Environmental Politics* 1, 4, pp. 65–86.

Breitmeier, Helmut, Marc A. Levy, Oran R. Young and Michael Zürn (1996) *The International Regimes Database as a Tool for the Study of International Cooperation*, Laxenburg, Austria: International Institute for Applied Systems Analysis, WP-96-160.

Bretherton, Charlotte (2003) 'Movements, Networks, Hierarchies: A Gender Perspective on Global Environmental Governance', *Global Environmental Politics* 3, 2, pp. 103–19.

Buttel, Frederick H. (2002) 'Environmental Sociology and the Classical Sociological Tradition: Some Observations on Current Controversies', in Riley E. Dunlap et al. (eds) *Sociological Theory and the Environment: Classical Foundations, Contemporary Insights*, Lanham, Md: Rowman and Littlefield.

Craig, W., T. Harris and D. Weiner (2002) *Community Participation and Geographic Information Systems*, London: Taylor and Francis.

Dimitrov, Radislav S. (2003) 'Knowledge, Power, and Interests in Environmental Regimes Formation', *International Studies Quarterly* 47, 1, pp. 123–50.

Dryzek, John (2001) 'Resistance is Fertile', *Global Environmental Politics* 1, 1, pp. 11–17.

Elster, Jon (1986) 'Introduction', in Jon Elster (ed.) *Rational Choice*, New York: New York University Press, pp. 1–33.

ESRI (2001) 'GIS for Homeland Security: An ESRI White Paper', <www.esri.com/library/whitepapers/ pdfs/homeland_security_wp.pdf>.

Fanow, Mary Margaret, and Judith A. Cook (eds) (1991) *Beyond Methodology: Feminist Scholarship as Lived Research*, Bloomington: Indiana University Press.

Fay, Brian (1975) *Social Theory and Political Practice*, London: George Allen and Unwin.

Fenno Richard E., Jr (1990) *Watching Politicians: Essays on Participant Observation*, Berkeley: University of California, Institute of Governmental Studies.

Fierke, Karin (2001) 'Critical Methodology and Constructivism', in Karin Fierke and K. E. Jørgensen (eds) *Constructing International Relations: The Next Generation*, Armonk, NY, and London: M. E. Sharpe, pp. 115–35.

Finnemore, Martha, and Kathryn Sikkink (2001) 'Taking Stock: The Constructivist Research Program in International Relations and Comparative Politics', *Annual Review of Political Science* 4, pp. 391–416.

Fiske, John (1998) 'Audiencing: Cultural Practice and Cultural Studies', in Norman K. Denzin and Yvonna S. Lincoln (eds) *The Landscape of Qualitative Research: Theories and Issues*, Thousand Oaks, Calif.: Sage Publications, pp. 359–78.

Frank, David John, Ann Hironaka and Evan Schofer (2000) 'The Nation-State and the Natural Environment over the Twentieth Century', *American Sociological Review* 25, 1, pp. 96–116.

George, Alexander L. (1979) 'Case Studies and Theory Development: The Method of Structured, Focused Comparison', in P. G. Lauren (ed.) *Diplomacy: New Approaches in History, Theory, and Policy*, New York: The Free Press, pp. 43–68.

Gibson, Clark C., Margaret A. McKean and Elinor Ostrom (2000) *People and Forests: Communities, Institutions, and Governance*, Cambridge, Mass.: MIT Press.
Gleditsch, Nils Petter (1998) 'Armed Conflict and the Environment: A Critique of the Literature', *Journal of Peace Research* 35, 3, pp. 381–400.
Goldfrank, Walter L., David Goodman and Andrew Szasz (eds) (1999) *Ecology and the World System*, Westport: Greenwood Press.
Goodchild, M. (1995) 'Geographic Information Systems and Geographic Research', in J. Pickles (ed.) *Ground Truth: The Social Implications of Geographic Information Systems*, New York: Guilford, pp. 31–50.
Green, Donald P., and Ian Shapiro (1994) *Pathologies of Rational Choice Theory: A Critique of Applications in Political Science*, New Haven and London: Yale University Press.
Greene, R. (2000) *GIS in Public Policy: Using Geographic Information for More Effective Government*, Redlands, Calif.: ESRI Press.
Grundig, Frank, Hugh Ward and Ethan P. Zorick (2001) 'Modeling Global Climate Negotiations', in Urs Luterbacher and Detlef Sprinz (eds) *International Relations and Global Climate Change*, Cambridge, Mass.: MIT Press, pp. 153–82.
Hardin, Garrett (1968) 'The Tragedy of the Commons', *Science* 162, pp. 1243–8.
Hill, Michael R. (1993) *Archival Strategies and Techniques*, Newbury Park, Calif., and London: Sage.
Homer-Dixon, Thomas (1996) 'Strategies for Studying Causation in Complex Ecological-Political Systems', *Journal of Environment and Development* 5, 2, pp. 132–48.
Kilgour, D. Marc, and Yael Wolinsky-Nahmias (2004) 'Game Theory and International Environmental Policy', in Detlef F. Sprinz and Yael Wolinsky-Nahmias (eds) *Models, Numbers, and Cases: Methods for Studying International Relations*, Ann Arbor: University of Michigan Press, pp. 317–43.
King, Gary, Robert O. Keohane and Sidney Verba (1994) *Designing Social Inquiry: Scientific Inference in Qualitative Research*, Princeton: Princeton University Press.
Litfin, Karen (1994) *Ozone Discourses: Science and Politics in Global Environmental Cooperation*, New York: Columbia University Press.
Lo, C., and A. Yeung (2002) *Concepts and Techniques of Geographic Information Systems*, Upper Saddle River, NJ: Prentice Hall.
McKeown, Timothy J. (1999) 'Case Studies and the Statistical Worldview: Review of King, Keohane, and Verba's *Designing Social Inquiry: Scientific Inference in Qualitative Research*', *International Organization* 53, 1, pp. 161–90.
Midlarsky, Manus I. (1998) 'Democracy and the Environment: An Empirical Assessment', *Journal of Peace Research* 35, 3, pp. 341–62.
Miles, Edward L., A. Underdal, S. Andresen, J. Wettestad, J. B. Skjærseth and E. M. Carlin (2002) *Environmental Regime Effectiveness – Confronting Theory with Evidence*, Cambridge, Mass.: MIT Press.
Miller, Marian (2001) 'Tragedy for the Commons: The Enclosure and Commodification of Knowledge', in Dimitris Stevis and Valerie Assetto (eds) *The International Political Economy of the Environment: Critical Perspectives*, Boulder: Lynne Rienner, pp. 111–34.
Milliken, Jennifer (2001) 'Discourse Study: Bringing Rigor to Critical Theory', in Karin Fierke and K. E. Jørgensen (eds) *Constructing International Relations: The Next Generation*, Armonk, NY, and London: M. E. Sharpe, pp. 136–59.

Mitchell, Ronald (2002) 'A Quantitative Approach to Evaluating International Environmental Regimes', *Global Environmental Politics* 2, 4, pp. 58–83.
Mitchell, Ronald, and Thomas Bernauer (1998) 'Empirical Research on International Environmental Policy: Designing Qualitative Case Studies', *Journal of Environmental and Development* 7, 1, pp. 4–31.
Monroe, Kristen Renwick (1991) 'The Theory of Rational Action: Its Origins and Usefulness for Political Science', in Kristen Renwick Monroe (ed.) *The Economic Approach to Politics: A Critical Reassessment of the Theory of Rational Action*, New York: HarperCollins, pp. 1–31.
Morrow, James D. (1994) *Game Theory for Political Scientists*, Princeton: Princeton University Press.
Morrow, Raymond A., with David D. Brown (1994) *Critical Theory and Methodology*, Thousand Oaks, Calif.: Sage Publications.
Ostrom, Elinor (1990) *Governing the Commons*, Cambridge: Cambridge University Press.
Ostrom, Elinor, R. N. Gardner and J. Walker (1994) *Rules Games and Common-Pool Resources*, Ann Arbor: University of Michigan Press.
Paehlke, Robert (2001) 'Environment, Equity and Globalization: Beyond Resistance', *Global Environmental Politics* 1, 1, pp. 1–10.
Peet, Richard, and Michael Watts (1996) *Liberation Ecologies: Environment, Development, Social Movements*, London and New York: Routledge.
Peluso, Nancy, and Michael Watts (eds) (2001) *Violent Environments*, Ithaca: Cornell University Press.
Peritore, N. Patrick (1990) 'Reflections on Dangerous Fieldwork', *American Sociologist* 21, 4, pp. 359–72.
Peters, B. Guy (1998) *Comparative Politics: Theory and Methods*, New York: New York University Press.
Punch, Maurice (1986) *The Politics and Ethics of Fieldwork*, Newbury Park, Calif., and London: Sage.
Reed, Rebecca, Julia Johnson-Barnard and William Baker (1996) 'Contribution of Roads to Forest Fragmentation in the Rocky Mountains', *Conservation Biology*, 10, 4 (August), pp. 1098–06.
Roberts, J. Timmons (1996) 'Predicting Participation in Environmental Treaties: A World-System Analysis', *Sociological Inquiry* 66, 1, pp. 38–57.
Roberts, J. Timmons and Peter E. Grimes (2002) 'World-System Theory and the Environment: Toward a New Synthesis', in Riley E. Dunlap et al. (eds) *Sociological Theory and the Environment: Classical Foundations, Contemporary Insights*, Lanham, Md: Rowman and Littlefield.
Schwartz, Daniel M., Tom Deligiannis and Thomas F. Homer-Dixon (2000) 'The Environment and Violent Conflict: A Response to Gleditsch's Critique and Some Suggestions for Future Research', *Environmental Change and Security Project Report* 6, pp. 77–94.
Shrader-Frechette, K. S., and E. D. McCoy (1993) *Method in Ecology: Strategies for Conservation*, Cambridge: Cambridge University Press.
Spivak, Gayatri Chakravorty (1985) 'The Rani of Sirmur: An Essay in Reading the Archives', *History and Theory* 24, pp. 347–72.
Sprinz, Detlef F. (1999a) 'Modeling Environmental Conflict', in Alexander Carius and Kurt M. Lietzman (eds) *Environmental Change and Security: A European Perspective*, Berlin and Heidelberg: Springer, pp. 183–94.

Sprinz, Detlef F. (1999b) 'Empirical-Quantitative Approaches to the Study of International Environmental Policy', in Stuart S. Nagel (ed.) *Policy Analysis Methods*, Commack, NY: Nova Science Publishers, pp. 41–64.

Sprinz, Detlef F. (2004) 'Environment Meets Statistics: Quantitative Analysis of International Environmental Policy', in Detlef Sprinz and Yael N. Wolinsky-Nahmias (eds) *Models, Numbers, and Cases: Methods for Studying International Relations*, Ann Arbor: University of Michigan Press, pp. 177–92.

Sprinz, Detlef F., and Yael N. Wolinsky-Nahmias (eds) (2004) *Models, Numbers, and Cases: Methods for Studying International Relations*, Ann Arbor: University of Michigan Press.

Stevis, Dimitris, and Valerie J. Assetto (2001a) 'Introduction: Problems and Solutions in the International Political Economy of the Environment', in Dimitris Stevis and Valerie J. Assetto (eds) *The International Political Economy of the Environment: Critical Perspectives*, Boulder: Lynne Rienner, pp. 1–12.

Stevis, Dimitris, and Valerie J. Assetto (eds) (2001b) *The International Political Economy of the Environment: Critical Perspectives*, Boulder: Lynne Rienner.

Taylor, Steven J., and Robert Bogdan (1998) *Introduction to Qualitative Research Methods: A Guidebook and Resource*, 3rd edn, New York: Wiley.

Toke, Dave (2002) 'Wind Power in UK and Denmark: Can Rational Choice Help Explain Different Outcomes?', *Environmental Politics* 11, 4, pp. 83–100.

Tsebelis, George (1990) *Nested Games: Rational Choice in Comparative Politics*, Berkeley: University of California Press.

University Consortium for Geographic Information Science (1996) 'Research Priorities for Geographic Information Science', *Cartography and Geographic Information Systems* 23, 3, <www.ncgia.ucsb.edu/other/ucgis/CAGIS.html>.

Veregin, Howard (1995) 'Computer Innovation and Adoption in Geography: A Critique of Conventional Technological Models', in John Pickles (ed.) *Ground Truth: The Social Implications of Geographic Information Systems*, New York: Guilford Press, pp. 88–112.

Weyland, Kurt (2002) 'Limitations of Rational-Choice Institutionalism for the Study of Latin American Politics', *Studies in Comparative International Development* 37, 1, pp. 57–85.

Wolinsky, Yael N. (1997) 'Two-Level Game Analysis of International Environmental Politics'. Paper presented at the Conference of the International Studies Association, Toronto.

Young, Oran R. (ed.) (1999) *The Effectiveness of International Environmental Regimes: Causal Connections and Behavioral Mechanisms*, Cambridge, Mass.: MIT Press.

annotated bibliography

Sprinz, Detlef, and Yael N. Wolinsky-Nahmias (eds) *Models, Numbers, and Cases: Methods for Studying International Relations*, Ann Arbor: University of Michigan Press. This volume provides a comparative analysis of positivist methodologies. There are chapters on qualitative, quantitative, and rational choice methodologies as general approaches to the study of international relations. Additional chapters discuss how these methodologies appear in the study of particular international relations topics, including international environmental politics. The parallel presentation of three different methodologies will help scholars to identify the most useful ones for their research agendas.

Mitchell, Ronald, and Thomas Bernauer (1998) 'Empirical Research on International Environmental Policy: Designing Qualitative Case Studies', *Journal of Environmental and Development* 7, 1, pp. 4–31. This article is an excellent introduction to positivist qualitative methodologies. In addition to describing the aims of the qualitative approach, the article includes discussions of process tracing and generalization. Perhaps most usefully, it lays out a series of concrete research design tasks for undertaking research on international environmental policy through qualitative case studies.

Fierke, Karin, and K. E. Jørgensen (eds) (2001) *Constructing International Relations: The Next Generation*, Armonk, NY, and London: M. E. Sharpe. The chapters by Fierke and Milliken in this volume provide a good starting point for identifying a critical or postpositivist approach to the empirical study of international relations. While the chapters do not focus on environmental politics in particular, they still provide useful guidance for projects of this kind.

Craig, W. J., Trevor M. Harris and Daniel Weiner (2002) *Community Participation and Geographic Information Systems*, New York: Taylor and Francis. This book is a collection of articles that detail the practice, methodology and application of geospatial information technologies and community action. It provides a description of the conceptual basis for the interaction of communities and technology for environmental action and empowerment. Several case studies are also included.

part ii
the forces that shape international environmental politics

5
the environment as a global issue
gabriela kütting and sandra rose

This chapter presents the debates involving globalization and the environment. A general introductory discussion of the phenomenon of globalization is needed before the relationship between globalization and environment can be examined. There is no single discourse on globalization, or on globalization and the environment. Rather, globalization is a multifaceted phenomenon and this is reflected in the literature on globalization as well as on globalization and ecology. In order to structure this chapter, we have used the traditional distinction between political, economic and sociocultural globalization as heuristic categories. However, we have added an introductory historical section in order to place our understanding of globalization in a historical context. We conclude by illustrating our analysis with a case study on trade and agriculture. Throughout this chapter, we address a variety of schools of thought and discuss their strengths and weaknesses. Our understanding is that most attempts at conceptualizing, or theorizing, about globalization from an international political economy (IPE) perspective tend to sideline the environmental and social consequences of globalization and that these issues are usually treated as part of an analysis of global civil society and new social movements. As a result, we find that only transnational actors representing social and environmental issues are incorporated into the analysis rather than the structural and systemic forces, as well as constraints, within which actors operate. Thus an integrated study of environment and economy eludes much of social science analysis and is the biggest challenge faced by environmental social science researchers – whether in the field of globalization or in general – at the beginning of the twenty-first century. This position necessarily influences the way we interpret existing writings on globalization.

globalization as structure: the historical dimension

Debate about the origins and historical beginnings of globalization can generally be traced back to the author's disciplinary focus. Whilst globalization is generally perceived to be a recent phenomenon, its starting date is often placed much earlier – be it the beginnings of trade, modernity, early capitalism, late capitalism, or the collapse of the Bretton Woods system. The structural origins of globalization are also contested: they are political, economic or sociocultural. Thus, a starting point for any text on globalization has to be an analysis of this literature and an explanation of the wide-ranging differences in the definition of both globalization and its historical origin. This needs to be followed by a discussion on the origins of environmental degradation in order to understand the relationship between environment and globalization.

Globalization both as a concept and as a process is a contested term – its usage has become generally accepted but there is no generally accepted definition of what constitutes globalization and/or its empirical features.

Having said that it is generally agreed that in the 1970s fundamental changes in the way the international political economy is organized led to a more global approach both in IPE/GPE (global political economy) but also environmental studies (Lipschutz and Mayer, 1996; Mittelman, 1997; Scholte, 1993; Strange, 1996). However, it is debatable whether these changes are deserving of the term 'globalization' that has been assigned to them. Amongst IPE/GPE scholars the age of globalization is taken to be the post-Fordist era that has engendered economics of flexibility, increasing trade liberalization, financial deregulation, an increasingly global division of labour and a transnational capitalist class (Lipiètz, 1997; Sklair, 2002; Strange, 1996). Although the phenomenon of globalization itself is contested, these changes in the international political economy and other globalizing tendencies are not. So in a way there are two parallel debates about globalization as a phenomenon: whether there is such a thing and whether changes in the global/international political economy during the past 30 years are a historically new phase or the continuation of a historically rooted phenomenon (Hirst and Thompson, 1996; Schwartz, 2000).

Some academics argue that globalization started with the formation of societies and the social relations between them and that we have now entered a higher stage of a linear, historically-determined process (Frank, 1998). Others would see globalization as coexisting with capitalism or modernity and again, depending on one's definition of capitalism and its beginnings, different start dates are given (Cox, 1996; Robertson, 1992; Wallerstein, 1995). This school of thought sees recent globalization as a higher or even the latest stage of capitalism – an acceleration of an already established phenomenon. Last, some researchers would suggest that globalization and the socioeconomic changes witnessed since the 1970s are an entirely new phenomenon that is separate from the other processes mentioned above (Cox, 1997; Mittelman, 1997).

Writers such as Mittelman and Cox focus on the economic side of globalization and see these economic changes as the driving factor behind other politically and culturally motivated global changes. These economic changes are empirically observable phenomena and have altered the shape of the global political economy, in fact transforming an international into a global political economy. In this view, these developments have also led to institutional developments, such as alterations in World Bank and International Monetary Fund (IMF) policies, changes in the General Agreement on Tariffs and Trade (GATT) and the establishment of the World Trade Organization (WTO), shifts in the way the United Nations is used as a political instrument by states and the rise of global civil society, as well as some change in the role of the state in the international system (Hoogvelt, 1997).

However, it cannot be argued that these emerging processes have led to a fundamental reorganization of the international system and to the emergence of a global system despite the wealth of literature suggesting so (Mittelman, 1997; Prakash and Hart, 1998). States are legally still the only sovereign actors in the international system and sanction fundamental changes, although the international system has become much more pluralistic in nature. The role of states has certainly changed in recent times and they are engaged in more consultation exercises and are more constrained in their choices, as are multilateral funding agencies (and incidentally also multinational corporations). Nevertheless, the structural changes experienced over the past 30 years in the international system have not made states the servants of multinational corporations as is often maintained, nor have civil society organizations moved into decision- and rule-making positions. There may be anecdotal evidence to the contrary but no case for wholesale change can be made.

Therefore the fashionable argument about globalization being an entirely new world order is not pursued as a viable argument here. This approach is largely based on cultural and social ideas of globalization and the reach of better transport and communication links which decrease physical and virtual distance between places and people (Robertson, 1992; Scholte, 1993). However, communication and travel methods have continuously improved over history and there is no reason to suggest that the changes in the past 30 years are so fundamentally different to what existed before that they are deserving of an entirely new term. In response, cultural globalization writers argue that the globalization process has been in motion for hundreds of years. This still does not address the point, however, that modern transport and communication means are only available to a relatively small elite of the world population and are by no means global in reach. Thus the *social relations* of transport and communication have not changed although the *spatial reach or speed* of these communications means has increased. Vast parts of developing countries do not have regular electricity supply or telephone access, therefore talking about a global village is an exaggeration.

globalization as structure: the environmental dimension

In this section we discuss the rise of a global consciousness of environmental problems. This is not a globalization of the environment but rather an account of the historical evolution of the concept of a global environment and how it impacts on and informs the subject of international/global environmental politics. Therefore, there is no environmental/ecological globalization as such but rather the globalization of the perception of the environment, which will be discussed in more detail in the following sections (see Stevis in this volume).

The global nature of environmental degradation can largely be linked to the rise of the fossil fuel economy and the decreasing distance of time and space in the relations between different parts of the globe (Daly, 1996). However, historically, the *understanding* of the global concept of 'one earth' can be traced back to developments in the 1960s and early 1970s (Conca et al., 1995). On the one hand, the doomsday feeling of *one* planet reaching its limits was reinforced by the first photographs of the planet from outer space, which showed the global rather than national nature of ecosystems at a time when the Club of Rome, in the *Limits to Growth* report, pointed out the limits to growth of existing consumptive patterns in industrialized countries (Meadows et al., 1972). On the other hand, seminal pieces of literature such as Rachel Carson's *Silent Spring* (1962) or Paul Ehrlich's *The Population Bomb* (1968), as well as Garrett Hardin's work (for example, 1968, 1974) on the tragedy of the commons and the lifeboat ethic (whose messages have been subject to violent criticism since), influenced the way environmental thought became a political priority in international politics from the 1970s onwards. Like James Lovelock's *The Ages of Gaia* (1988), these texts influenced the study of international and/or global environmental politics with their holistic and thus also global view of ecology or environment and social interaction.

Traditionally, political economy analysis in international relations (IR) goes back to the beginnings of modern capitalism and the social relations that evolve in this period and then develop and change throughout modernity (Gill, 1996; Hoogvelt, 1997). Although the relationship between environmental degradation and a global economy was identified relatively early on in the environmental thought literature, the environment has not formed part of mainstream political economy analysis within IR. It has entered the field through radical/political/historical ecology and types of ecological economics analysis (Daly, 1996; Dryzek, 1997; Eckersley, 1995; Merchant, 1992). These approaches usually define the rise of modern capitalism as the point in history when society became increasingly alienated from its physical environment but perceived itself to be mastering, or harnessing it – a process that became more intense as modern capitalism became more sophisticated. Basically, the rise of modern capitalism, the Enlightenment, Newtonian science and the industrial revolution acted in concert to bring about a change in society–environment relations as humans in the core economies saw themselves as increasingly mastering nature rather than being dependent on and dominated by it (Merchant, 1992). This in turn led to the perceived notion of decreasing dependency on the environment which resulted in its neglect through lack of understanding of ecological processes and their significance for life on the planet. There is an underlying if unstated concept of a global environment and a global outlook in this literature.

Perhaps the best way to describe how economic organization affects the environment is to have recourse to the idea of Daly's steady-state economy. He describes the two visions of the economy, that of standard economics and that of the steady-state economy:

> For standard economics, ... the economy is an isolated system in which exchange value circulates between firms and households. Nothing enters from the environment, nothing exits to the environment. It does not matter how big the economy is relative to its environment. For all practical purposes an isolated system has no environment. For steady-state economics, the preanalytic vision is that the economy is an open subsystem of a finite and non-growing ecosystem (the environment). The economy lives by importing low-entropy matter-energy (raw materials) – and exporting high-entropy matter-energy (waste). Any subsystem of a finite non-growing system must itself at some point also become non-growing. (Daly 1992, p. xiii)

The implication of the steady-state economy approach is that it is physically impossible to continue extracting resources and creating waste while expecting unlimited economic growth. In addition, current economic organization of society disregards the first two laws of thermodynamics which determine the existence of energy on the planet. The first law states that the amount of existing energy and matter is constant, that is, cannot be changed. The second law of thermodynamics argues that the state and quality of existing energy can change. In industrial society existing energy gets transformed into 'waste', that is, a form of energy that cannot be reused, thus in effect diminishing the amount of energy available. In addition to this physical side of environmental change, ecological economics writers such as Martinez-Alier (2002) and Daly argue that conventional economics neglects the moral side of environmental exploitation, being too fixated on markets and efficiency rather than on connections. This links up to the environmental justice literature (Martinez-Alier, 2002; see Parks and Roberts in this volume).

This argument combines well with the ecological world-systems theory literature, which also focuses on global structures and their relations with environmental degradation (Chew 2001; Goldfrank et al., 1999; Hornborg, 1998). The main argumentative thrust of this type of analysis suggests that the rise and fall of world civilizations can be traced to environmental degradation as a main contributory factor of the decline of empires or large powers. Thus the nature of capitalism can be understood through the social relations of production, labour and the environment. Ponting (1991) in his environmental history of the world advances a similar argument, although not couched in theoretical terms. These are views of history that integrate an environmental or ecological perspective into predominantly social historical accounts.

The main argument of Chew's thesis, for example, is that different phases in world history and the rise and fall of trading relations can be analysed from a historical materialist perspective as done by Wallerstein or alternatively, Frank and Gills, and focus on the social relations of production. However, these approaches neglect the relationship between nature/natural resources and the material basis of production. In fact, the demise of most empires or large powers also coincides with a decline in the natural resource base through overexploitation or other exhaustion. In fact, forensic research suggests that even the two historical periods of dark ages are linked to the depletion of the natural resource base and this has been documented in carbon testing from these areas (Chew, 2001).

This type of approach integrates the environment into political economy in a holistic manner combining social with environmental analysis. Environmental degradation has always existed under systems of mass production and modern capitalism is a qualitatively different phase of this problem, while environmental degradation under globalization is qualitatively different again.

The points made so far have focused on the structural dimension of globalization and the relationship between globalization and environment. There are a variety of ways in which structure in the international or global system can be perceived and few of them take account of the environmental dimension. The literature on environmental thought often has much more pertinent contributions to make to the understanding of a global environment. The following sections will bring together the structural points made with the issues of agency more usually associated with globalization. This is done by making a heuristic distinction between political, economic and sociocultural phenomena.

globalization as a political process

In IR, the main dimension of political globalization is usually identified as the changing role and nature of actors in the international system as well as the increasing institutionalization of the international system and rise of global institutions. This part of the discussion will be rather short so as not to reproduce material from the chapter on global governance (see Biermann in this volume). Instead, this section will focus on a discussion of the effects of a more global (meaning in a geographical but also transnational sense) participation in the institutional architecture and ensuing equity issues from a North–South perspective. This means that the role of the state as well as the rise of transnational actors will be discussed here. To quote Prakash and Hart:

> Ipso facto, globalization refers to processes that potentially encompass the whole globe. The process does not have to have actually encompassed the whole globe to be associated with the phenomenon of globalization but there has to be at least a potential for its omnipresence. Thus, one should be able to identify the degree to which a particular globalization process has actually attained globality. (1998, p. 3)

global governance and the state

Political globalization does not necessarily mean that any intergovernmental or transnational organization needs to operate globally, but rather that new processes and new agency and structural developments have a global impact. A large part of the academic debate on global governance focuses on the changing role of the state in the international system and its potential replacement by other actors and the decline of sovereignty (Baker, 2002). In the words of Lipschutz:

> One of the central issues facing human civilisation at the end of the 20th century is governance: Who rules? Whose rules? What rules? What kind of rules? At what level? In what form? Who decides? On what basis? Many of the problems that give rise to questions such as these are transnational and transboundary in nature, with the result that the notion of global 'management' has acquired increasing currency in some circles. This is especially true given that economic globalization seems to point toward a single integrated world economy in which the sovereign state appears to be losing much of its authority and control over domestic and foreign affairs. (1999, p. 259)

The debate about the loss of sovereignty of the state is one of the cornerstones of political globalization studies although from a critical global political economy perspective, it makes more sense to talk of a transfer of power or political division of labour (Mittelman, 2000). Although it may seem that states are losing power, they are the founders and funders of the very institutions which are supposed to challenge the power of the state. It seems that rather than declining, the power of the Northern or industrialized state is actually fortified through the global economic governance institutions, which, at the end of the day, represent its interests. It is actually the power of the developing country state that is being undermined by global governance or rather prevented from developing in the first place as most developing countries have never been in a position of structural power. Therefore it could be argued that the global economic institutions are a form of structural power in Lukes' terms rather than the decline of the power of the state (Lukes, 1974).

The global politico-economic framework legitimized by states and global institutions provides a formidable system for the efficient transfer of resources from the periphery to the core (Cox, 1996; Mittelman, 2000; Saurin, 1996b). Any environmental governance efforts are subordinated to this goal and do not generally form part of IPE or GPE analyses of globalization practices. Instead, global environmental governance is analysed through the institutional literature (Bernstein, 2001; Young, 1997, 2002; see Biermann in this volume). Despite the increasing environmental rhetoric in the form of the sustainable development discourse (Redclift, 1987 – despite its age, still one of the best books on the subject), there has been no real attempt to take on board the strained nature of environment–society relations and consequently there has been no real effort to accommodate environmental with social needs – with a few notable exceptions (Gillespie, 2001; Lipschutz, 1999; Saurin, 1996a; see Bruyninckx in this volume). However, the core of the literature dealing with political globalization and environment can be found in the civil society literature and the regime literature (see Betsill and Wettestad in this volume).

global civil society and transnational actors

The increasing disappointment with state-sponsored policies and international organizations has led to the rise of new actors in the form of transnational protest movements and the rise of non-governmental actors in both the civil society and corporate fields (although a strict definition of civil society includes the corporate sector, modern usage of the term suggests a distinction (Betsill in this volume). These civil society actors have been busy creating additional and alternative forms of global governance which have become part of the global network of regulations, norms and ethics (Ford, 2004; Wapner, 1995; Willetts 2002). Non-governmental organizations (NGOs) are ever more important participants in international environmental institutions which were previously state-only activities (Breitmeier and Rittberger, 1998; Raustiala, 1997). In line with this argument, environmental NGOs (ENGOs) have gained more influence on states as well as public affairs by working within and across societies themselves (Wapner, 1995). Wapner argues that ENGOs are political actors in their own right and that transnational activist societal efforts should be seen through the concept of 'world civic politics'. The activists today do not target their efforts directly at the state level, but work through transnational economic, social and cultural networks to achieve their aims, which might include the empowerment of local communities. Rohrschneider and Dalton (2002), for instance, found empirical evidence of a relatively dense network of international action by green groups and a substantial resource transfer from green groups in the OECD countries to those in the developing countries.

In some cases, civil society actors contribute to and help shape international governance, and sometimes transnational governance exists in addition to international governance (Keck and Sikkink, 1998; Princen and Finger, 1994). This governance is institutionalized in the form of hundreds of international environmental agreements and voluntary arrangements, covering all sorts of regional and global issues ranging from the Climate Change Convention to forest stewardship councils.

These form the main subject matter of the study of the environment in international and global politics. However, it could be argued that the heart of the matter is not environmental governance per se but the relationship between economic and environmental governance and the lack of environmental provisions in the economic sphere or the precedence that economic institutions and regulations take over environmental ones (Conca, 2000; Gillespie, 2001; Jeong, 2001). This status quo determines that environmental governance can only ever be a sideshow of limited environmental effectiveness.

Both environmental and non-environmental organizations play a role in global environmental governance. There are a number of global governance organizations which are closely related to global environmental governance, such as the environmental elements of the UN. More importantly, non-environmental organizations such as the WTO, the IMF and the World Bank have a strong impact on environmental governance through their economic, trade, investment and development policies (Clapp, 2001; Dauvergne, 2001). We will not cover the role of international environmental agreements here, which is mostly put into practice through UN agencies, as this has been done in great detail elsewhere (Young, 1997, 2002) and it has been demonstrated that these agreements are fairly marginal to global environmental governance from an ecocentric perspective (Kütting, 2000).

Global economic and political governance, which structurally determines environmental governance, leads to the sidelining of ecological considerations and a lack of understanding of environment–society relations. This means that global governance policies do not take into account the social dependence on ecological foundations. Thus, the absence of environmental priorities in the WTO, the main organization of global economic governance, is more indicative of global environmental governance than the various international environmental agreements on particular issue areas which are negotiated under the constraints of this global institutional economic framework (Williams, 2001). Likewise, the structural adjustment policies of the World Bank and the IMF have a strong environmental impact as a result of the limited attention apportioned to environmental considerations. Although the World Bank has put environmental policy high on its agenda, this has been done in a sustainable development framework which assumes unlimited growth and denies the basic realities of environmental equity and resource access (Miller, 1995).

Neoliberal institutions, such as the World Bank or the IMF, may be committed to the alleviation of poverty and environmental degradation.

However, there is an unspoken assumption that this can be done without structural change (we do not count structural adjustment policies as structural change). In fact, social justice and equity are quite deliberately not major priorities in neoliberal thinking because of the importance of the competition principle. It could well be argued that an excessive pursuit of equity or social justice could be pereceived as a hindrance to competition. However, the implications of this are economic just as much as political and will be discussed in the next section.

globalization as an economic process

Since studying economic globalization from an IR perspective in many ways replicates the political dimensions of globalization, this section necessarily deals with a similar subject matter as the previous section, albeit from a different perspective. International economic integration institutions such as the World Bank, the IMF, the WTO and regional economic integration organizations have changed the political and economic landscape from the 1970s onwards. Most of these institutions have, of course, existed since the end of World War II and were architects of the postwar political and economic order. With the collapse of the Bretton Woods system in the 1970s and the ensuing changes in the production structure, these institutions have also gradually experienced the changes in the international economic order to which they have contributed.

The field of trade is perhaps the area in which most of the globalizing change has taken place and will be discussed in detail in the chapter on international political economy (see Clapp in this volume). Although there have been historical periods of trade liberalization before, notably in the nineteenth century, the degree of institutionalization of the present trade liberalization era is quite unprecedented (Hirst and Thompson, 1996; Schwartz, 2000). The WTO is perceived as one of the main culprits of the negative effects of economic globalization by the media and the general public, largely because of the attention it has received from protest movements. Both GATT and the WTO have changed the landscape of international trade in an accelerated movement.

These developments show that the liberal and neoliberal approaches to globalization are very much the hegemonic approach to the global political economy as they are so much ingrained in actual practice and are the embedded value structure of the global institutional architecture. But where does that leave the environment? Liberal approaches have a strong environmental component but this is related to wealth creation. For example, the Brundtland Report (WCED, 1987) equates poverty with environmental degradation and sees the solution to environmental problems in the increase of wealth within a society, which will then give society the financial means to put regulatory structures in place. Of course this approach denies the essential link between environmental degradation and wealth creation as it is after all excessive consumption and use of resources and sinks that cause most environmental degradation. Thus liberalism and an advanced economic society generate the financial resources necessary to manage the environmental problems they generated in the first place with their excessive wealth generation.

This interpretation of structural change in the political economy relates to the discussion of environment–society relations in a global environment, earlier in this chapter. This structural dimension will be elaborated with a brief overview of the global accounts of economic globalization and environment, namely the concept of ecological modernization after the exploration of the impact of capital on the environmental system.

In line with the above, Sachs (2000) argues that since the 1980s the image of the planet has turned into an emblem of transnational business which offers the world unrestricted movement, promises access in every direction, and seems to present no obstacle to expansionism other than the limits of the globe itself. The image of the globe symbolizes the limitation in the physical sense and expansion in the political sense which makes it a valuable symbol that can be used by environmental groups and transnational corporations alike. This is at the heart of the basic conflict in our epoch, according to Sachs, which includes, on one side, the ecological limits of the earth and on the other side the dynamics of economic globalization that push for the removal of all boundaries associated with political and cultural space. The flows of capital and investment, as well as the removal of boundaries to these flows through organizations such as the WTO, are of prime importance for a positive development of economic globalization. According to the utopian model of economic globalization, only supply and demand are supposed to speed up or slow down these flows. Efficiency gains (in economic terms) can be achieved, according to Sachs, through technical advances which allow movement of capital from one place to the other without delay. (The most extensive globalized market is the one that deals with the least physical of all commodities – money.) The flow of capital and investment in countries enable the economies there to grow but at the same time causes a higher usage of resources in that particular country; and more resources in general if the volume of economic activity expands. Sachs argues that monetary growth is always intertwined with material growth whereby the favoured targets for investments are raw material extraction or energy infrastructure, all of which use up natural resources extensively. It has been observed that the absolute volume of flow of resources has been increased where the flows of capital have increased. However, other approaches to environment and the global economy are more optimistic about the possibilities of ecological improvement.

The ecological modernization school of thought (Mol, 2001) recognizes that the technological advancements of modernity and globalization have led to a global environmental crisis. In this respect technology is perceived to be a good starting point for dealing with this crisis (Schuurman, 2001; Spaargaren et al., 2000). It sees technological solutions to problems created by technology as the way forward and cites the examples of Germany and the Netherlands as cases where ecological modernization has been successful in moving towards a more ecologically conscious society. However, it does not address the root causes underlying technologically harmful industries and the equity/justice argument (Blowers, 2003) nor does it offer any solutions as to how ecological modernization could become a viable model for developing countries. Thus it is predominantly still an environmental management approach.

The overview of the study of environment and globalization from a political and economic perspective has demonstrated quite clearly that all political aspects are influenced by economic aspects and that all economic aspects of globalization are heavily influenced by political decisions. The next section will superimpose a sociocultural perspective.

globalization as a social and cultural process

The cultural aspect of globalization has a historical and material perspective in that transnational cultural influences can be traced back to the evolution of empires, the spread of organized religion, forces such as colonialism, ideological diffusion, and so on (Jameson and Miyoshi, 1999; Said, 1994). These social forces complement economic processes of globalization and are intrinsically linked with them. The nature and form of capitalism are culturally as well as economically and politically informed. An emphasis on culture is predominantly found in reflectivist and particularly postmodern approaches to social science rather than in rationalist theories. These approaches are geared towards exploring the socially constructed nature of social practices and are therefore very open towards the concept of culture as an influential explanatory variable as culture can ideally explain different social values across spatial or temporal boundaries.

On the subject of culture and environment, the literature in the field of IR and globalization has been extremely limited and the subject has mostly received attention in a sociological or postmodern/poststructuralist context (Conley, 1997), or also through the construction of global NGO networking as 'global culture' (Boli and Thomas, 1999). However, for an environmentally informed study of globalization processes, an awareness of the culturally perceived notion of environment is very necessary. Definitions of environment, environmental change and environmental degradation are culturally dependent as environmental values and environment–society relations are culturally, historically and geographically specific. Likewise, the social construction of environment is not a global but rather a culturally specific undertaking which is linked to the direct experience of environmental degradation. At the moment, both the cultural and environmental dimensions of globalization studied from a political science perspective are underdeveloped and need to be integrated further. However, this also means that the integration of culture and environment from an environmental–global perspective is the least developed of all.

Culture is a fuzzy concept to define and mainly refers to social practices and beliefs that are rooted in particular forms of religious, social, economic and political practices predominant in a society. It is also a predominantly Western concept and can be found in Western literature on sociology and social and political theory. In fact, cultural globalization is often equated with the spread of Western social and cultural practices at the global level, often through multinational corporations such as 'McDonaldization' or the spread of Western media and entertainment culture. However, such a concept of cultural globalization would not constitute globalization but rather Westernization. In fact, in the West/North consumption is seen as a cultural phenomenon as it is the key instrument through which culture is represented and reproduced (Miles, 1998, p. 3; Sklair, 2002: chs 5, 7). This section will consider consumptive and postmaterial issues as manifestations of global culture and environment.

culture, consumption and environment

Another aspect of cultural globalization is the increasing popularity and spread of particular fashions that lead to a global consumer class. It is manifested by the spread of brand names which are globally known and recognizable (Klein, 2000). This is both an economic and a cultural phenomenon and a form of manufactured culture. By this I mean it is an attempt to manufacture a particular type of 'global citizen' to which it is desirable to aspire. This is not an evolved form of culture but an attempt at artificially creating a transnational cultural type which is predominantly characterized by the consumption of global brand names and a particular form of news and entertainment (Sklair, 2002; Klein, 2000). However, as noted, this is not global culture but, rather, the spread of one particular type of Western consumption pattern. It is also not a cultural but an economic phenomenon. This point illustrates very clearly how interlinked economy and culture are.

The social and structural origins of environmental degradation can be found in the excessive consumption of the planet's resources. The dominant neoliberal or even liberal approach in global management institutions is based on the assumption that the current standard of living enjoyed by the richest 20 per cent of the world population can be extended to the whole globe (Sachs et al., 1998). In terms of resource availability, this is clearly problematic. However, the rhetoric at the recent World Summit for Sustainable Development in Johannesburg in 2002 suggested that economic globalization is the way out of poverty and environmental degradation for the world's poorer countries (Wapner, 2003, p. 6).

The argument that excessive consumption leads to environmental degradation is not a new argument and dates back to the late 1960s and early 1970s and the beginnings of the environmental movement and the *Limits to Growth* report by the Club of Rome (Meadows et al., 1972). It is based on the 'need not want' philosophy. The early environmental movement in the 1970s questioned the ideology of consumerism in the period of unlimited expectations of the late 1960s and argued that the ideology of wanting more and more was fundamentally flawed and would lead to the ecological collapse of the planet. Rather, there should be an ideological shift to considering what people actually *needed* for a fulfilled life rather than *wanted*, that is, a questioning of the ideology of unlimited economic growth and of an expected rise in the standard of living of those who had already achieved a high level. This movement coincided with the first oil crisis and the first United Nations Conference for the Human Environment in Stockholm in 1972. The idea that there are insufficient resources has often been discredited with the discovery of new oil fields and the introduction of more energy-efficient technology and the perceived success of ecological modernization. These are problems of distribution and access to resources rather than availability. Therefore the concern about running out of resources and the *need not want* campaign have lost their immediate urgency.

Given the intrinsic linkage between political economy and cultural phenomena, it is very difficult to make a case for either factor as a singular engine for world development. Since economic, political and social practices are culturally informed and as culture is determined by the economic and social make-up of a society, the two can only heuristically be separate. For this reason, many historical materialists or political economists have tended to ignore the subject of culture as it is seen as being covered by the focus on political economy. However, there is more to culture than that and global culture is not the spread of Starbucks coffee and Nike trainers, the global popularity of Harry Potter or the use of English as a global language. Likewise, global culture is not the availability of Ethiopian food in US restaurants or the global appeal of reggae or salsa music. The spread of Western consumer goods cannot be described as global culture and the use of English as a global language is historically determined. Culture refers to social practices and how they influence belief systems and therefore political and economic practices.

An alternative view of culture and environment, similar to the ecological modernization view, is Ronald Inglehart's idea of a postmaterialist society (Abramson and Inglehart, 1995; Inglehart, 1977). Postmaterialist societies are those that do not need to worry about their economic well-being, having achieved a high level of economic security and they now turn to postmaterialist values such as environment, feminism and well-being, focusing on 'the project of the self'. With respect to environmental globalization as a cultural and social process, norms can be amplified and should include postmaterial values. However, ultimately these postmaterialist values manifest themselves in very materialist consumptive patterns in a small part of the industrialized world.

culture and global environmental change

The cultural dimension of environmental change or degradation is seldom discussed in global institutions or academic texts. Thus a universal or global understanding of what constitutes 'the environment' is assumed. The economic tools used under modern capitalism and especially the production and financial patterns practised under globalization have a large-scale systemic and often irreversible impact on the environment, leading to probing questions on the ethics of such practices in environmental (and of course also social) terms. Under globalization the increasing time–space distanciation of rights and responsibilities for environmental degradation through the dependence on participation in the global political economy is unique. One example of this is the role of indigenous knowledge and use of indigenous plants for economic purposes:

The rights to environmental knowledge developed and used by indigenous peoples and rural farmers have become a highly contested issue as a result of the growth of multinational biotechnology firms and their search for scientifically unknown, highly valuable plants, which has taken them to remote parts of the globe and placed them in contact with the local people. One response by local groups has been to issue calls for payment of royalties for use of their knowledge, and a more anthropological one has called into question the clash of cosmovisions whereby western legal concepts of originality and innovation embedded in intellectual property law are not only sharply at odds with their indigenous counterparts, but are primed to serve the interests of biocolonialism. (Little, 1999, p. 267)

Traditionally, governments have taken over the role of safeguarding natural resources and sinks and protecting their citizens from environmental harm. Nationally this happens through the rule of law. Internationally, this has been effected through the role of international environmental agreements and through other more private forms of regulation. As governments are the appropriate legal channels through which such interests can be represented at the international level, there does not seem to be an ethical problem with this form of organization. However, under globalization there have been practical (but not de jure) changes in the role of governments at the international level.

Thus it can be argued that the social and cultural aspects of globalization have led to a universal concept of environmental degradation and environmental institutionalization determined by a Western/Northern understanding of environment. This becomes very clear in the case study.

case study: agriculture and trade

Trade in agriculture is a resource issue that addresses all of the dimensions of globalization and environment discussed above. Developing countries are brought into the globalizing political and economic frameworks through agricultural trade. This has a resource issue dimension approached from an environmental point of view but also is a problem of distribution and access to resources rather than availability. The study of the agricultural sector and trade shows the local–global linkages between Northern consumption and its social and environmental consequences in developing countries dependent on agricultural exports. Many African and Caribbean states have drastically increased their agricultural production in the hope of using the income from these cash crops for debt servicing or wealth generation in general. These agricultural strategies have serious social and environmental consequences for the region and are intricately linked to higher demand in the Northern hemisphere where the consumption of fashion is ever increasing in velocity.

There are serious consequences for food security and environmental degradation as well as general equity questions. The historical dimension will be discussed through a comparative analysis with agricultural trade under colonialism and then the current global trade framework with its political and economic manifestations will be used to illustrate the structural and transnational dimensions of North–South relations under globalization.

Agriculture has always been an ideal economic activity for trade because geographical limitations obviously put natural limits to what can be grown where. It has become even more important as a trade area with the rise of technological innovations that increase the speed and improve the cooling methods available, thus making it viable to transport, for example, apples or bananas halfway around the world. Agriculture was the profit-making activity that sustained colonialism, through, for example, cotton and sugar plantations (Isaacman and Roberts, 1995; Mintz, 1985). Since World War II the velocity of increasing yields and the variety in the industrialized countries' shops and kitchens has increased exponentially (Miles, 1998). This has also had an enormous impact on the environment. The amount of land used for agriculture also increases steadily every year and the methods of agriculture have changed. Although none of this can be directly reduced to the globalization process, all the issues identified here as the roots and symptoms of globalization can also be found in the changing nature of the agricultural process. Thus, before entering into a detailed discussion, a short list of the environmental issues relating to the agricultural sector will be highlighted here. They include environmental problems on site and the problems relating to distribution of produce.

agriculture and ecology

Environmental problems associated with agriculture can generally be divided into four categories: chemical use (pesticides, fertilizers and defoliants), water (shortages due to irrigation projects and water pollution from agricultural runoff) and soil (degradation through sterilization or erosion ultimately leading to desertification). Thus soil degradation is the consequence of excessive chemical use but also general excessive use.

Fourth, a more recent problem is the genetic modification of seeds and crops and the proliferation of non-traditional crops.

Along the product chain, agricultural problems relating to environment and globalization are the increasingly global nature of markets and the increasing importance of transportation (McMichael, 1995). More transport means more environmental degradation due to more energy being extended to bring product and consumer together. Another environmental dimension is the changing modern diet and the increased consumption of the individual consumer in the industrialized countries. For example, the average North American or European consumes several times the amount of meat their ancestors only a couple of generations back did. In agricultural terms, this means that more and more land is used for the production of livestock feed. The consequences of this are summarized very much to the point by the World Environment Atlas:

> Meat eating significantly affects global food production, contributing to food scarcity for many and glut for the few. There has been a fundamental shift in world agriculture this century from food grains to feed grains, and cattle now compete with people for food. A third of the world's fish catch and more than a third of the world's total grain output is fed to livestock. In the USA, the world's premiere meat eating country, 80 per cent of the corn grown and 95 per cent of the oats grown are fed to livestock. On average, over 70 per cent of grain produced in the USA is fed to livestock. Cattle are inefficient converters of food, however. Thus the greater the human consumption of animal products, the fewer people can be fed. (Seager, 1995, p. 103)

Thus, taking this need for animal feed into consideration whilst looking, for example, at the debt crisis, it becomes clear that there are very obvious global dimensions in not only agricultural policy but also everyday social practices. However, such dimensions are not exactly new. A global aspect of agriculture was also discernible during the colonial period. In fact, colonialism was exactly about that: global reach of resources.

trade and agriculture: then and now

When comparing the colonial period with today's globalizing or globalized agricultural systems, several differences can be pointed out. First of all there is a question of scope. In the colonial period, economic organization of agriculture was about making available what was not naturally available to the colonial powers on their own territories. Second, colonialism was very much a state-driven activity (despite the importance of the forerunners of the multinational corporations) while today's agriculture is dominated by large corporations as well as states (Goodman and Watts, 1997). Third, today's agricultural policies are dominated by international institutions such as the WTO or regional economic integration organizations (Hines, 2000). Fourth, technological improvements and speed make a massive difference to the way agricultural practices have evolved. Fifth, increasing migration has changed demand for certain foods and likewise, Western consumption and taste preferences are being taken up in other parts of the world. What has essentially stayed the same is that global agriculture is organized in such a way that it benefits the dominant states (Hines, 2000).

The main environmental dimension is the increased demand on the soil or other medium and how this increased demand is coped with through chemical, biological or genetic help as well as the increased energy demand through transport and distribution. However, there is also a social and environmental justice dimension.

political and economic governance

This becomes nowhere clearer than in the world trade field: a principle of the neoliberal world economy that has a huge structural impact on agriculture is that of international trade rules. Produce in many African countries, for example, is undercut on the local markets by imports subsidized by their states of origin, thus pushing farmers even more toward cash crops (Oxfam, 2003). At the same time, industrialized countries subsidize their own produce on world markets, thus making it difficult for developing countries to remain competitive despite lower production costs (Oxfam, 2003). Thus the combination of a free market with no import levies and subsidies in other parts of the world are having disastrous social as well as environmental consequences in many developing countries, particularly in Africa. These problems can only be overcome through fairer, regulated prices, most obviously through taxation of incoming agricultural produce. However, this is not an option under the current world trade framework. In a state with no industry, going beyond subsistence agriculture to cash crop production is the only way to generate economic surplus for much-needed imports and the existing international economic structures do not consider the needs of the rural populations and their environment in such areas. This is only one example of the equity dimension of globalization and global trade rules in agriculture and one that is increasingly being highlighted by transnational actors at protests, for example at WTO ministerial meetings.

The argument that the issues of social justice and environmental degradation cannot be separated means that there are serious implications for the uneven pattern of consumption globally and this is particularly relevant in the agricultural field. If the current pattern of consumption in developed countries cannot be extended, at least hypothetically, to the global population, then clearly a redistribution of income is called for in order to share the existing resources more equitably in order to be in harmony with the principles of embedded liberalism (George, 1994). However, for debates on this subject to become pertinent, the myths of unlimited economic growth and of wealth for all need to be discredited first (see 'Globalization as Structure' section, above). In fact, the global agricultural economy is not only organized in order to achieve the cheapest/most efficient possible production process but also to guarantee the inflow of produce to fit the consuming elite of the planet. Although neoliberalism pushes the policy of free trade, for example, the areas where trade liberalization has largely been achieved are those that benefit Northern/Western consumers, such as capital goods. In the agricultural sector, the farmers in the industrialized countries continue to be protected from the free market, and this has dramatic consequences for many Southern farmers. In fact, the structural basis of the current trade regime is geared towards supplying industrialized countries' markets to the most efficient and cheapest extent. Such policies can even be found in the policy advice given to applicants for structural adjustment policies as they are advised to grow cash crops which then leads to depressed prices for developed countries' consumers, if applied across the board (IMF, 2002).

The impact of such policies is illustrated with this example of cotton farmers in West Africa – a case that took centre stage at the WTO Ministerial Conference in Cancun in September 2003. West African farmers are dependent on cotton export income for debt repayment. At a time when world market prices for cotton are in decline, US production and exports continue to be on the increase. For example, between 1998 and 2001, the volume of US cotton exports nearly doubled (Oxfam, 2003, p.11). At the same time, in 2001/02 the value of US cotton production came to US$3 billion at world market prices and the value of outlays in terms of subsidies to cotton farmers was US$3.9 billion (Oxfam, 2003, p. 12). Thus US cotton is not produced under free market conditions and does not reflect real production costs. It effectively undercuts cotton farmers who produce cotton at a very efficient cost which is what is happening to West African farmers.

Only subsidized cotton producers (in the US and southern Europe) are making a profit. The consequences for the West African farmers and the state are potentially horrific: not able to afford subsidies themselves (and subsidies being in a legally grey area anyway) and overly dependent on cotton as an export crop, their outlook is extremely bleak. In fact, according to Oxfam, the losses in the cotton industries due to US subsidies offset the debt relief granted through the Heavily Indebted Poor Countries (HIPC) initiative (2003, p. 17). In the fall of 2002, Brazil launched a complaint in the WTO contending that US cotton subsidies have caused overproduction, increased exports and thus have contributed to the fall of world cotton prices. However, its first panel request at the WTO dispute settlement body was blocked by the US in March 2003. In early May 2003, West African Trade Ministers met in Ouagadougou and in an official declaration announced that they would table the issue of cotton subsidies at the WTO ministerial conference in September and also demand compensation for damages suffered and to be incurred during the period it takes to dismantle these subsidies (F. K. Ametepe, 5 May 2003, *Le Pays*, Burkinabé national newspaper). The tabling of this motion at Cancun fits in with wider developing–developed country chasms that have arisen in preparatory meetings and have resulted in deadlock. This example of cotton can also be extrapolated to other commodities such as sugar, coffee, tea, and so on.

consumption and agriculture

The social and power relations underlying the structure of the global political economy are a very important subject that is particularly visible when looking at the agricultural sector. There is no question that there is a globalizing production economy in many (but not all) economic sectors but this is not matched by a global consumption economy (Kütting, 2004). Globalized production is very clear in the agricultural sector but consumption is limited and geared to the consumers in the industrialized states. Thus the globalizing aspects of the political economy benefit certain actors whilst other actors are unable to improve their position in the globalization process.

Although historically, there have been global aspects to agriculture for hundreds of years, the recent changes of the rise of agribusiness and the growing importance of international institutions in the regulation of trade are new and symptomatic of globalization (McMichael, 1995). Thus, although reports of the decline of the state are exaggerated, as the role of the US in the regulation of subsidies shows, other important actors in the shape of international institutions, multinational corporations and non-governmental organizations as well as social movements (the landless movement, for example) are clearly gaining in importance. These are both political and economic phenomena.

The sociocultural aspects of agricultural globalization become obvious in the changes in diets worldwide. Not only do Western consumers expand the scope of their diets, but also diets in developing countries are changing as a result of the increasing global nature of agriculture. For example, in West Africa, more and more people are resorting to a wheat-based diet although wheat is not grown in the region. This is the influence of Western tastes, leading to an import dependence on wheat and a decline in markets for local produce such as maize, sorghum, millet, and so on. Here, the distinction between economic and sociocultural aspects of this phenomenon is increasingly blurred.

These changes have profound environmental implications as they increase the stress on soils due to the demand for increased yields. This takes us back to the steady-state economy of the ecological economists and the question about the possibility of limited and unlimited growth – the view of a cornucopia leading towards policies of ecological modernization and environmental management and the view of the limits to growth leading towards warnings about technological determinism. Agriculture as a case study illustrates the multidimensional nature of the subject of globalization.

conclusion

This chapter has discussed the various dimensions of the relationship between globalization and environment. Environmental degradation can be understood as a structural issue which is directly related to the emergence of a global economy. Here, both the structural origins as well as the consequences of environmental degradation are global in reach. Environmental degradation can also be studied from an agency perspective where the increasingly global nature of the environmental phenomenon manifests itself through the rise of transnational activity in the form of new actors, new forms of political and economic governance and also an increasing awareness of the involvement of the individual citizen in this process – as part of civil society but also as consumer. These linkages between structure and agency, between local and global and between the social, political and economic can be traced through the case study which shows the limitations of global political and economic governance when it comes to social and environmental justice but also shows that the sociocultural, the economic and the ecological/environmental dimensions are intrinsically linked.

Future research agendas in the field of global change/globalization and the environment are likely to continue in the areas singled out here as these are very much emerging research areas. It is particularly the areas of global governance and the interplay between political and economic globalization that are likely to capture the interest of environmental researchers in the near future. Important issue areas that have not yet fully found their place on research agendas in global environmental studies are the subject of equity and global environmental politics, both as conceptual issues but also as empirical studies of governance procedures and institutional workings. However, these are some of the biggest challenges of the twenty-first century.

bibliography

Abramson, Paul, and Ronald Inglehart (1995) *Value Change in Global Perspective*, Ann Arbor: University of Michigan Press.
Baker, Gideon (2002) 'Problems in the Theorisation of Global Civil Society', *Political Studies*, 50, 5, pp. 928–43.
Bernstein, Steven (2001) *The Compromise of Liberal Environmentalism*, New York: Columbia University Press.
Blowers, Andrew (2003) 'Inequality and Community and the Challenge to Modernization', in Julian Agyeman, Robert Bullard and Bob Evans (eds) *Just Sustainabilities: Development in an Unequal World*, Cambridge, Mass.: MIT Press.
Boli, John, and George Thomas (eds) (1999) *Constructing World Culture*, Stanford: Stanford University Press.
Breitmeier, Helmut, and Volker Rittberger (1998) *Environmental NGOs in an Emerging Civil Society*, Tuebingen: Tuebinger Arbeitspapiere zur Internationalen Politik und Friedensforschung, no. 32.
Carson, Rachel (1962) *Silent Spring*, Boston: Houghton Mifflin Company.
Chew, Sing (2001) *World Ecological Degradation: Accumulation, Urbanization and Deforestation 3000 BC–AD 2000*, Walnut Creek, Calif.: Altamira.
Clapp, Jennifer (2001) *Toxic Exports: The Transfer of Hazardous Wastes from Rich to Poor Countries*, Ithaca: Cornell University Press.
Conca, Ken (2000) 'The WTO and the Undermining of Global Environmental Governance', *Review of International Political Economy* 7, 3, pp. 484–94.
Conca, Ken, Michael Alberty and Geoffrey Dabelko (eds) (1995) *Green Planet Blues*, Boulder: Westview Press.
Conley, Verena (1997) *Ecopolitics: The Environment in Poststructuralist Thought*, London: Routledge.
Cox, Robert (1997) 'A Perspective on Globalization', in James Mittelman (ed.) *Globalisation: Critical Reflections*, Boulder: Lynne Rienner, pp. 21–33.
Cox, Robert, with Timothy Shaw (eds) (1996) *Approaches to World Order*, Cambridge: Cambridge University Press.
Daly, Herman (1996) *Beyond Growth: The Economics of Sustainable Development*, Boston: Beacon Press.

Daly, Herman (1992) *Steady State Economics*, 2nd edn, Washington, DC: Island Press.
Dauvergne, Peter (2001) *Loggers and Degradation in the Asia-Pacific*, Cambridge: Cambridge University Press.
Dryzek, John (1997) *The Politics of the Earth: Environmental Discourses*, Oxford: Oxford University Press.
Eckersley, Robin (ed.) (1995) *Markets, the State and the Environment*, London: Routledge.
Ehrlich, Paul (1968) *The Population Bomb*, New York: Ballantine Books.
Frank, Andre Gunther (1998) *ReOrient: Global Economy in the Asian Age*, London: University of California Press.
Ford, Lucy (2004) 'Challenging Global Environmental Governance: Social Movement Agency and Global Civil Society', *Global Environmental Politics* 3, 2, pp. 120–34.
George, Susan (1994) *A Fate Worse than Debt*, London: Penguin.
Gill, Steven, and James Mittelman (eds) (1997) *Innovation and Transformation in International Studies*, Cambridge: Cambridge University Press.
Gillespie, Alexander (2001) *The Illusion of Progress*, London: Earthscan.
Goldfrank,Walter, David Goodman and Andrew Szasz (eds) (1999) *Ecology and the World-System*, London: Greenwood Press.
Goodman, David, and Michael Watts (eds) (1997) *Globalising Food*, London: Routledge.
Hardin, Garrett (1968) 'The Tragedy of the Commons', *Science*, 162, 3859 (13 December), pp. 1243–8.
Hardin, Garrett (1974) 'Living on a Lifeboat', *BioScience*, 23, 10 (October), pp. 561–8.
Hines, Colin (2000) *Localization: A Global Manifesto*, London: Earthscan.
Hirst, Paul, and Graheme Thompson, (1996) *Globalization in Question*, London: Polity Press.
Hoogvelt, Ankie (1997) *Globalisation and the Postcolonial World*, Basingstoke: Macmillan.
Hornborg, Alf (1998) 'Ecosystems and World Systems: Accumulation as an Ecological Process', *Journal of World-Systems Research* 4, 2, pp. 169–77.
IMF (2002) Public Information Notice No. 02/2, 9 January.
Inglehart, Ronald (1977) *The Silent Revolution: Changing Values and Political Styles Amongst Western Publics*, Princeton: Princeton University Press.
Isaacman, Allen, and Richard Roberts (eds.) (1995) *Cotton, Colonialism and Social History in Sub-Saharan Africa*, Portsmouth, NH: Heinemann.
Jameson, Fredric, and Masao Miyoshi (eds) (1999) *The Cultures of Globalization*, Durham: Duke University Press.
Jeong, Ho-Won (ed.) (2001) *Global Environmental Policies*, New York: Palgrave.
Keck, Margaret, and Kathryn Sikkink (1998) *Activists Beyond Borders*, Ithaca: Cornell University Press.
Klein, Naomi (2000) *No Logo*, London: Flamingo Press.
Kütting, Gabriela (2000) *Environment, Society and International Relations*, London: Routledge.
Kütting, Gabriela (2004) *Globalization and Environment: Greening Global Political Economy*, Albany: SUNY Press.

Lipiètz, Alain (1997) 'The Post-Fordist World: Labour Relations, International Hierarchy and Global Ecology', *Review of International Political Economy* 4, 1, pp. 1–41.
Lipschutz, Ronnie (1999) 'From Local Knowledge and Practice to Global Environmental Governance', in M. Hewson and T. Sinclair (eds) *Approaches to Global Governance Theory*, Albany: SUNY Press, pp. 259–83.
Lipschutz, Ronnie, with Judith Mayer (1996) *Global Civil Society and Global Environmental Governance*, Albany: SUNY Press.
Little, P. E. (1999) 'Environments and Environmentalisms in Anthropological Research: Facing a New Millennium', *Annual Review of Anthropology* 28, pp. 253–84.
Lovelock, James (1988) *The Ages of Gaia: A Biography of our Living Earth*, Oxford: Oxford University Press.
Lukes, Steven (1974) *Power: A Radical View*, Basingstoke: Macmillan.
Martinez-Alier, Joan (2002) *The Environmentalism of the Poor: A Study of Ecological Conflicts and Valuation*, Cheltenham: Edward Elgar.
McMichael, Philip (1995) *Food and Agrarian Orders in the World Economy*, Westport: Praeger.
Meadows, Donella, et al. (eds) (1972) *The Limits to Growth: A Report on the Club of Rome's Project on the Predicament of Mankind*, New York: Universe Books.
Merchant, Carolyn (1992) *Radical Ecology: The Search for a Livable World*, London: Routledge.
Miles, Steven (1998) *Consumerism as a Way of Life*, London: Sage.
Miller, Marian (1995) *The Third World in Global Environmental Politics*, Boulder: Lynne Rienner.
Mintz, Sidney (1985) *Sweetness and Power: The Place of Sugar in Modern History*, London: Penguin.
Mittelman, James (2000) *The Globalization Syndrome*, Princeton: Princeton University Press.
Mittelman, James (ed.) (1997) *Globalization: Critical Reflections*, Boulder: Lynne Rienner.
Mol, Arthur (2001) *Globalization and Environmental Reform*, Cambridge, Mass.: MIT Press.
Oxfam (2003) *Cultivating Poverty*, Oxfam Briefing Paper No. 30, Oxford: Oxfam.
Ponting, Clive (1991) *A Green History of the World*, London: Penguin.
Prakash, Aseem, and Jeffrey A. Hart (eds) (1998) *Globalization and Governance*, London: Routledge.
Princen, Thomas, and Mathias Finger (eds) (1994) *Environmental NGOs in World Politics*, London: Routledge.
Raustiala, Kal (1997) 'States, NGOs and International Environmental Institutions', *International Studies Quarterly*, 41, 4, pp. 719–40.
Redclift, Michael (1987) *Sustainable Development: Exploring the Contradictions*, London: Routledge.
Robertson, Roland (1992) *Globalization: Social Theory and Global Culture*, London: Sage.
Rohrschneider, Robert, and Russell Dalton (2002) 'A Global Network? Transnational Cooperation by Environmental Groups', *Journal of Politics*, 64, 2, pp. 510–33.
Sachs, Wolfgang (2000) *Planet Dialectics: Explorations in Environment and Development*, Basingstoke: Palgrave.

Sachs, Wolfgang, Reinhard Loske and Manfred Linz, (1998) *Greening the North: A Post-industrial Blueprint for Ecology and Equity*, London: Zed Books.
Said, Edward (1994) *Culture and Imperialism*, London: Vintage Books.
Saurin, Julian (1996a) 'International Relations, Social Ecology and the Globalisation of Environmental Change', in John Vogler and Mark Imber (eds) *The Environment and International Relations*, London: Routledge, pp. 77–98.
Saurin, Julian (1996b) 'Globalization, Poverty and the Promise of Modernity', *Millennium* 25, 3, pp. 657–80.
Scholte, Jan Aart (1993) *International Relations of Social Change*, Milton Keynes: Open University Press.
Schuurman, Frans (ed.) (2001) *Globalization and Development Studies*, London: Sage.
Schwartz, Herman (2000) *States versus Markets*, 2nd edn, Basingstoke: Macmillan.
Seager, Joni (1995) *The State of the Environment Atlas*, London: Penguin Books.
Sklair, Leslie (ed.) (1994) *Capitalism and Development*, London: Routledge.
Sklair, Leslie (2002) *Globalization, Capitalism and its Alternatives*, Oxford: Oxford University Press.
Spaargaren, Gert, Arthur Mol and Fred Buttel (eds) (2000) *Environment and Global Modernity*, London: Sage.
Strange, Susan (1996) *The Retreat of the State*, Cambridge: Cambridge University Press.
Wallerstein, Immanuel (1986) *Africa and the Modern World*, Trenton, NJ: Africa World Press.
Wallerstein, Immanuel (1995) *After Liberalism*, New York: The New Press.
Wapner, Paul (1995) 'Politics Beyond the State: Environmental Activism and World Civic Politics', *World Politics* 47, 3, pp. 311–40.
Wapner, Paul (2003) 'World Summit on Sustainable Development: Toward a Post Jo'Burg Environmentalism' *Global Environmental Politics* 3, 1, pp. 1–10.
WCED (World Commission on Environment and Development) (1987) *Our Common Future*, Oxford: Oxford University Press.
Willetts, Peter (2002) 'Non-Governmental Organizations, Article 1.44.3.7', in *UNESCO Encyclopedia of Life Support Systems*, Section 1, Institutional and Infrastructure Resource Issues, Paris: UNESCO.
Williams, Marc (2001) 'In Search of Global Standards: The Political Economy of Trade and the Environment', in Dimitris Stevis and Valerie Assetto (eds) *The International Political Economy of the Environment: Critical Perspectives*, Boulder: Lynne Rienner, pp. 39–62.
Young, Oran (ed.) (1997) *Global Governance: Drawing Insights from the Environmental Experience*, Cambridge, Mass.: MIT Press.
Young, Oran (2002) *The Institutional Dimensions of Environmental Change*, Cambridge, Mass.: MIT Press.

annotated bibliography

Lipschutz, Ronnie, with Judith Mayer (1996) *Global Civil Society and Global Environmental Governance: The Politics of Nature from People to Planet*, Albany: SUNY Press. Lipschutz and Mayer want us to rethink environmental action by 'global civil society' to see how it as fundamentally grounded in local efforts to preserve livelihoods and to expand control over a place that is home. If our experiments in global governance can appreciate this rootedness of effective action by civil society, they argue, global action may be able to take on more effective forms and protect local environments as well as the livelihoods that depend on them. The most important ideas they pursue in this book are: (a) environmental change is a social process; (b) environmental governance is most fruitfully approached locally; and (c) in policy terms, local resource regimes are creating new and potentially more effective means of managing resources through combinations of institutional innovation, market forces, and local accountability mechanisms. Lipschutz's reconceptualization of environmental protection and civil society is in many ways a satisfying framework for thinking about the development of a system of global environmental governance.

Miller, Marian A. L. (1995) *The Third World in Global Environmental Politics*, Boulder: Lynne Rienner. Marian Miller's work fills a valuable niche with *The Third World in Global Environmental Politics* since there was and is relatively little written about the role of Third World countries within the arena of international environmental politics. Miller brings together existing information and perspectives in a way that is clearly organized. She provides an overview of the conceptual category 'Third World' and reviews literature by international relations theorists on the concept of an international *regime*, its formation and development. The concept of a regime then becomes central to several case studies of international environmental regimes – those dealing with ozone depletion, hazardous waste trading, and biodiversity. She offers some brief, but wide-ranging, comments on the Third World, structural inequality and sustainable development. Miller argues that while Third World countries are vastly disadvantaged in international power politics, the distinctive character of environmental concerns can give these countries a bit more leverage than they have in other contexts.

Princen, Thomas, and Mathias Finger (eds) (1994) *Environmental NGOs in World Politics: Linking the Local and the Global*, London: Routledge. *Environmental NGOs in World Politics* examines the importance of NGOs in world environmental politics. This pioneering book explains how NGOs perform key roles in an emerging world of environmental politics. It shows how they act as independent bargainers and as agents of social learning, to link biophysical conditions to the political realm at both the local and global levels. The authors argue that NGOs are able to appropriate those environmental issues irresolvable by traditional politics, building their own bargaining assets to negotiate with other international actors. Four major case studies illustrate the richness of NGO activity and the geographic and substantive diversity of their politics.

Young, Oran (ed.) (1997) *Global Governance: Drawing Insights from the Environmental Experience*, Cambridge, Mass.: MIT Press. The contributors to this volume draw upon the experiences of environmental regimes to analyse the problems of international governance in the absence of a world government. In the process they address a number of central points. One is whether regime analysis has produced a distinct conception of governance that can be applied to the solution of collective action problems at the international level. Another important question they address is whether or not it is possible to identify the conditions necessary for international 'governance without government' to succeed. Further, they examine the correlation of the emergence of regimes in specific issue areas and broader consequences for the future of international society. The contributors address the issue regarding the possibility of generalizing the experience with environmental issues to a broader range of international governance problems.

6
international political economy and the environment
jennifer clapp

In this era of economic globalization, there has been remarkable growth in the volume and value of global trade, investment and finance. These international economic relationships have important implications for the natural environment, as all have been identified as having some linkage to environmental quality. The extent to which these international economic relationships contribute to environmental problems or to solutions for environmental problems is the subject of extensive debate (see, for example, Stevis and Assetto, 2001). Some see the relationship as largely positive, with environmental benefits being attached to the economic growth that global economic transactions seek to facilitate. While there may be some cases where the linkages have negative outcomes for the environment, for these thinkers environmental policies are seen to be able to address the situation in ways that do not harm global economic transactions. Others, however, see mainly negative implications arising from global economic relationships and the economic growth that is associated with it. For them, it is important that environmental policies do restrict global economic transactions. A third view is also gaining prominence which seeks to bridge the divide, arguing that while there are some potential negative aspects of global economic relations for the environment, management of the global economy can bring both economic and environmental benefits.[1]

This chapter outlines the relationship between the international political economy (IPE) and the global environment, as well as the debates that surround that relationship. The IPE–environment interface is multifaceted and complex. It encompasses the linkages between global economic interactions and the emergence of environmental problems on the one hand, as well as the global economic institutions and actors that take action to address those problems through international governance mechanisms. In some cases, some of the same actors (for example, transnational corporations (TNCs)) are involved in both economic activities that may contribute to environmental harm as well as in the governance of global economic institutions that seek to regulate those activities. This can lead to conflicts of interest and also highlights the importance of power relationships in the study of the interface between IPE and the environment.

tracing the roots of debates over ipe and the environment

It was in the early 1990s that debates over the global political economy and the environment erupted in full force. This was in large part a product of concerns over trade policy and its interface with environmental issues at that time, most particularly the negotiation of the North American Free Trade Agreement (NAFTA) and the General Agreement on Tariffs and Trade (GATT) Tuna–Dolphin challenge (Esty, 1994; Williams, 2001). There was, at this time, concern that trade and investment issues would override environmental considerations and these cases brought the question to the fore. The result was a large volume of literature on the topic, from a variety of viewpoints.

The debate over the relationship between the international political economy and the environment overlaps to some extent but is not entirely encompassed within traditional debates in the field of International Relations (IR). In other words, the roots of various positions in both policy and theory on the links between trade, investment and finance on the one hand, and environment on the other, originate in a number of fields apart from IR. These include, most importantly, ecology and economics. Paterson (see Chapter 3 in this volume) recognizes the need to look beyond the three 'traditional' camps in IR, which is a step in the right direction. When examining debates about IPE and the environment, we could also add the neoclassical economic view which has been a very vocal participant. I argue that there are three main 'camps' in the debate over IPE and the environment. I explain these below using Paterson's categories with the addition of neoclassical economics.

Neoclassical economists are a dominant voice in both policy and academic debates, and tend to see the expansion of global trade, investment and finance as on the whole positive for the quality of the natural environment. This position derives from a fundamental assumption taken by neoclassical economic thinking, comparative advantage, which dates back to the writings of David Ricardo, and assumes that international trade benefits all partners in material terms. In today's global economy, neoclassical economists also argue that transnational investment and financing are positive for the material gain of nations, as they are seen mainly as activities to facilitate trade and enhance growth. The neoclassical economists have accommodated environmental concerns within this view by arguing that material gain that arises from global economic interactions can be used to finance environmental improvements. To back up this argument these thinkers rely on evidence from studies which show that an inverted 'U' relationship exists in Organization for Economic Cooperation and Development (OECD) countries between income growth and a cleaner environment (also known as the Environmental Kuznets Curve, or EKC). In other words, as incomes rise, environmental problems may initially get worse, but then improve as incomes rise further. The rationale is that a wealthier population will demand a cleaner environment, that governments will respond by enacting stricter environmental laws, and that firms react by introducing 'greener' products (Grossman and Krueger, 1995; World Bank, 1992). From this perspective, the global political economy and the environment are mutually supportive, such that growth can go on indefinitely, and environmental improvements will result. For these neoclassical economic thinkers the liberalization of trade, investment and finance are all seen to be increasing global economic integration, and in turn generating more wealth with which to protect the environment.

In opposition to this viewpoint we see other views in coalition which have argued vigorously that the global political economy and environment are not mutually supportive. Ecological economists and many radical thinkers have been very vocal critics of the neoclassical economic position. This critical camp includes thinkers drawing on the ecoauthoritarian, sustainable analysis, structural conflict and accumulation schools outlined by Paterson. The thinkers from each of these schools have a deep scepticism about the impact of growth on the environment, as well as scepticism of the actors promoting it, and thus do not buy into the EKC argument presented by neoclassical economists. They argue instead that the liberalization of trade, investment and finance, all aimed to increase economic activity, will have disastrous results for the natural environment. Growth is based on nature, for both resources as well as sinks for the wastes that result, and this is not accounted for in neoclassical economics. Economic growth, those in the critical camp argue, cannot go on indefinitely. For many of these thinkers it has in fact gone beyond its sustainable limit. Ecological economists use the laws of thermodynamics to demonstrate that these limits are real and are being surpassed (Daly, 1996; Georgescu-Roegen, 1971). Others in this camp also focus on the ways in which global economic relationships create or perpetuate inequalities that have negative environmental implications.

While these two camps have tended to dominate the debate over the global political economy and the environment over the past decade, a third view is increasingly being expressed. This third view originates in large part from what Paterson calls the liberal institutionalist school within traditional IR. These thinkers argue that in some cases common ground can be found between neoclassical economists and their critics when it comes to the environment (Neumayer, 2001). Following their focus on structured cooperation between states, they advocate strong rules to govern the global economy in ways that ensure that the environment does not suffer. In making this argument, institutionalists largely agree with neoclassical economists about the possibility that the growth of the global economy can have positive impacts on the environment, and they also agree with the critics that in some instances they are not mutually supportive. Thus their policy advice is to create global rules to avoid the cases where harm occurs, and encourage those where positive outcomes are likely.

Each of these camps in the broader debate over the global political economy and the environment has also engaged in the more specific debates over trade, investment and finance and their relationship to the environment.[2] The theoretical and policy debates in each of these areas of the global political economy are discussed below.

trade and the environment

International trade is an extremely important part of the global political economy. World trade has expanded greatly in recent years, growing from 25 per cent of global GDP in 1960 to 58 per cent in 2001. During the 1950–2001 period, there was a 20-fold increase in global exports of goods (World Trade Organization (WTO) statistics online, <http://www.wto.org/english/res_e/statis_e/statis_e.htm>). There has also been a massive increase in the value of world trade, from US$58 million in 1948 to over US$6 trillion in 2000 (WTO data, see <www.wto.org>). Much of the literature on trade and the environment tends to engage either with the debates over the environmental merits and demerits of international trade or with debates over the way in which environmental issues should or should not be incorporated into international trade governance and vice versa.

the impact of trade on the environment

There has been heated debate over the impact of trade on the environment in recent years (Esty, 1994; Neumayer, 2001). Those who argue that trade's impact on the environment is overall positive tend to fall within the neoclassical economic camp. These thinkers have put forward several arguments to back their views. A principal argument made by those taking this view hinges on the relationship between trade and economic growth, as noted above.

A further argument of those who stress that trade has positive environmental implications is that free trade allocates resources most efficiently, resulting in less resource use and less wastage. Specialization of production based on comparative advantage results in more efficient allocation of resources than would be the case if countries pursued self-sufficiency. This means in practice less wastage of scarce resources. Further, trade barriers create distortions that result in inefficiencies (Bhagwati, 1993). Trade restrictions such as tariffs, quotas and export subsidies, it is argued, can lead to the underpricing of resources domestically, which encourages their overuse. Again, this works against the goals of efficiency and conservation. From this perspective, then, the key is better environmental policy, not restriction of trade.

Taking issue with these arguments are ecological economists and other critical thinkers, who argue that the assumptions on which the liberal economic view is based are fundamentally flawed. From their perspective, economic growth that should result from trade is itself a large part of the environmental problem. This is because it results in more 'throughput' in the economy, ultimately resulting in more environmental degradation. In other words, more physical materials from the earth are used in production, and more waste is created (Daly, 1993, 1996). Efficiency gains are seen to be outstripped by this growth, ultimately destroying the planet (Sachs, 1999).

There are a host of other problems with trade identified by its critics. Specialization for trade purposes may distribute pollution unevenly. Pockets of the developing world, for example, are seen to produce more toxic products and export their natural resources more than would be the case without trade (Karliner, 1997; Sachs, 1999). Related to this is the problem of a 'race to the bottom'. Rather than environmental standards rising with trade liberalization, critics argue that countries fear raising standards, and in fact may lower them, in a bid to attract investment and improve their trade competitiveness (Daly, 1993; Porter, 1999). Additionally, growth in trade inevitably means more pollution from transportation of goods around the world (Conca, 2000, p. 485). Restriction of trade on environmental grounds is a perfectly legitimate policy response for these thinkers.

While the debate between free traders and anti free trade critics with respect to the environment has been polarized for years, the institutionalist view is increasingly seen in the literature that seeks to forge a middle ground. This view points out the commonality between the two extremes to map out scenarios when managed trade can be beneficial for the environment. It also proposes global governance mechanisms to achieve this goal (for example, Esty, 2001; Neumayer, 2001; Weinstein and Charnovitz, 2001). These thinkers argue, for example, that there are certain obvious cases where free trade should not be encouraged, such as the trade in toxic substances and dangerous chemicals. These problems, they argue, can be managed effectively through trade measures incorporated into environmental agreements and environmental measures incorporated into trade agreements. Yet at the same time they do see a role for trade liberalization when clear efficiency gains will result, for example, the removal of subsidies, such that resources are managed more effectively.

global governance for trade and environment

The theoretical debate over trade and environment as discussed above is extremely important because it underpins debates on global governance over trade and environment issues. The mechanisms of global governance that deal with trade and environment issues include international agreements on trade as well as international environmental agreements. Each is discussed below.

International trade agreements have historically paid little attention to the interface between trade and environment. The General Agreement on Tariffs and Trade, first negotiated in 1947, made no explicit mention of 'environment' and none has been added since. In 1995 the WTO replaced the GATT Secretariat, and the new organization now oversees a series of trade agreements, of which GATT is just one. Though the WTO preamble does stress the need to promote 'sustainable development', there is no specific language in the WTO agreements that allows countries to relax trade rules in the name of environmental protection.

GATT rules do, however, stress that countries cannot discriminate against products based on country of origin or how they are produced. What this means is that countries cannot apply trade restrictions on goods based on production and processing methods (PPMs) if the products are otherwise identical. This rule keeps countries from applying trade restrictions on goods that are produced in ways that are known to be environmentally damaging (Esty, 1994, pp. 49–51).

Only Article XX of the GATT has the potential to allow for trade restrictions on environmental grounds. Article XX sets out circumstances where states are eligible for exceptions to the GATT rules. These include measures undertaken to protect human, animal or plant life or health, or to ensure natural resources conservation. But these exceptions are qualified. Measures taken to protect human, animal or plant life or health must be shown to be 'necessary' before they can be exempted from GATT rules. And measures taken to ensure conservation of natural resources must apply strictly to depletion of natural resources and must be taken in conjunction with domestic measures to protect that resource. Further, if the article is applied a country must prove that it was not invoked in an arbitrary or unjustified way. Moreover, Article XX says little about global environmental issues. It is thus very difficult for states to be exempted from GATT rules for measures taken to address environmental issues outside of their borders or that affect the global commons (see Esty, 1994; Neumayer, 2001).

Given all of these qualifications to Article XX, it should not be surprising that it is extremely difficult for states to successfully gain exemptions to GATT rules for environmental reasons. Several states have tried to justify trade restrictions for environmental purposes by claiming exemption under Article XX (for example, the Tuna–Dolphin disputes of 1991 and 1994), but these were struck down by the GATT dispute panels. In this 1991 dispute, the US claimed an exemption under Article XX for its restrictions against imports of tuna that were caught in dolphin unfriendly ways, in accordance with the US Marine Mammal Protection Act. It claimed that such actions were necessary both to protect animal life and to conserve natural resources. Mexico disputed the US action, and the GATT dispute panel ruled that the US restrictions were not eligible for exemption under Article XX because they were seen to be applied unilaterally and they were extra-jurisdictional, such that they clearly discriminated against like products based on the way in which they were produced. The panel ruled that the US could have tried to solve the problem in a multilateral way (Vogel, 2000).

In 1994 a second Tuna–Dolphin dispute panel had to be formed. In this case, the EU challenged a secondary embargo that the US imposed on tuna from third-party sellers that did not meet US regulations. Again the GATT dispute panel ruled against the US, this time with the argument that the US application of the Marine Mammal Protection Act was arbitrary and unilateral, making it ineligible for an exemption under Article XX (Perkins, 1998). Other cases of attempted use of Article XX that have been challenged by other parties have since been brought forward and nearly all have been struck down (Neumayer, 2001). The various rulings by the GATT and subsequently WTO dispute resolution panels indicate that the trade body prefers countries to deal with environmental issues via multilateral efforts rather than via unilateral trade restrictions. Some say that the WTO is not inherently anti-environment, however, and that recent rulings show that it is making efforts to incorporate environmental issues more fully (DeSombre and Barkin, 2002).

The other main governance area in which the interface between trade and environment is prominent is in multilateral environmental agreements (MEAs). About 10 per cent of the 200 or so MEAs incorporate trade provisions, indicating that parties to those agreements felt that the best way to address those environmental issues was with trade restrictions of one sort or another. These include, for example, the Basel Convention (hazardous wastes), the Cartagena Protocol (biosafety), the Kyoto Protocol (climate change), the Montreal Protocol (ozone layer), the Rotterdam Convention (pesticides) and the Stockholm Convention (persistent organic pollutants, or POPs, chemicals). The rules that restrict trade vary according to the agreement and the issue at hand. Some restrict trade in dangerous items, while others restrict trade between parties and non-parties of items they seek to control as a mechanism to encourage countries to sign on to those agreements. Many of these MEAs also include other control measures that may potentially affect trade, such as provisions for technology transfer and prior informed consent (see Stilwell and Tarasofsky, 2001).

While these agreements have incorporated trade provisions, it is not clear in international law which type of agreement takes precedence, trade law or environmental law. In 1994 the GATT/WTO constituted a trade and environment committee to discuss, among other issues, the relationship of global trade rules to MEAs. At the launch of the most recent talks, the Doha Round, the WTO ministers agreed to undertake talks to clarify the relationship between WTO rules and MEAs. The need to clear up this issue was reinforced at the World Summit on Sustainable Development in 2002, and this work is being initiated at the WTO via its Committee on Trade and Environment (WTO website: <www.wto.org/english/tratop_e/dda_e/dohaexplained_e.htm#environment>). The United Nations Environment Programme (UNEP) is also studying this issue. Some would like to see a World Environment Organization (WEO) to counter the power of the WTO (Biermann, 2000; also in this volume). Regional trade agreements and organizations are also important mechanisms of governance that touch on trade and environment issues. Unlike the GATT and WTO, the NAFTA, negotiated in the early 1990s, does explicitly attempt to incorporate environmental concerns not just in its preamble but directly into the main text of the agreement. It does this by mentioning specific international environmental treaties that should take precedence over trade rules incorporated into the agreement, provided they are carried out in the least trade-distorting manner. The treaties mentioned include the Convention on the International Trade in Endangered Species, the Basel Convention and the Montreal Protocol, as well as four bilateral treaties (Soloway, 2002). The NAFTA also established an environmental side agreement, the North American Agreement on Environmental Cooperation (NAAEC). This side-agreement aims to ensure that states comply with and enforce their own national environmental laws and it establishes a mechanism for settling environmental disputes. A Commission on Environmental Cooperation (CEC) was also established to oversee the NAAEC. This body allows citizen input into reporting of international environmental violations (Hufbauer et al., 2000). It is still early to assess the full environmental implications of the NAFTA in practice, and arguments have been put forth both praising its environmental performance as well as criticizing it (Hufbauer et al., 2000; Logsdon and Husted, 2000).

The European Union (EU) is another important regional organization grappling with both trade and environmental issues. The EU is different from the NAFTA in that it promotes both political and economic integration, including the harmonization of environmental laws (Geradin, 2002). Since the 1980s the EU has developed and adopted EU-wide policies for environmental protection, including those regarding waste, air pollution, water, nature protection and climate change. These laws were developed in a parallel process to the EU's economic integration, and are thus not widely seen to be explicitly tied to trade performance (Stevis and Mumme, 2000, p. 31). Some EU states, however, primarily those with weaker environmental standards to begin with, such as Greece, Portugal, Spain and Italy, were more hesitant than others to adopt EU-wide environmental policies because of their concern over the economic implications of more stringent environmental regulations. It is for this reason that there is flexibility built into the EU's environmental regulations to allow for some degree of difference in requirements according to member states' economic and environmental situations (Geradin, 2002, pp. 128–9). Despite these differentials, however, the EU is widely seen to be a successful case of upward harmonization of environmental laws within the context of enhanced economic integration.

The Asia Pacific Economic Cooperation (APEC) is a third example of a regional trade organization which has had to grapple with trade and environment issues. Although APEC had 'sustainable development' as a key policy goal from very early on, in practice very little to date has been done with respect to environmental cooperation in the region. As a broad grouping covering economic cooperation across and within the Asia and Pacific region, it includes advanced industrial countries (for example, Japan, US, Canada, Australia) as well as a number of newly industrializing countries (for example, the Philippines, Indonesia, Mexico, Chile) and countries in transition (such as China, Russia).[3] The central aims of APEC are to facilitate trade, including trade liberalization, as well as to promote economic and technical cooperation, the latter of which includes environmental cooperation. These goals, however, have been pursued along separate diplomatic tracks. Unlike the case of the EU, this has hampered rather than improved environmental cooperation. This is because APEC is not seeking to harmonize environmental regulations, and its particularly wide diversity in terms of environmental and economic conditions has made agreement in this area difficult. As a result, there has been very little integration of trade and environmental issues (Zarsky, 2002).

transnational investment and the environment

There has been extensive growth in both the number of transnational corporations and the amount of foreign direct investment (FDI) in the past few decades. Today there are over 65,000 TNC parent firms globally, up from just 7,000 in 1970. There are, in addition, some 850,000 foreign affiliate firms – that is, corporations affiliated with a TNC – operating around the world. Together, these firms make up one-tenth of world GDP and one-third of world exports (UNCTAD, 2001, p. 9, 2002, pp. xv, 272). FDI flows have also grown in this period. In 1970 the level of FDI inflows stood at US$9.2 billion, and by 2001 it stood at US$735 billion (UNCTAD, 2001, p. 1; World Bank, 2003). With such a significant weight in the global economy, it is important to examine the impact of TNCs and FDI for the environment. It is recognized that TNCs tend to invest in the most environmentally damaging industries (UN, 1992, p. 226). The literature on transnational investment and the environment has thus far focused mainly on debates over whether firms relocate to take advantage of differential environmental standards, and also on debates over whether firms' voluntary 'greening' has been effective.

environmental standards and transnational investment

There is much debate over whether international investment, particularly foreign direct investment undertaken by TNCs, is negative or positive for the environment. Those from the critical camp argue that TNCs invest most heavily in jurisdictions where environmental regulations are weakest (Karliner, 1997; Korten, 1995). This kind of behaviour leads to several phenomena. First, it can trigger 'industrial flight' from countries which raise environmental standards. Second, it can lead to 'pollution havens' when some countries, primarily developing ones, set out to lower their environmental standards in a deliberate attempt to attract foreign investment. Not unrelated to the first two, we may see 'double standards' appear where different branches of the same TNC have different sets of environmental standards depending on where they are operating (Clapp, 2001). For these thinkers, this only contributes to the race to the bottom whereby states lower standards not just to improve trade competitiveness but also to attract FDI.

Most neoclassical economists do not see the environmental impact of TNCs as being a significant enough issue to worry about. They argue that different environmental standards in different countries are normal and part of a country's own capacity to absorb pollution. Proof that there is industrial flight and pollution havens for them is elusive, as statistical studies based on aggregate data on pollution costs and investment patterns have failed to conclusively show that they exist (Mani and Wheeler, 1998; Pearson, 1987). The main reason cited for this is that environmental costs make up a low percentage of firms' operating costs (typically 2–3 per cent of their sales). Because these costs are low, changes to regulations which affect those costs are not significant enough to cause firms to flee from jurisdictions with more stringent regulations or to set up shop in locations with more lax regulations. If firms move, it is likely for other reasons, such as labour costs, which can be up to 30 per cent of a firm's costs (Leonard, 1988). TNCs, these thinkers argue, in fact help to raise environmental performance in developing countries by transferring cleaner, state-of-the-art technologies, compared with local firms that use outdated and more polluting methods.

This debate over pollution havens and industrial flight has been ongoing since the 1970s. While it is agreed by most that the share of pollution intensive industry is rising in the developing world (Low and Yeats, 1992), there has been little agreement on why it is occurring.

Critical thinkers see this trend as a clear case of corporate abuse of the environment, while others say it is due to changes in domestic demand, and not linked to global investment decisions on the part of TNCs. Because each side of the debate uses a different methodology and sources of information, it is hard to compare their data.

Recent literature on these issues has stressed that we need to open up this debate and move beyond the narrow focus on whether it can be proven statistically that pollution havens do or do not exist. Some, for example, point out that the neoclassical economic argument ignores the pollution haven tendencies in the natural resources sectors. This is because their studies are based on data from manufacturing firms only, and thus do not provide enough evidence that the phenomenon does not occur in other areas, such as mining and forestry (for example, Clapp, 2002; Hall, 2002).

There has also been some suggestion from the liberal institutionalist camp that policy actions, even in the absence of widely agreed evidence that pollution havens exist, would be beneficial, and could be acceptable from both sides of this debate (Neumayer, 2001). A focus on a different concept, the 'regulatory chill', is helpful here. This refers to situations where states fail to raise environmental standards for fear of losing investment and weakening trade competitiveness. Whether such outcomes would result from raising standards is not the point. What matters is that states act in certain ways based on the belief that they may lose investment if they impose stricter standards (Porter, 1999). And existing trade agreements with investment clauses, such as Chapter 11 of NAFTA (and proposed investment provisions in the WTO) appear to be keeping governments from strengthening regulations for fear of being sued (Mann, 2001). If environmental regulations are not going to become any more stringent than the status quo, then we will see a continuation of poor quality regulations and an entrenchment of regulatory differences between countries. Aiming policy initiatives to ward against such a regulatory chill would be an easy start toward improving the situation.

tncs and global environmental governance

While TNCs are the subject of debate in terms of their performance with respect to the environment, there is also much discussion on the role TNCs play in the formation of global environmental governance. Moreover, there is debate over the idea of a treaty on corporate accountability that would incorporate environmental performance standards.

TNCs have operated through a number of channels in their bid to take part in the negotiation of global environmental governance (Levy and Newell, 2005). Lobbying the state at the domestic level before state delegations head to international environmental negotiations has traditionally been a key strategy for industry. In this way they are able to exert significant influence over governments' positions from behind the scenes through lobbying activities at the national level (for example, Susskind, 1994; Gleckman, 1995). While this is still an important strategy, business players are increasingly lobbying at the international level as well, via industry associations and industry representatives who gain observer status at such meetings. The presence of these actors at global negotiations is now a regular feature, as we have seen in the case of the waste trade, climate change, ozone depletion, biosafety and POPs.

Corporate lobby groups have also been active in other global forums for sustainable development, such as the Rio Earth Summit in 1992 and the World Summit on Sustainable Development (WSSD) in 2002. At Rio, industry groups formed the Business Council on Sustainable Development as a coordinated voice for industry (Chatterjee and Finger, 1994). Similarly, at WSSD, industry groups established the Business Action for Sustainable Development (BASD) to form a single industry voice. At both of these meetings, the main message from industry was that voluntary initiatives, rather than outside regulation of TNCs, would be the most effective and efficient way to promote sustainable development (Corporate Europe Observer, 2001).

A more diffuse but no less important way that TNCs have influenced global environmental governance is via their 'structural power', stressed by the critical camp. This structural power has been exerted in two important ways. First is the role they have carved out for themselves in terms of defining 'sustainable development' (Sklair, 2001). In defining the mainstream version of sustainable development in a way that enables them to maintain goals of economic globalization, including continued openness to global investment and faith in industry efforts to save the environment, they can escape much regulation. Second, the structural power of capital – nationally and globally – has been seen by some to be key in terms of explaining the influence that industry does indeed seem to have over government policy more broadly (Levy and Newell, 2005; Newell and Paterson, 1998). Some argue that the current era of increased global economic competition has meant that many states have pursued domestic policy outcomes which would be acceptable to corporations in order to keep or attract investment in their country (Barnet and Cavanagh, 1994). The mere threat or potential threat of relocation by global firms could be keeping governments from tightening or enforcing domestic environmental regulations (Korten, 1995).

Transnational corporate actors also influence global environmental governance through other international forums. Private firms have been involved in recent years in the establishment of private forms of governance such as voluntary codes of environmental conduct at both the national and international levels (Nash and Ehrenfeld, 1996). This includes participation in establishing industry-based environmental codes of conduct, such as the ISO 14000 and Responsible Care. The ISO 14000 environmental management standards is an interesting case in this regard, as it is the most widely recognized set of global standards of this sort. But debates are emerging as to whether these standards will really make a difference to firms' environmental behaviour, whether they will help improve environmental practice amongst firms in developing countries, and whether the drafting process was democratic (see Finger and Tamiotti, 1999; Krut and Gleckman, 1998). For example, the ISO 14000 standards only ask firms to abide by the environmental laws in the country in which they operate, which does little to raise standards. Moreover, the standards do not include performance criteria, but only changes to management practices. Thus, there is little direct pressure to improve performance. While they may not be the strongest in terms of leading to environmental improvements, the ISO 14000 environmental management standards have been recognized as legitimate standards by the WTO. In this way they are part of global structures of environmental governance. Some have expressed concern about this, as the drafting process was dominated by TNCs with very little input from governments and environmental groups (Clapp, 1998a; Roht-Arriaza, 1995).

It is often assumed that business players generally oppose strong global environmental rules because they impose costs on firms, but deciphering the business position on a particular environmental issue is not always so straight forward. In some cases corporate actors push for weak global environmental rules, but in some cases they are content to go along with strong rules pushed for by NGOs and states. Economic considerations are often the key to understanding the positions taken by corporate actors, though uncovering these motivations is often complex (Levy and Newell, 2000). In the cases of ozone-depleting substances and POPs, industry actors have largely been united in favor of strict rules calling for a ban on the production and trade of these harmful substances, largely because these industries can gain economically from the sale of substitutes. But in the case of waste recycling and biosafety, the entrenched industries' chances at gaining from substitutes are slim, so they have a much stronger stake in opposing strong rules which they see as harming the very core of their industry (Clapp, 2001, 2003; Levy, 1997). In the case of climate change, industry response has been varied among different firms, indicating that individual firms have different economic and political interests in this issue (Levy, 1997; Rowlands, 2000).

Recent years have seen some multilateral efforts geared toward promoting corporate accountability from the outside to complement the voluntary measures taken by industry itself. These include the UN's *Global Compact* and the OECD *Guidelines on Multinational Enterprises*, both of which include environmental goals and aspirations. To date these codes have been voluntary.

Some environmental NGOs are promoting a binding global treaty on corporate accountability. The idea of a global treaty on corporate accountability was put forward by several NGOs in the run-up to the WSSD in 2002, including proposals by Friends of the Earth International, the World Development Movement, Christian Aid and the Alliance for a Corporate-Free UN (Corporate Europe Observer, 2001, p. 6). Greenpeace, for example, unveiled its 'Bhopal Principles' at Johannesburg. The idea with these principles is to introduce them first as voluntary measures, but use them as a basis for a binding treaty. These proposals for a treaty on corporate accountability stress the importance of assigning liability to corporations for environmental damage that they cause, as well as requiring them to consult with and compensate affected communities. They also call for corporations to fully report their social and environmental impacts. NGOs would also like to see citizen and community rights as well as minimum environmental, social, labour and human rights standards incorporated into such an agreement (Friends of the Earth International, 2002; Greenpeace International, 2002).

Not surprisingly, industry is not at all keen on the idea of a legally binding treaty on corporate accountability, especially one that places a strong emphasis on the need to extend corporate liability for damages their operations cause (Moody Stuart, 2002). Though the issue was discussed at the WSSD, no binding commitments were made in this direction.

global development finance and the environment

The linkages between global finance and the environment have many facets. They may be less 'visible' than those of trade and investment, but are no less important. Global financing comprises many types of global transactions, including international borrowing and development assistance. Here the focus will be on sources of development assistance and lending. The largest development lending agency is the World Bank, which lent some US$19.5 billion dollars to developing countries in 2002

(World Bank website: <www.worldbank.org>). Both World Bank project lending to developing countries and accumulated developing country debt along with accompanying adjustment programmes have been subject to critique over their environmental implications. Export credit agencies have also become important sources of funding over the past decade, and like World Bank lending have drawn fire from NGOs over their environmental practices.

world bank project lending

Today some two-thirds of World Bank lending is for projects in developing countries, down from higher levels in previous decades. World Bank project lending has been the focus of much criticism from environmental groups since the 1980s. These groups targeted the World Bank because of its visibility and power to influence other sources of lending. There was widespread concern that the projects being funded were not taking environmental concerns adequately into account and resulted in a number of environmentally unsustainable projects. The World Bank's procedures for project design and evaluation had few if any references to the natural environment. This approach is not surprising given that the Bank's main approach to environmental issues at the time was to assume that economic growth would be beneficial not just in financial terms, but also in environmental terms, such that any problems were 'externalities' (Reed, 1997, pp. 229–30).

But the World Bank focus on lending for large-scale infrastructure projects – such as dams, power projects and roads – as well as migration schemes and industrial agricultural projects was seen by critics to be causing a great deal of environmental harm. Although the World Bank had some environmental policies in place by the 1980s, often these policies were not followed (Wade, 1997, pp. 634–7). Environmental groups followed the World Bank's environmental record closely, assisted in many cases by local groups in developing countries. Cases such as the Polonoreste road project in Brazil, which was linked by environmental groups to massive deforestation, as well as the Narmada Dam scheme in India which led to resettlement of thousands of people, were targeted by these campaigns (Rich, 1994). The Environmental Defense Fund (now Environmental Defense) played a key role in the US in terms of exposing the World Bank's record in an attempt to force it to adopt more environmentally friendly lending policies.

When the Environmental Defense Fund testified in the US Congress in 1985, detailing the deforestation in Brazil as a result of the Polonoreste project, as well as the fact that the World Bank was not following its own environmental policies, the US government threatened to withhold funds from the Bank. This in turn led to the cessation of the project on the part of the Bank. By 1987 the World Bank began to make a specific effort to 'green' its project lending policies (Wade, 1997, p. 680). It undertook a major restructuring at this time and created a new Environment Department and increased its environment staff over the course of the early 1990s (Reed, 1997). After 1991 the World Bank also began to require environmental impact assessments of all projects. With its new policies in place, the Bank conducted an independent review of the Narmada Valley projects. The 1992 report that came out of this review highlighted the severe problems associated with the project (Morse, 1992). The following year the World Bank along with the Indian government agreed to halt the loan for the Narmada Dam scheme, though the Indian government carried on with the project without international financing (Caufield, 1996, p. 28).

Though the Bank began to 'green' itself in the early part of the 1990s, critics noted that the newfound enthusiasm for the environment began to fade at the Bank in the late 1990s. While environmental project lending from the World Bank had increased by a factor of 30 between 1990 and 1995, the trend did not continue. Environmental lending fell from 3.6 per cent of total lending in 1994 to just 1.02 per cent of total lending by 1998 (Friends of the Earth et al., n.d.). Moreover, there were also continuing complaints that the World Bank did not engage in adequate consultation with NGOs or affected communities regarding the environmental impacts of loans. Many, including some within the Bank, now see that while the World Bank has begun to pay more attention to the environment in its project lending, there is still a long way to go (World Bank, 2001, p. 20).

The World Bank is also involved in the Global Environment Facility (GEF), a multilateral source of funding for developing countries specifically geared toward projects with environmental benefits. The Bank, along with the United Nations Development Programme (UNDP) and the UNEP administer this fund. The World Bank, however, has acted as the lead agency and has handled the financing of the projects. The aim of the GEF is to provide grant funding for developing countries that covers the 'incremental costs' of meeting their global environmental obligations under international environmental agreements. These 'incremental costs' are in effect the 'extra' costs incurred by developing countries to undertake projects that have global benefits (Streck, 2001, p. 73). The GEF initially only granted funds for projects with global benefits for international waters, climate change, ozone depletion and biodiversity. It has since added efforts to reduce POPs and land degradation to its list (<http://lnweb18.worldbank.org/ESSD/envext.nsf/45ByDocName/Themes>). Environmental NGOs and developing countries were very critical of the GEF in its early years (Fairman, 1996). With the World Bank as lead institution, it soon came to dominate the GEF's operations, resulting in charges of little consultation on project design and implementation with NGOs and affected local communities. Discussions on restructuring the GEF began in the early 1990s, and resulted in a more democratic decision-making procedure and more involvement of NGOs and affected communities (Streck, 2001). Critics, however, remain sceptical of the organization (Horta et al., 2002).

lending for structural adjustment

In addition to critiques that have been launched against the environmental impact of project lending, there is also a growing debate over the environmental implications of lending for structural adjustment programmes (SAPs). SAP loans are typically coordinated between the IMF and the World Bank. Today, about one-third of World Bank lending supports programs of structural adjustment, a figure which has grown since the loans were first introduced in the early 1980s. These loans from the IMF and World Bank are general balance of payments support in return for macroeconomic policy changes, designed to help countries improve economic growth and ultimately their ability to repay their external debts. These types of reforms originate from neoliberal economic thinking, and are typically those that open up a country's economy to become more integrated into the global economy. They include currency devaluation, trade liberalization, removal of subsidies, liberalization of investment policies and cuts to government expenditures.

The IMF and World Bank did not take into consideration the environmental implications of the early structural adjustment loans, as both institutions assumed that policies promoting growth would be positive for the environment (Reed, 1996). However, some were quick to disagree with this assessment. Critics argued that currency devaluation, trade liberalization and cuts to government spending in particular in many developing countries had direct impacts on the quality of the environment. Such effects are seen, for example, in the rising rates of deforestation in a number of adjusting countries, including countries such as Brazil, Ghana, Cameroon, the Philippines and Zambia, as exports of timber were encouraged by these policy shifts (for example, George, 1992; Hogg, 1994; Toye, 1991). There have also been charges that mining has increased in some adjusting countries as a result of currency devaluation and the liberalization of trade and investment policies. The Philippines for example was asked to rewrite its mining code, which led to increased foreign direct investment and more intensive mining activities (Friends of the Earth, 1999, p. 10). Cuts to government expenditures have also been criticized for reducing governments' environmental budgets. Friends of the Earth, for example, claims that Thailand's budget for pollution control was slashed under its SAP by 80 per cent between 1997 and 1999 (Friends of the Earth, 1999, p. 7).

While the IMF has tended not to engage in the debate, the World Bank has defended its adjustment policies on the environment front. It has claimed that while it is difficult to generalize about the impact of adjustment on the environment, in most cases the effect has probably been neutral or positive (see Glover, 1995; Pearce, et al., 1995; World Bank, 1994). This is because removal of subsidies and other inefficiencies in the economy as a result of adjustment measures is seen to lead to prices for natural resources that more accurately reflect their scarcity, and should lead to improved conservation. Cuts to subsidies for fuel and pesticides, for example, should lead to less use of these environmentally harmful substances. Benefits from trade liberalization and currency devaluation also encourage the production of export crops like coffee, rubber, palm oil and cocoa – all crops with strong root systems which can help prevent soil erosion. While it has defended adjustment lending on environmental grounds, the World Bank in recent years has undertaken studies on ways to better understand the linkages between this type of lending and the environment (World Bank, 2001).

export credit agencies

An increasingly important source of finance for developing countries are export credit agencies (ECAs). Based in developed countries, ECAs are public agencies that provide support or credit – in the form of government-backed loans, investment guarantees and insurance – to developing countries. The credits from ECAs are specifically tied to business contracts with companies based in the lending country (Rich, 2000, p. 34). The developing country borrower that takes the credit from the ECA, sometimes a government, although sometimes a private company, is responsible for repaying the loan.

There are ECAs based in most industrialized countries, including, for example, the Export-Import Bank of the United States (the Ex-Im Bank), the US Overseas Private Investment Corporation (OPIC), the Japan Bank for International Cooperation (JBIC)/International Financial Corporation (formerly JEXIM), the Canadian Export Development Corporation, the

German Hermes Kreditversicherung-AG (Hermes), the UK Export Credit Guarantee Department (ECGD), and the Australian Export Finance and Insurance Corporation (EFIC) (ECA Watch: <www.eca-watch.org/eca/directory.html>). These ECAs now provide US$100–200 billion in lending to developing countries every year, subsidizing nearly 10 per cent of global exports (Goldzimer, 2003, p. 2; UNEP webpage: <www.uneptie.org/energy/act/fin/ECA/>). This makes these agencies extremely important global financial actors for developing countries. For example, they account for around 40 per cent of developing country debt owed to official creditors (Goldzimer, 2003, p. 4).

While they account for a growing amount of project financing to developing countries, until recently little attention was paid to their operations or their environmental impacts. But in the past few years ECAs have come under attack by environmental groups. Export credits have typically been given for very risky ventures, including financing and insurance for oil, gas, logging and mining projects, as well as large-scale dams, nuclear power, chemical facilities and road building in remote areas. Environmental groups have charged ECAs with failing to consider the environmental implications of the projects they fund. This is not surprising, as unlike the World Bank, most ECAs are not bound to any environmental or social policies or standards. Critics charge that they are also very secretive, making it extremely difficult to get full details on the projects they finance and their environmental impact (Rich, 2000, p. 35).

Interestingly, the US ECAs are the only ones to be required to undergo environmental assessments for the projects they finance. This is largely because the ECAs in the US have strong links to the US Agency for International Development (USAID), which adheres to strict environmental standards (Rich, 2000, pp. 35–6). But while the US has taken a lead role on the environmental front when it comes to ECAs, this kind of pressure is lacking in other countries. If a US ECA rejects a project on environmental grounds, there is little stopping another ECA from another country financing it instead. The Three Gorges Dam in China is seen as a classic case of this problem. When the US Ex-Im Bank turned it down for environmental reasons, it was instead funded by the Canadian, German, Swedish and Swiss ECAs (Friends of the Earth et al., n.d.).

Recent years have seen an effort on the part of NGOs and the US and UK governments to work toward a global set of guidelines for ECAs that would include both environmental and social standards. Following on this push, the OECD has initiated talks on a set of common standards for ECAs. Thus far, however, little concrete has come out of this effort.

Parallel to the OECD efforts, environmental NGOs presented the Jakarta Declaration for Reform of Official Export Credit and Investment Insurance Agencies in 2000.[4] It calls not only for greater transparency of ECAs, but also for binding social and environmental guidelines and standards that are at least equal to those of the World Bank and the OECD. It also calls for the cancellation of ECA debt of the poorest countries. Thus far, while many NGOs have signed onto it, the Jakarta Declaration has received little attention from governments.

case study: international transfer of hazardous wastes from rich to poor countries

The debates regarding the global political economy and the environment are large and multifaceted. Trade, investment and financial issues overlap in many ways, and in looking at a specific environmental problem, it is important to keep this in mind. I have chosen here to focus on the problem of the international transfer of hazardous waste from rich to poor countries. This case study is useful in that it is closely linked to global trade, investment and finance, in different ways (Clapp, 2001). However, while a complex problem, it does not necessarily illustrate all of the points made in the discussion above. In particular, it highlights questions of trade, investment, and to a lesser extent financing issues.

Trade is the most explicit link to the transfer of hazardous wastes from rich to poor countries. It became apparent in the 1980s that toxic wastes from industrial countries were being shipped to a number of developing countries. Africa, Asia and Latin America were all affected by this practice. So the problem was identified as a trade issue. But the problem has ties to global finance, as it was the indebtedness of these countries in the 1980s that led them to be targets for waste, as accepting it brought desperately needed sources of foreign exchange. But while it was initially seen by some in these countries as a means to earn their way out of debt, it was soon recognized that the price was simply not worth it. Moreover, public outcry in both rich and poor countries over this practice exploded in the 1980s.

In the face of public criticism once this practice came to light, the international community negotiated the Basel Convention on the Transboundary Movement of Hazardous Wastes and their Disposal in 1989. This multilateral environmental agreement seeks to control the international trade in hazardous waste by only allowing the trade to occur with prior notification and consent. But it did not ban the trade in hazardous wastes outright. Regional and national level laws also emerged, many of which were more stringent than the provisions laid out in the Basel Convention. The rules of the Basel Convention enabled many developing countries to refuse waste imports, and by the mid-1990s the blatant dumping of the 1980s had dwindled to a trickle. But a new problem appeared on the scene with respect to the waste trade. This was the export of wastes destined for recycling operations in the developing world. Although wastes exported for recycling are covered by the rules of the Basel Convention, it was difficult to enforce the prior notification and consent procedure when the wastes were labelled as 'products' to be recycled rather than as wastes to be dumped. But the recycling of hazardous wastes in poor countries is more often than not carried out in environmentally unsound conditions, which many consider to be as bad if not worse than simple disposal of the wastes (Puckett, 1994).

The surfacing of the recycling problem in the early 1990s prompted environmental NGOs to call for changes to the rules of the Basel Convention to ban the trade in wastes, for both disposal and recycling, between rich countries (explicitly the OECD countries) and poor countries. Developing countries were largely in agreement that the Basel Convention should incorporate more explicit measures to halt this practice. However, industry lobby groups, along with several key OECD countries, were opposed. They did not want to see stricter trade rules in an MEA like the Basel Convention, and even argued that it contravened GATT/WTO rules to ban the waste trade (Krueger, 1999; O'Neill, 2000). Despite their protests, the conference of parties to the Basel Convention adopted an amendment in 1995 which bans the export of toxic wastes from OECD to non-OECD countries for both disposal and recycling. Since the amendment was adopted, environmental groups have lobbied hard for states to ratify it, as three quarters of the parties must ratify it before it comes into force. Industry groups have focused their energies on encouraging countries not to ratify the agreement. At the time of writing there were 56 ratifications, six short of the 62 ratifications required for the amendment to come into force.

The transfer of hazards between rich and poor countries also has important links to investment. While much of the literature on pollution havens and industry flight argues that firms do not relocate to developing countries for environmental reasons, most of these studies do make an exception to this argument when it comes to the most highly hazardous industries (Leonard, 1988). In other words, it appears that the dirtiest and most toxic industries do tend to move to jurisdictions with weak environmental regulations. Those from a more radical perspective have been making this argument for years. They have argued that hazards were moving both as a result of higher regulation being imposed in rich countries (industry flight) as well as due to weaker rules in poor countries (pollution havens) (Castleman, 1985). Moreover, double standards in these industries were blatant. The accident at the Union Carbide pesticide plant in Bhopal in 1984 is a prime example (Morehouse, 1994; Rajan, 2001). The problem has been present since at least the 1970s, but the concern over it is heightened in recent years. Since the Basel Convention amendment which bans the export of waste to developing countries, many are worried that the problem of hazardous industry transfer could only get worse. If firms that produce toxic wastes are unable to send their wastes to developing countries, they may be even more encouraged to move their entire production process to these countries (Clapp, 1998b). Evidence of this type of hazard transfer has been seen in a number of case studies, most prominently the *maquiladora* firms in Mexico (Frey, 2003).

Hazardous waste transfers from rich to poor countries via exports and transnational investment demonstrate clearly the linkages between the global political economy and the environment. The globalization of the production process and the footloose nature of transnational investment is a key factor behind movement of hazards around the world. Developing countries have been particularly vulnerable to this type of trade and investment due to weaker environmental regulations and/or lack of enforcement of such regulations. The liberalization of trade and investment policies in those countries pursuing SAPs has played a role in opening up these countries to new trade and investment of this sort.

conclusion

This chapter has outlined the theoretical debates and policy dilemmas that currently dominate the field of international political economy and the environment. While there have been strongly opposing views expressed by neoclassical economists and critical thinkers over the past decade regarding the environmental benefits and downsides of international trade, investment and financing, liberal institutionalists have begun to advocate a middle ground. This middle view has been expressed especially regarding environmental policy issues linked to the global economy. Advocated by this view is enhanced state (and in some cases private sector) management of international trade, investment and finance where clear environmental harm results from these activities. Such an approach, these advocates argue, is necessary in the face of a deadlock between the opposing sides in the theoretical debate, which they see as hampering progress in the policy realm. This institutionalist approach was very much present at the WSSD in 2002. Whether this approach will make significant improvements to the state of the world's environment remains to be seen. Assessing the influence of this theoretical development is an important area for future research on the links between the international political economy and the environment. Careful study of the effectiveness of the various approaches in practice will not only provide an empirical base to judge their merits, but also inform the theoretical debates and help them to move forward.

In addition to the rich theoretical avenues for further research, there are several policy areas that merit further investigation. I will highlight three areas here that I feel deserve more detailed analysis in the literature. The first is the relationship between multilateral environmental agreements and trade agreements. While a body of work on the interface between WTO rules and MEAs has emerged, further work is necessary on this front as the WTO begins to tackle this issue. Moreover, a number of countries are now pursuing bilateral trade agreements, and there is a need to study the impact of these regional agreements on states' environmental obligations through MEAs. Second, there is a need for further work on the environmental impacts of ECAs. A number of NGOs have begun to examine this issue and a small set of case studies have been publicized, but there is ample room for more detailed and systematic case studies by academics of the impacts of this type of lending on the environment. Finally, the area of corporate accountability is ripe for further study. With the promotion of the idea of a corporate accountability treaty at the WSSD by a number of NGOs, it will be important for researchers to watch and investigate future developments in this area.

notes

1. For a much more detailed look at these debates, see Clapp and Dauvergne (2005).
2. In Clapp and Dauvergne (2005) these various camps of thinkers are defined as market liberals, institutionalists, bioenvironmentalists, and social greens. Here the neoclassical economic thinkers equate to the market liberals; the liberal institutionalists equate to the institutionalists; and the critical camp equates to the social greens and bioenvironmentalists.
3. APEC's members include: Australia, Brunei Darussalam, Canada, Chile, People's Republic of China, Hong Kong, China, Indonesia, Japan, Republic of Korea, Malaysia, Mexico, New Zealand, Papua New Guinea, Peru, Philippines, Russia, Singapore, Chinese Taipei, Thailand, United States, and Vietnam.
4. The Jakarta Declaration is available at <www.eca-watch.org/jakarta_english.html>.

bibliography

Barnet, Richard J., and John Cavanagh (1994) *Global Dreams: Imperial Corporations and the New World Order*, New York: Simon & Schuster.
Bhagwati, Jagdish (1993) 'The Case for Free Trade', *Scientific American* 269, 5, pp. 42–9.
Biermann, Frank (2000) 'The Case for a World Environment Organization', *Environment* 42, 9, pp. 22–31.
Castleman, Barry (1979) 'The Export of Hazardous Factories to Developing Nations', *International Journal of Health Services* 19, 4, pp. 569–606.
Castleman, Barry (1985) 'Double Standards in Industrial Hazards', in Jane Ives (ed.) *The Export of Hazard*, London: Routledge.
Caufield, Catherine (1996) *Masters of Illusion: The World Bank and the Poverty of Nations*, London: Macmillan.
Chatterjee, Pratap, and Matthias Finger (1994) *The Earth Brokers: Power, Politics and World Development*, London: Routledge.
Clapp, Jennifer (1998a) 'The Privatization of Global Environmental Governance: ISO 14000 and the Developing World', *Global Governance*, 4, 3, pp. 295–316.
Clapp, Jennifer (1998b) 'Foreign Direct Investment in Hazardous Industries in Developing Countries: Rethinking the Debate', *Environmental Politics* 7, 4, pp. 92–113.
Clapp, Jennifer (2001) *Toxic Exports: The Transfer of Hazardous Wastes from Rich to Poor Countries*, Ithaca: Cornell University Press.
Clapp, Jennifer (2002) 'What the Pollution Havens Debate Overlooks', *Global Environmental Politics* 2, 2, pp. 11–19.
Clapp, Jennifer (2003) 'Transnational Corporate Interests and Global Environmental Governance: Negotiating Rules for Agricultural Biotechnology and Chemicals', *Environmental Politics* 12, 4, pp. 1–23.
Clapp, Jennifer, and Peter Dauvergne (2005) *Paths to a Green World: The Political Economy of the Global Environment*, Cambridge, Mass.: MIT Press.
Conca, Ken (2000) 'The WTO and the Undermining of Global Environmental Governance', *Review of International Political Economy* 7, 3, pp. 484–94.
Corporate Europe Observer (2001) 'Industry's Rio+10 Strategy: Banking on Feelgood PR', *CEO Quarterly Newsletter*, 10.
Daly, Herman (1993) 'The Perils of Free Trade', *Scientific American*, 269, 5, pp. 50–7.
Daly, Herman (1996) *Beyond Growth: The Economics of Sustainable Development*, Boston: Beacon Press.
DeSombre, Elizabeth R., and J. Samuel Barkin (2002) 'Turtles and Trade: The WTO's Acceptance of Environmental Trade Restrictions', *Global Environmental Politics* 2, 1, pp. 12–18.
Esty, Daniel (1994) *Greening the GATT: Trade, Environment and the Future*, Washington, DC: Institute for International Economics.
Esty, Daniel (2001) 'Bridging the Trade–Environment Divide', *Journal of Economic Perspectives* 15, 3, pp. 113–30.
Fairman, David (1996) 'The Global Environment Facility: Haunted by the Shadow of the Future', in Robert Keohane and Marc Levy (eds) *Institutions for Environmental Aid*, Cambridge, Mass.: MIT Press, pp. 55–87.

Finger, Matthias, and Ludivine Tamiotti (1999) 'The Emerging Linkage Between the WTO and the ISO: Implications for Developing Countries', *IDS Bulletin, Globalisation and the Governance of the Environment*, 30, 3, pp. 8–16.

Frey, R. Scott (2003) 'The Transfer of Core-Based Hazardous Production Processes to the Export Processing Zones of the Periphery: The Maquiladora Centers of Northern Mexico', *Journal of World-Systems Research*, 9, 2, pp. 317–54.

Friends of the Earth (1999) *The IMF: Selling the Environment Short*. Washington, DC: Friends of the Earth. Available at <www.foe.org/imf/index.html>.

Friends of the Earth et al. (n.d.) 'A Race to the Bottom: Creating Risk, Generating Debt, and Guaranteeing Environmental Destruction', available at <www.foe.org/international/ecareport.pdf>.

Friends of the Earth International (2002) 'Towards Binding Corporate Accountability', available at <www.foei.org/publications/corporates/accountability.html>.

George, Susan (1992) *The Debt Boomerang*, London: Pluto Press.

Georgescu-Roegen, Nicholas (1971) *The Entropy Law and the Economic Process*, Cambridge, Mass.: Harvard University Press.

Geradin, Damien (2002) 'The European Community: Environmental Issues in an Integrated Market', in Richard Steinberg (ed.) *The Greening of Trade Law: International Trade Organizations and Environmental Issues*, Lanham, Md: Rowman and Littlefield, pp. 117–54.

Gleckman, Harris (1995) Transnational Corporations' Strategic Responses to 'Sustainable Development', in Helge Ole Bergesen and Georg Parmann (eds) *Green Globe Yearbook,* Oxford: Oxford University Press, pp. 93–106.

Glover, David (1995) Structural Adjustment and the Environment, *Journal of International Development* 7, 2, pp. 285–9.

Goldzimer, Aaron (2003) 'Worse than the World Bank? Export Credit Agencies – The Secret Engine of Globalization', *Food First Backgrounder* 9, 1, available at <www.foodfirst.org/pubs/backgrdrs/2003/w03v9n1.pdf>.

Greenpeace International (2002) *Corporate Crimes: The Need for an International Instrument on Corporate Accountability and Liability*, Amsterdam: Greenpeace.

Grossman, Gene, and Alan Krueger (1995) 'Economic Growth and the Environment', *Quarterly Journal of Economics* 110, 2, pp. 353–77.

Hall, Derek (2002) 'Environmental Change, Protest and Havens of Environmental Degradation: Evidence from Asia', *Global Environmental Politics* 2, 2, pp. 20–8.

Hogg, Dominic (1994) *The SAP in the Forest,* London: Friends of the Earth.

Horta, Korinna, Robin Round and Zoe Young (2002) 'The Global Environmental Facility: The First Ten Years – Growing Pains or Inherent Flaws?', Report for Environmental Defense and the Halifax Initiative.

Hufbauer, Gary, Daniel Esty, Diana Orejas, Luis Rubio and Jeffrey Schott (2000) *NAFTA and the Environment: Seven Years Later*, Washington, DC: IIE.

Karliner, Joshua (1997) *The Corporate Planet, Ecology and Politics in the Age of Globalization*, San Francisco: Sierra Club.

Korten, David C. (1995) *When Corporations Rule the World*, West Hartford and San Francisco: Kumarian Press and Berrett-Koehler Publishers.

Krueger, Jonathan (1999) *International Trade and the Basel Convention*, London: Earthscan.

Krut, Riva, and Harris Gleckman (1998) *ISO 14001: A Missed Opportunity for Global Sustainable Industrial Development*, London: Earthscan.

Leonard, H. Jeffrey (1988) *Pollution and the Struggle for the World Product*, Cambridge: Cambridge University Press.

Levy, David (1997) 'Business and International Environmental Treaties: Ozone Depletion and Climate Change', *California Management Review* 39, 3, pp. 54–71.

Levy, David, and Peter Newell (2000) 'Oceans Apart? Business Responses to Global Environmental Issues in Europe and the United States', *Environment* 42, 9, pp. 8–20.

Levy, David, and Peter Newell (eds) (2005) *The Business of Global Environmental Governance*, Cambridge, Mass.: MIT Press.

Logsdon, Jeanne, and Bryan Husted (2000) 'Mexico's Environmental Performance under NAFTA: The First Five Years', *Journal of Environment and Development*, 9, 4, pp. 370–83.

Low, Patrick, and Alexander Yeats (1992) 'Do "Dirty" Industries Migrate?' (World Bank Discussion Paper No. 159), in Patrick Low (ed.) *International Trade and the Environment*, Washington, DC: World Bank, pp. 89–103.

Mani, Muthukumara, and David Wheeler (1998) 'In Search of Pollution Havens? Dirty Industry in the World Economy, 1960–1993', *Journal of Environment and Development* 7, 3, pp. 215–47.

Mann, Howard (2001) *Private Rights, Public Problems: A Guide to NAFTA's Chapter on Investor Rights*, Winnipeg: IISD. Available at <www.iisd.org/pdf/trade_citizensguide.pdf>.

Moody Stuart, Mark (2002) 'Globalization in the 21st Century: An Economic Basis for Development', *Corporate Environmental Strategy* 9, 2, pp. 115–21.

Morehouse, Ward (1994) 'Unfinished Business: Bhopal Ten Years After', *The Ecologist* 24, 5, pp. 164–8.

Morse, Bradford (1992) *Sardar Sarovar: Report of the Independent Review*, Ottawa: Resource Futures International Inc.

Nash, Jennifer, and John Ehrenfeld (1996) 'Code Green,' *Environment* 37, 1, pp. 16–45.

Neumayer, Eric (2001) *Greening Trade and Investment: Environmental Protection Without Protectionism*, London: Earthscan.

Newell, Peter, and Matthew Paterson (1998) 'A Climate for Business: Global Warming, the State and Capital', *Review of International Political Economy*, 5, 4, pp. 679–703.

O'Neill, Kate (2000) *Waste Trading among Rich Nations: Building A New Theory of Environmental Regulation*, Cambridge, Mass.: MIT.

Pearce, David, Neil Adger, David Maddison and Dominic Moran (1995) 'Debt and the Environment', *Scientific American*, June, pp. 52–6.

Pearson, Charles (ed.) (1987) *Multinational Corporations, the Environment, and the Third World*, Durham, NC: Duke University Press.

Perkins, Nancy (1998) 'World Trade Organization: United States – Import Prohibition of Certain Shrimp and Shrimp Products'. Available at American Society for International Law website: <www.asil.org/ilm/nancy.htm>.

Porter, Gareth (1999) 'Trade Competition and Pollution Standards: "Race to the Bottom" or "Stuck at the Bottom"?', *Journal of Environment and Development* 8, 2, pp. 133–51.

Puckett, Jim (1994) 'Disposing of the Waste Trade: Closing the Recycling Loophole', *The Ecologist* 24, 2, pp. 53–8.

Rajan, S. Ravi (2001) 'Toward a Metaphysic of Environmental Violence: The Case of the Bhopal Gas Disaster', in Nancy Lee Peluso and Michael Watts (eds) *Violent Environments*, Ithaca: Cornell University Press, pp. 380–98.

Reed, David (ed.) (1996) *Structural Adjustment, the Environment and Sustainable Development*, London: Earthscan.

Reed, David (1997) 'The Environmental Legacy of the Bretton Woods: The World Bank', in Oran Young (ed.) *Global Governance: Drawing Insights from the Environmental Experience*, Cambridge, Mass.: MIT Press, pp. 227–45.

Rich, Bruce (1994) *Mortgaging the Earth: The World Bank, Environmental Impoverishment, and the Crisis of Development*, London: Earthscan.

Rich, Bruce (2000) 'Exporting Destruction', *Environmental Forum* September–October, pp. 32–40.

Roht Arriaza, Naomi (1995) 'Shifting the Point of Regulation: The International Organization for Standardization and Global Law-Making on Trade and the Environment', *Ecology Law Quarterly* 22, 3, pp. 502–15.

Rowlands, Ian (2000) 'Beauty and the Beast? BP's and Exxon's Positions on Global Climate Change', *Environment and Planning C: Government and Policy* 18, 3, pp. 339–54.

Sachs, Wolfgang (1999) *Planet Dialectics: Explorations in Environment and Development*, London: Zed Books.

Sklair, Leslie (2001) *The Transnational Capitalist Class*, London: Blackwell.

Soloway, Julie (2002) 'The North American Free Trade Agreement: Alternative Models of Managing Trade and the Environment', in Richard Steinberg (ed.) *The Greening of Trade Law: International Trade Organizations and Environmental Issues*, Lanham, Md: Rowman and Littlefield, pp. 155–88.

Stevis, Dimitris, and Valerie Assetto (eds) (2001) *The International Political Economy of the Environment: Critical Perspectives*, Boulder: Lynne Rienner.

Stevis, Dimitris, and Stephen Mumme (2000) 'Rules and Politics in International Integration: Environmental Regulation in NAFTA and the EU', *Environmental Politics* 9, 4, pp. 20–42.

Stilwell, Matthew, and Richard Tarasofsky (2001) *Towards Coherent Environmental and Economic Governance: Legal and Practical Approaches to MEA-WTO Linkages*, Gland: WWF. Available at <www.ciel.org/Publications/Coherent_EnvirEco_Governance.pdf>.

Streck, Charlotte (2001) 'The Global Environment Facility – a Role Model for International Governance?', *Global Environmental Politics* 1, 2, pp. 71–94.

Susskind, Lawrence E. (1994) *Environmental Diplomacy: Negotiating More Effective Global Agreements*, Oxford: Oxford University Press.

Toye, John (1991) 'Ghana', in Paul Mosley, Jane Harrigan and John Toye (eds) *Aid and Power*, Vol. 2, London: Routledge, pp 151–200.

UNCTAD (2001) *World Investment Report 2001: Promoting Linkages*, New York and Geneva: UN.

UNCTAD (2002) *World Investment Report 2002: Transnational Corporations and Export Competitiveness*, New York and Geneva: UN.

UN Transnational Corporations and Management Division, Department of Economic and Social Development (1992) *World Investment Report 1992*, New York: United Nations.

Vogel, David (2000) 'International Trade and Environmental Regulation', in Norman J. Vig and Michael E. Kraft (eds) *Environmental Policy: New Directions for the Twenty-First Century*, 4th edn, Washington, DC: CQ Press.

Wade, Robert (1997) 'Greening the Bank: The Struggle over the Environment, 1970–1995', in John Lewis and Richard Web (eds) *The World Bank: Its First Half Century*, Vol. 2, Washington, DC: Brookings Institute, pp. 611–734.

Weinstein, Michael, and Steve Charnovitz (2001) 'The Greening of the WTO', *Foreign Affairs* 80, 6, pp. 147–56.

Williams, Marc (2001) 'In Search of Global Standards: The Political Economy of Trade and the Environment', in Dimitris Stevis and Valerie Assetto (eds) *The International Political Economy of the Environment: Critical Perspectives*, Boulder: Lynne Rienner, pp. 39–61.

World Bank (1992) *World Development Report 1992*, New York: Oxford University Press.

World Bank (1994) *Adjustment in Africa: Reforms, Results and the Road Ahead*, Oxford: Oxford University Press.

World Bank (2001) 'Adjustment Lending Retrospective: Final Report, Operations Policy and Country Services', available at <http://lnweb18.worldbank.org/ext/language.nsf/0a70c4735d4fc79e852566390063b35c/eec155910a2720a885256ae30076a40c/$FILE/alretro.pdf>.

World Bank (2003) *World Development Indicators* (online). Available at <www.worldbank.org>

Zarsky, Lyuba (2002) 'APEC: The "Sustainable Development" Agenda', in Richard Steinberg (ed.) *The Greening of Trade Law: International Trade Organizations and Environmental Issues*, Lanham, Md: Rowman and Littlefield, pp. 221–47.

annotated bibliography

Neumayer, Eric (2001) *Greening Trade and Investment: Environmental Protection Without Protectionism*, London: Earthscan. This book provides a comprehensive and detailed analysis of trade and investment issues as they relate to the environment. It covers the environmental implications of WTO trade rules as well as the links between the NAFTA investment provisions and the environment. The book also suggests areas for reform of international trade and investment regimes which would take the environment more seriously into account.

Levy, David, and Peter Newell (eds) (2005) *The Business of Global Environmental Governance*, Cambridge, Mass.: MIT Press. This book is a collection of articles on the role of business actors in global environmental governance. The contributions, from a range of authors and disciplines, provide a comparative examination of the response of business to a variety of environmental issues, as well as industry 'greening' strategies.

Stevis, Dimitris, and Valerie Assetto (eds) (2001) *The International Political Economy of the Environment: Critical Perspectives*, Boulder: Lynne Rienner. This book is a collection of critical essays on the interface between the global economy and the environment. Its authors examine a variety of issues, including problems at both theoretical and policy levels. The book also includes both Southern and Northern perspectives on the problems.

Wade, Robert (1997) 'Greening the Bank: The Struggle over the Environment, 1970–1995', in John Lewis and Richard Web (eds) *The World Bank: Its First Half Century*, Vol. 2, Washington, DC: Brookings Institute, pp. 611–734. This article provides a definitive history of the World Bank's environmental record to the mid-1990s. It covers the challenges and campaigns against the Bank's poor environmental record up to the 1980s, as well as the subsequent 'greening' of the World Bank. It also assesses the effectiveness of the Bank's environmental reforms.

Clapp, Jennifer, and Peter Dauvergne (2005) *Paths to a Green World: The Political Economy of the Global Environment*, Cambridge, Mass.: MIT Press. This book systematically examines the linkages between the global economy and the environment through the lenses of four worldviews: market liberals, institutionalists, bioenvironmentalists and social greens. The book explains the contrasting viewpoints and policy dilemmas that dominate the field, covering globalization, growth, trade, investment and finance.

7
transnational actors in international environmental politics

michele m. betsill

This chapter examines scholarship on transnational actors in international environmental politics (IEP). Transnational actors engage in interactions across national boundaries but do 'not operate on behalf of a national government or an intergovernmental organization' (Risse-Kappen, 1995, p. 3).[1] While realists dismiss claims about the significance of transnational actors in world politics, scholars of IEP have long recognized their importance in processes of global governance. Virtually every study of IEP acknowledges their presence in political processes related to the environment and such actors increasingly are the primary focus of analysis. This largely reflects the fact that transnational actors have a stronger presence in the environmental issue area than in many other areas of concern to international relations scholars (such as security and trade).

The chapter begins with a discussion of the field's theoretical roots, followed by consideration of 'who' participates in transnational environmental politics. The field lacks a clear consensus on the nature (or name) of its basic unit of analysis. The third section presents findings on how transnational actors engage in IEP, the effects of their participation, and issues related to their internal dynamics. In this section, I also discuss some of the methodological challenges encountered by scholars of transnational environmental politics. The fourth section consists of a brief case study of the Climate Action Network, a transnational advocacy network involved in the international politics of climate change. In the conclusion, I offer some thoughts about the future direction of this field of research.

theoretical roots

The study of transnational environmental politics came of age in the early 1990s, motivated in large part by the 1992 United Nations Conference on Environment and Development held in Rio de Janeiro, Brazil (the Rio Conference). Tens of thousands of citizens from around the world gathered to interact with government delegates and to participate in their own parallel conference. The Rio Conference raised awareness that governments alone cannot manage global environmental threats and created a new research agenda focused on understanding what transnational actors do, how they participate in international environmental politics and with what effect.

However, the field's roots are located in earlier debates about the state-centrism of international relations theories and state–society relations. The study of transnational actors in world politics first emerged in the context of challenges to the state-centrism of the dominant theoretical approaches in international relations, neorealism and neoliberalism. Risse (2002) claims the first wave of challenge came from integration theory in the 1950s (for example, Deutsch, 1957; Haas, 1958; Mitrany, 1966), followed by a second wave beginning in the 1970s centered on debates about transnational relations (for example, Keohane and Nye, 1977; Skjelsbaek, 1971; Willetts, 1982). These two waves have been differentiated in terms of pluralism versus transnationalism (Reinalda, 2001) as well as types of institutional considerations, with the integration literature primarily concerned with supranational organizations such as the European Community and the transnational relations literature particularly interested in multinational corporations (Risse, 2002; Wapner, 1996).

More recently, the interest in 'global governance' has created a new space for highlighting the role of transnational actors (Edwards, 2001; Falkner, 2003; Khagram et al., 2002; Rosenau and Czempiel, 1992; Young, 1997). Scholars working in this area suggest that states are increasingly sharing responsibility for addressing 'the world's growing agenda of border-crossing problems' with transnational actors (Florini, 2000, p. 3; see also Mathews, 1997). Here it should be noted that research on transnational actors in IEP has shaped the wider discussion of global governance in the field of international relations (Haas, 1992; Keck and Sikkink, 1998; Lipschutz, 1992; Princen and Finger, 1994; Wapner, 1996).

The study of transnational environmental politics also engages debates about state–society relations and has links to work on the concept of civil society in the fields of political theory and comparative politics. In those fields, civil society is generally conceptualized as a domain separate from the state in which individuals engage in voluntary association (Cohen and Arato, 1992; Hall, 1995; Seligman, 1992). The concept of civil society has been extended into the international arena as scholars recognized that such voluntary association increasingly takes place across national boundaries (Edwards and Gaventa, 2001; Florini, 2000; Lipschutz, 1992; Wapner, 1996). Wapner (1997, p. 66) defines global civil society as 'the domain that exists above the individual and below the state but also across state boundaries, where people voluntarily organize themselves to pursue various aims'.

For some scholars, the significance of global civil society lies in its challenge to the notion that governments monopolize political space. Wapner (1996, p. 3) argues: 'there are other arenas for carrying out efforts that are separate from the realm of the government. These other arenas can be found in what is called global civil society, and the attempt to use them for environmental protection purposes is a form of world civic politics.' While some scholars contend that environmental politics is characterized by struggles between states and transnational actors (Clark et al., 1998; Close, 1998; Newell, 2000; Willetts, 1996c), others suggest that the relationship may be complementary (Falkner, 2003). In contrast, Lipschutz (1992; Lipschutz and Mayer, 1996) portrays global civil society as a force usurping the authority of states. The development of governance regimes outside the intergovernmental realm is seen as indicative of declining state power when it comes to managing the global environment (Gereffi et al., 2001; Mathews, 1997). Alternatively, some scholars argue that distinguishing between the intergovernmental arena and global civil society does not reflect the reality of world politics where transnational actors and states exist within a broader structure, dominated by the interplay of state interests or the logic of global capitalism (Chartier and Deleage, 1998; Kellow, 2000; Pasha and Blaney, 1998).

transnational actors: who are they?

Transnational actors are typically defined in the negative, characterized by what they are *not* (for example, *non*-governmental organizations or *non*-state actors). Scholars lack a consensus on what types of actors should be studied and what they should be called. Broadly, transnational actors can include grassroots organizations, scientific associations, special interest groups (national and international), academics, businesses, trade associations, environmentalists, individuals, the media, churches and religious organizations, independence movements, subnational governments, political parties, foundations and consumer groups. However, most scholars explicitly narrow the concept, and three general categories of actors have emerged as the primary units of analysis in the study of transnational environmental politics: non-governmental organizations (NGOs), transnational networks and multinational corporations (MNCs).

ngos

The majority of IEP scholars refer to transnational actors in environmental politics as NGOs. According to Caldwell (1990, p. 111), NGOs are 'the most diversified and least easily classified' component of the institutional architecture for environmental policy making. A few scholars use the term in reference to virtually any non-state actor seeking to influence decision-making at the global level, although most reserve the term for non-profit organizations that have not been established by a government. This is consistent with the United Nations (UN) definition of NGOs, which also excludes organizations that advocate violence, are political parties, and/or do not support UN objectives (Oberthür et al., 2002; Willetts, 1996c).

There is some debate about whether to include organizations with commercial interests as NGOs. While individual MNCs are generally treated separately (see below), non-profit associations representing commercial interests (for example, trade associations and/or coalitions whose members are MNCs) are often referred to as NGOs. This is consistent with the UN guidelines mentioned above and emphasizes the commonalities among actors that operate in distinction to the state. Alternatively, some scholars contend that NGOs represent broader societal concerns rather than narrow commercial interests (Biliouri, 1999; Fox and Brown, 1998; Mol, 2000). This approach, which assumes three spheres of human activity where NGOs are distinct from state and market actors, risks romanticizing NGOs by suggesting that they alone represent what is good for society and that they do so in all instances.

NGOs are often differentiated in terms of geographic scope, substantive interest and/or type of activity. In the UN, the term NGO was originally limited to those organizations working in at least three countries, referred to today as *international* NGOs (INGOs) (UIA, 2003; Willetts, 1996c). While the UN system now also accredits *national* NGOs, provided they have international interests, INGOs, defined as a subcategory of NGOs, remain the central focus of many studies (for example, Arts, 2001; Chartier and Deleage, 1998; Frank et al., 1999; McCormick, 1993).

Still others concentrate their research on organizations with particular substantive interests, distinguishing between *environmental* NGOs (ENGOs) (for example, Betsill, 2002; Close, 1998; Kellow, 2000; Princen, 1994; Rowlands, 1995; Wright, 2000) and other types of NGOs that may also engage in the transnational politics of the environment, such as *human rights* NGOs (for example, Clark, 1995) and *business/industry* NGOs (for example, Arts, 2001; Broadhurst and Ledgerwood, 1998; Skodvin and Andresen, 2003). Finally, some scholars differentiate between NGOs based on the character of their primary activities: *advocacy* NGOs promote specific policies, *programmatic* NGOs engage in specific projects, think tanks or *scientific* organizations conduct research, and *educational* NGOs focus on outreach (Caldwell, 1990; Charnovitz, 1997; Fox and Brown, 1998; Gough and Shackley, 2001; O'Brien et al., 2000).

Today, the major international environmental NGOs include the World Wide Fund for Nature, Greenpeace and Friends of the Earth. In 1982, the Environment Liaison Centre identified 2,230 national environmental NGOs located in developing countries and 13,000 in industrialized countries (Caldwell, 1990, p. 314). Wapner (1996, p. 2) estimates that nearly 100,000 NGOs work as advocates for environmental protection. All of these estimates have a normative bias in that they primarily include those organizations seen to be working for the common 'good' of enhancing environmental protection while excluding those organizations working against such protection.

transnational networks

Narrowing the scope of one's analysis has clear logistical advantages. At the same time, doing so runs the risk of oversimplification. One alternative is to focus on transnational networks of actors, such as transnational advocacy networks (Keck and Sikkink, 1998), epistemic communities (Haas, 1992)[2] and social movements (Hochstetler, 2002; Mol, 2000; O'Brien et al., 2000). Transnational networks are typically defined by the ideas that bind members together rather than specific characteristics of the individual members. For example, transnational advocacy networks are linked by shared principled beliefs while epistemic communities are held together by shared causal beliefs as well. Keck and Sikkink (1998) contend that transnational networks of corporate actors are held together by shared instrumental goals.[3] Social movements, also held together by shared principled beliefs, are distinguished by the fact that they tend to mobilize their constituencies through protest or disruptive action and provide the opportunity for mass participation (Khagram et al., 2002; Yearley, 1994). The transnational networks approach naturally allows for the diversity of actors engaged in international environmental politics, does not force researchers to draw fine distinctions between them, and more fully recognizes the complexity of international life where different types of actors operating in a variety of political spheres are engaged in a common enterprise.

In IEP, many types of transnational networks have emerged. Advocacy networks linking international and national environmental NGOs include the Asia Pacific People's Environmental Network; Environment Liaison Centre, the Climate Action Network, the Stakeholder Forum, the Pesticide Action Network, and the Third World Network. While Smith et al. (1997) note that such networks can be seen as part of a transnational environmental movement, Rootes (1999) cautions that there does not as yet exist a truly *global* environmental movement. Other types of transnational networks include the Cities for Climate Protection campaign, a transnational network linking local authorities engaged in the governance of global climate change (Betsill and Bulkeley, 2004). Prominent transnational networks involving corporate actors include the World Business Council for Sustainable Development, the Global Climate Coalition (now defunct), and the International Chamber of Commerce (Charnovitz, 1997; Kolk, 2001; Newell, 2000; Rowlands, 2001).

Scholars face two potential difficulties when using the transnational networks approach. First, it is likely to mask very real differences in the sources of leverage, internal dynamics and motivations of different types of actors involved in a network (Chatterjee and Finger, 1994; Jordan and Van Tuijl, 2000). Second, it can be difficult to conduct empirical research. Scholars frequently focus on the formal NGOs that are prominent in transnational networks in their analyses, thus raising many of the definitional concerns discussed above (for example, Duwe, 2001; Williams and Ford, 1999). Alternatively, Keck and Sikkink (1998) study transnational networks through analyses of specific campaigns rather than particular organizations.

mncs

Finally, a growing number of scholars focus on the role of MNCs in transnational environmental politics (for example, Chatterjee and Finger, 1994; Clapp, 1998a; Garcia-Johnson, 2000; Kolk, 2001; Levy and Newell, 2005; Newell, 2001b; Rowlands, 2000; Skodvin and Skjærseth, 2001). MNCs are easily distinguished from NGOs in that they are profit-seeking entities. While recognizing that MNCs may engage in IEP through transnational networks and/or non-profit organizations that promote commercial interests (NGOs), this body of research focuses on the ways that individual MNCs participate in and shape international policy-making processes. As was the case in studies focused on NGOs, scholars of MNCs differ in terms of how they view the relationship between MNCs and civil society.

transnational actors in iep

The vast majority of scholarship on transnational environmental politics is in the form of qualitative case studies whereby researchers draw upon data obtained through interviews, archival research, participant observation and in some cases questionnaires. Early case studies provided a rich history of transnational actors' engagement in IEP. Scholars then moved on to consider how transnational actors participate in specific political processes and with what effect. Below I note several methodological challenges related to this area of research. As it has become clear that transnational actors matter in IEP, a body of literature has emerged that looks inside transnational actors to better understand their internal dynamics and consider questions about their legitimacy, representation and accountability.

historical overview

Although studies of transnational environmental politics are a relatively recent phenomenon, transnational actors have engaged in IEP for more than a century. Research about them has focused on how *international environmental* organizations have engaged in IEP, especially within the UN system. More recently, attention has turned to the involvement of the environmental community within international economic institutions. Of course, other types of transnational actors participate in IEP as well, however their historical involvement is less well documented and thus not presented here.

The first international environmental NGOs appeared in the late nineteenth century, with early examples including the International Union of Forestry Research Organizations (1891) and the International Friends of Nature (1895) (Frank et al., 1999). While their numbers have increased over the last century, estimates differ considerably. One study found that INGOs focused on the environment increased from two in 1953 to 90 by 1993, accounting for 14.3 per cent of all 'social change organizations' identified (second only to human rights groups) (Keck and Sikkink, 1998, p. 11). Another study reports the founding of 173 international environmental NGOs between 1882 and 1990 (Frank et al., 1999, p. 84).

In a detailed study using event history analysis, Frank et al. (1999) identified several interesting facets related to the evolution of international environmental NGOs between 1875 and 1990. There have been three 'spurts' in the growth of international environmental NGOs, coinciding with the establishment of the UN following World War II, the 1972 UN Conference on Human Development (Stockholm Conference) and the 1992 Rio Conference (see also Caldwell, 1990; Smith, 1972). In 1925, 77 per cent of INGO members hailed from European countries; that number dropped to 43 per cent in 1970 and 31 per cent in 1990, suggesting increased participation from individuals in the Americas, Africa and Asia (Frank et al., 1999, p. 86). Moreover, the sheer number of individual memberships in international environmental NGOs has increased dramatically. Finally, Frank et al. found substantial increases in organizational resources (budget and staff) between 1968 and 1990 as well as a growing number of interorganizational linkages, indicating the development of transnational networks.

Today, the history of transnational environmental politics is often organized around the three major UN-sponsored global environmental conferences: the 1972 Stockholm Conference, the 1992 Rio Conference and the 2002 World Summit on Sustainable Development (Johannesburg Conference). Stockholm is seen to be a watershed in terms of NGO participation in global governance, marking the beginning of a 'slow yet steady liberalization of the NGO system occurring over the following two decades' (Willetts, 1996c, p. 57; see also Feraru, 1974; Johnson, 1972; Morphet, 1996). The Rio Conference recognized the role transnational actors play as partners with states in the global struggle to promote sustainable development (Chatterjee and Finger, 1994; Dodds, 2001; Kakabadse, 1994; Kolk, 1996; Morphet, 1996; Willetts, 1996c). More recently, transnational actors were central to the creation of 'partnerships' for sustainable development at the 2002 Johannesburg Summit (Gutman, 2003; Speth, 2003; United Nations, 2002; Wapner, 2003). In between these conferences, ENGOs have participated in countless treaty negotiations, typically organized under the UN umbrella.

Beginning in the 1980s, many ENGOs turned their attention to the international economic institutions, recognizing that global economic forces are often drivers of environmental degradation. The first campaigns were directed at reforming the lending practices of multilateral development banks such as the World Bank (Bramble and Porter, 1992; Conca, 1996; Keck and Sikkink, 1998; Kolk, 1996; Young, 1999; see also Clapp in this volume). In the 1990s, prompted by the 'tuna–dolphin' case, the international environmental community began calling for environmental protection within institutions related to international trade (for example, the World Trade Organization and the North American Free Trade Agreement) (Hogenboom, 2001; O'Brien et al., 2000; Williams and Ford, 1999; Wright, 2000). More recently, a number of transnational actors participated in the larger anti-globalization movement, seeking to challenge neoliberal economic ideas seen to dominate the international system (Williams and Ford, 1999; Yamin, 2001; Young, 1999).

effects on iep

Research on the impact of transnational actors in IEP can be differentiated along two dimensions: the location of political activity and the types of effects. By far the largest body of work examines the influence of transnational actors in *institutionalized global politics*.[4] These studies focus on the ways that transnational actors interact with governments and intergovernmental organizations in national and international policy processes. However, a growing body of research recognizes that transnational actors also work outside this institutionalized dimension of politics, in the realm of *global civil society*. In terms of effects, many scholars, particularly those taking a pluralist approach to world politics, view transnational actors as pressure groups, advocating particular *policies and practices*. Others, such as those scholars coming from a constructivist or sociological perspective, emphasize the role of transnational actors in shaping *ideas* about the environment. Finally, some scholars contend that transnational actors impact the *broader structures* (for example, notions of sovereignty and global capitalism) within which IEP takes place. Embedded in these claims are different assumptions about which strategies and sources of leverage are most significant.

institutionalized global politics

Scholars working in this area document the strategies used by transnational actors, identify their sources of leverage vis-à-vis states and make claims about the impact of transnational actors on policy-making processes. On the issue of strategies, scholars commonly differentiate between direct and indirect strategies; the former targeted directly at states and the latter targeting states through secondary channels such as the media and the public (Kakabadse, 1994; Wright, 2000; Young, 1999). A direct strategy may consist of providing technical information or policy advice, participating in working groups, serving on national delegations to negotiations or conferences, lobbying national governments or IGO officials, distributing printed materials, and/or drafting proposals and treaty text. An indirect strategy may involve holding parallel forums during intergovernmental meetings, conducting public awareness, advertising or educational campaigns organized around a governmental decision-making process, shaming states that seek to block negotiations or violate existing rules and/or interacting with the media. The literature suggests that while environmentalists employ both strategies regularly, business/industry groups tend to rely more heavily (though not exclusively) on direct strategies.

Alternatively, Young (1999) distinguishes between insider and outsider strategies (see also Williams and Ford, 1999). Insider strategies involve working within the international system as it currently exists. One might argue that the direct and indirect strategies discussed above both fall within the category of an insider strategy. In contrast, outsider strategies aim at prompting changes in the larger structures of the global system (for example, notions of democracy and sovereignty, capitalism) and often involve rejecting, rather than working within, existing intergovernmental institutions.

Networking is another common strategy employed in the realm of institutionalized global politics (Biliouri, 1999; Caldwell, 1990; Carpenter, 2001; Close, 1998; Skodvin and Andresen, 2003).[5] Much of the literature on environmentalists focuses on networks operating across scales, creating links between INGOs and grassroots organizations (Fox and Brown, 1998; Princen and Finger, 1994). NGOs often use these cross-scale networks to invoke the 'boomerang' strategy; by working at both the international and domestic levels, they seek to pressure states from above and below (Hochstetler, 2002; Keck and Sikkink, 1998; O'Brien et al., 2000). Other transnational networks create connections between groups with similar substantive interests, with the expectation that such networks result in greater efficiency (Arts, 2001; Betsill, 2002; Betsill and Bulkeley, 2004; Duwe, 2001). An increasing number of transnational networks link transnational actors with different substantive interests and/or transnational actors with states (DeSombre, 2000; Newell, 2000; O'Brien et al., 2000; Rowlands, 2001).

While different types of transnational actors are seen to employ similar strategies in international environmental politics, most scholars differentiate between environmentalists and business/industry groups when talking about their sources of leverage vis-à-vis states. Business/industry actors, it is argued, have 'tacit power' over states based on 'the role they play in the creation of economic growth and the production of energy' (Newell, 2000, p. 159). According to this line of reasoning, it is not their economic resources per se that make business/industry actors powerful but their central position in national economies and the international political economy (Levy and Newell, 2000; Rowlands, 2001). Alternatively, Chatterjee and Finger (1994) argue that business/industry has a privileged position in international environmental policy-making simply because 'money talks'. For environmentalists, their leverage in international environmental politics is seen to be related to their perceived legitimacy, derived from a particular type of expertise (for example, Corell, 1999; Gemmill and Bamidele-Izu, 2002; Keck and Sikkink, 1998; Yamin, 2001) or their claim to represent civil society (Anderson, 2001; Gough and Shackley, 2001). Such legitimacy claims are usually associated with the ability of the environmental community to *persuade* states to adopt particular policies and practices. However, Skodvin and Andresen (2003) note that ENGOs also use their resources to *coerce* states into making such changes. For environmentalists and the business/industry community, but perhaps more so for environmentalists, the ability to use the media is also an important source of leverage, particularly when used for 'shaming' what are seen to be uncooperative states.

When examining the impact of transnational actors on specific policy processes, most scholars distinguish between three distinct (but overlapping) phases: agenda-setting, policy formulation and implementation. During the agenda-setting phase, transnational actors catalyse policy action, particularly by identifying problems and calling upon states to do something (for example, Bilouri, 1999; Charnovitz, 1997; Gemmill and Bamidele-Izu, 2002; Morphet, 1996; Newell, 2000; Raustiala, 2001; Yamin, 2001). In the policy formulation phase, transnational actors convince national governments and intergovernmental organizations to change their policies and practices (for example, Bramble and Porter, 1992; Close, 1998; O'Brien et al., 2000; Wapner, 2001). When policy formulation takes place in the context of treaty negotiations, transnational actors influence debates about particular proposals, shape the positions that states take in the negotiations, and/or affect the final outcome of the negotiations (the treaty text) (for example, Arts, 1998, 2001; Corell and Betsill, 2001; Levy and Newell, 2000; Newell, 2000; Raustiala, 2001).[6] During the implementation phase, transnational actors can help states monitor compliance with international agreements and/or carry out projects (for example, Grundbrandsen and Andresen, 2004; Jasanoff, 1997; Raustiala, 2001; Yamin, 2001). Zürn (1998) argues that transnational actors, through their participation, can enhance the legitimacy of governmental and intergovernmental policies, making them more likely to be implemented. Newell (2000) adds that corporate actors may impede implementation through non-cooperative strategies. During all phases of the policy process, transnational actors work through both domestic and international channels (Skodvin and Andresen, 2003).

By engaging in institutionalized global environmental politics, transnational actors open up political space for future transnational relations in this arena (Dodds, 2001; Keck and Sikkink, 1998; Willetts, 1996a). Today, transnational actors can generally expect to be included in inter-governmental policy processes related to the environment. However, the rules for how transnational actors can participate differ between policy arenas and states routinely invoke their sovereign privilege to restrict NGO participation in such processes (Clark et al., 1998; Oberthür et al., 2002; UNEP, 2002).

Transnational actors can also shape ideas about the environment. Borrowing a concept from sociology, scholars argue that transnational actors frame (or reframe) environmental problems and in the process, establish the boundaries within which states must formulate their responses (Betsill, 2002; Chatterjee and Finger, 1994; Humphreys, 2004; Jasanoff, 1997; Keck and Sikkink, 1998; Williams and Ford, 1999).[7] This is most commonly seen as an important role for environmentalists (and scientists), although Newell (2000) contends that the media also frames understandings of global environmental problems.

Some scholars claim that transnational actors shape the broader structures within which institutionalized environmental politics take place. Newell (2000) and Rowlands (2001) contend that business/industry groups shape the context of IEP by engaging in institutionalized global politics outside of, but related to, the environment, such as the formation of trade and investment rules. Others argue that the involvement of transnational actors in institutionalized policy processes contributes to the democratization of world politics (Princen, 1994; Raustiala, 1997; Willetts, 1996c). Such claims, however, raise questions about the nature of accountability and representation among transnational actors (Chartier and Deleage, 1998; Friedman et al., 2005; Held, 1999; Jordan and Van Tuijl, 2000), and at least one study has found that NGO participation in global UN conferences has not resulted in such democratization (Clark et al., 1998). Finally, in the process of interacting with states in institutionalized policy processes, scholars claim that transnational actors have reoriented (though not necessarily diminished) the notion of state sovereignty (Clark et al., 1998; Friedman et al., 2005; Wapner, 1998).

global civil society

A growing number of scholars focus on how transnational actors interact in the realm of global civil society and the subsequent implications for world politics. The first wave of research in this area primarily examined interactions between members of the environmental community and between environmentalists and society (Chatterjee and Finger, 1994; Clark, 1995; Lipschutz, 1992; Lipschutz and Mayer, 1996; Wapner, 1996). Transnational actors working in the realm of global civil society create dense networks of communication among themselves, which in turn are seen to facilitate mutual learning and open up new opportunities for shaping global environmental governance (Betsill and Bulkeley, 2004). Even when transnational actors participate in intergovernmental conferences and treaty negotiations, some authors claim the significance of these activities lies not in their impact on governments but in the networks transnational actors create among themselves (Chatterjee and Finger, 1994; Clark, 1995; Clark et al., 1998).

Today, scholars also focus on transnational relations among business/ industry actors and between environmentalists and business/industry (Carpenter, 2001; Clapp, 1998b; Falkner, 2003; Garcia-Johnson, 2000; Gereffi et al., 2001; Newell, 2001a, 2001b; Rowlands, 2001). While the latter type of interaction is not new, it is occurring with increasing frequency in the area of IEP. Many environmentalists have chosen to target business/industry directly, bypassing states, which they believe are either unable or unwilling to regulate corporate practices (Biliouri, 1999; Falkner, 2003; Mol, 2000; Newell, 2001a). The relationship is often adversarial, with environmentalists adopting a confrontational strategy and engaging in activities such as corporate boycotts, public relations wars, creation of MNC monitoring groups, and shareholder activism to pressure change in business/industry policies and practices. However, environmentalists and business/industry increasingly work cooperatively, through activities such as ecoconsumerism, project collaboration, creation of codes of conduct and stewardship regimes (Newell, 2001a). Some observers express concern that these close relationships threaten the legitimacy of the environmental community and may have implications for power relations within the sphere of global civil society (Bernstein, 2002; Falkner, 2003; Mol, 2001).

The activities of transnational actors in global civil society are seen to have a range of effects on environmental politics. In some cases, strategies adopted by environmentalists have been successful in reshaping corporate practices (Gereffi et al., 2001). Some scholars argue that the rules of corporate environmental governance are being rewritten in the realm of global civil society (Carpenter, 2001; Clapp, 1998b; Falkner, 2003; Gough and Shackley, 2001; Kolk, 2001; Rowlands, 2001). For many scholars, the significance of world civic politics is that it provides a mechanism through which environmental ideas and values are spread (for example, Garcia-Johnson, 2000; Lipschutz and Mayer, 1996; Princen and Finger, 1994; Wapner, 1996, 2001). Such values, which can be used to establish standards of good conduct, are seen to provide a foundation for subsequent changes in destructive environmental practices by individuals, local communities, business/industry and society (as well as governments).

Lipschutz (1992) makes some of the strongest claims about the impact of world civic politics. He argues that through networks of economic, cultural and social relations, 'a politics of collective identity is developing around the world' (Lipschutz, 1992, p. 398). The emergence of global civil society is seen to have the potential to 'remap' world politics by challenging the legitimacy of a society dominated by states and opening up an alternative sphere in which transnational actors engage one another directly (Friedman et al., 2005; Lipschutz, 1992). For some, this alternative venue is part of the democratization of world politics (Lipschutz and Mayer, 1996; Wapner, 1996) and the shift toward global governance (Rosenau, 2003).

methodological challenges[8]

As noted above, most research on transnational actors in IEP consists of qualitative case studies. Collectively, this body of research supports the conclusion that transnational actors 'matter' in IEP. However, qualitative case studies as employed thus far have limitations in making claims about transnational actor influence in any given policy process and in considering the conditions under which transnational actors impact IEP (Arts, 1998; Betsill and Corell, 2001; Newell, 2000; Yamin, 2001; Zürn, 1998). One problem in the current literature is a tendency to treat all studies related to transnational actors in the environmental issue area as a single body of research without differentiating between the types of transnational actors being studied or the political arenas in which they operate. A second problem is a surprising lack of specification about what is meant by 'influence' and how to identify influence in any given political arena. Third, scholars often confuse correlation with causation and fail to specify the causal mechanisms linking transnational actors' activities to the observed effects.

Two initiatives have sought to make the study of NGO influence in the context of international environmental negotiations more systematic (Arts, 1998; Betsill and Corell, 2001). More systematic research strategies not only would strengthen claims of transnational actor influence in any particular process but could also provide the foundation for comparison across cases (Zürn, 1998). Some scholars have acknowledged variation in the levels of transnational actor influence and that sometimes transnational actors fail in their attempts to shape IEP (Arts, 1998; Clark, 1995; Keck and Sikkink, 1998; O'Brien et al., 2000; Raustiala, 2001; Rowlands, 1995; Willetts, 1996b). However, so long as scholars use different types of evidence and criteria for assessing influence, it is difficult to make reliable comparisons about levels of influence across cases and to consider the factors that shape transnational actor influence. To do so, it will also be necessary to expand the range of cases examined to include those involving low levels of influence and/or cases where transnational actors do not participate at all.

Despite these methodological difficulties, some scholars have begun to identify conditioning factors, particularly in the realm of institutionalized global politics. Broadly, it appears that there are three general categories of factors that shape the ability of transnational actors to influence IEP: (1) the institutional context (for example, Arts, 2001; Corell and Betsill, 2001; Hochstetler, 2002; Kakabadse and Burns, 1994; Keck and Sikkink, 1998; Newell, 2000; Skodvin and Andresen, 2003; Williams and Ford, 1999); (2) the characteristics of the transnational actors (for example, Biliouri, 1999; Chatterjee and Finger, 1994; Corell and Betsill, 2001; Dodds, 2001; Falkner, 2003; Gereffi et al., 2001; Kakabadse and Burns, 1994; Newell, 2000; Rowlands, 2001); and (3) the nature of the issue at hand (for example, Clark et al., 1998; Corell and Betsill, 2001; Keck and Sikkink, 1998; Rowlands, 2001). Of course, these must be viewed as working hypotheses since, as discussed above, few scholars have taken a systematic comparative approach to the study of transnational actors.

explaining transnational actors

To date, the majority of the literature considers transnational actors as independent variables. However, as it becomes clear that transnational actors are shaping environmental politics in a variety of ways, scholars have returned to the question of who these actors are, looking more closely at their origins and internal operations in order to better explain why these actors form and behave in particular ways. This research has been particularly significant in highlighting the extraordinary diversity within various transnational communities (Kellow, 2000; Kolk, 1996; O'Brien et al., 2000; Raustiala, 2001; Rowlands, 2001; Wapner, 1996; Willetts, 1996a; Williams and Ford, 1999). It is also addressing concerns that transnational actors (especially environmentalists) have been romanticized in much of the literature and considering questions about their legitimacy, accountability and representation (Anderson, 2001;

Chartier and Deleage, 1998; Jordan and Van Tuijl, 2000; Kellow, 2000; Pasha and Blaney, 1998; Tesh, 2000). In their study of the emergence of international environmental NGOs between 1875 and 1990, Frank et al. (1999) found that the emergence of NGOs is linked to the rise of a rationalized discourse around nature and the creation of intergovernmental environmental organizations. As nature became rationalized, international environmental NGOs served as 'expert bodies' helping states and societies protect (and exploit) nature for human needs. Changes in the institutional architecture of global governance (such as the formation of the UN) provided a forum through which transnational actors could actively engage in international politics. Some scholars argue that these new political spaces merely reflect the needs of states and international governance regimes (Dodds, 2001; Feraru, 1974; Kellow, 2000). In other words, states create spaces for transnational actors when it is in their interest to do so. If true, then as state capacity to address global environmental problems increases, there should be declining demand for transnational actors and thus reduced incentive for their formation. Indeed, Frank et al. (1999) found that the formation of international environmental NGOs has slowed recently due to consolidation of the intergovernmental arena.

Keck and Sikkink (1998) offer a slightly different argument about political space and mobilization of transnational actors. They contend that as 'channels between the state and its domestic actors are blocked, the boomerang pattern of influence may occur: domestic NGOs bypass their state and directly search out international allies to try to bring pressure on their states from outside' (Keck and Sikkink, 1998, p. 12). Rather than responding to openings in political space, transnational networks emerge as political space is closed at the national level.

Characteristics of the international political economy appear to shape the types of strategies that transnational actors employ in environmental politics. According to Mol (2000), ecological modernization discourses have prompted a transformation in environmental NGOs in the US and Europe, moving radical tactics to the periphery. Not surprisingly several studies have found that economic considerations are central in shaping the environmental strategies of corporate actors (Kolk, 2001; Levy and Newell, 2000). Newell (2001a) suggests that environmental NGOs target MNCs directly with increasing frequency because they believe the pressures of globalization give MNCs significant leverage over states and render governments unable and unwilling to regulate corporate behaviour themselves.

The strategies employed by transnational actors are also shaped by their internal structures. Some transnational networks have developed highly sophisticated institutional structures, which enable network members to coordinate activities, share information, identify priorities and specialize in particular aspects of an issue or policy process (Dodds, 2001; Duwe, 2001). However, these structures can also reveal tensions within transnational communities, reminding us that transnational actors, like states, are themselves political entities with their own power relations (Charnovitz, 1997; Clark, Friedman, and Hochstetler, 1998; Dodds, 2001; Duwe, 2001; Jordan and Van Tuijl, 2000; Kellow, 2000; Kolk, 1996; Rowlands, 2001). These tensions are often related to the structure of the international system in which transnational actors exist – North–South issues, for example – and may negatively impact the strategies employed by transnational actors and/or their effects.

Moreover, such tensions raise questions about the legitimacy, representation and accountability of transnational actors. How *global* is global civil society if members from the South are systematically disadvantaged (Chartier and Deleage, 1998; Duwe, 2001)? Can transnational networks hope to promote democratization in institutions of global governance if they 'reflect as much inequality as they are trying to undo?' (Jordan and Van Tuijl, 2000, p. 2061). What are the implications for representation when NGOs receive significant funding from state-based institutions (Kellow, 2000; Yamin, 2001)? Suggestions that some members of the environmental community employ questionable tactics, such as manipulating scientific findings, in their quest to achieve the common 'good' raise further questions about accountability (Harper, 2001; Jordan, 2001; Skodvin and Andresen, 2003; Tesh, 2000; Wapner, 2002).

the climate action network

In sum, the literature on transnational actors in IEP offers a rich picture of how such actors participate in environmental governance and a growing list of propositions about their impacts. Emerging work on the internal dynamics of transnational actors promises to provide a better sense of who these actors are and their implications for our understandings of world politics. In this section, I take some of the concepts introduced above and apply them in a case study of the Climate Action Network (CAN), a transnational advocacy network engaged in the institutionalized global politics of climate change. Data for this case study was collected between 1997 and 2001 in the form of semistructured interviews with CAN members and participant observation at the third Conference of the Parties to the Framework Convention on Climate Change in Kyoto, Japan in December 1997 and the resumed sixth Conference of the Parties in Bonn, Germany in June 2001.[9]

CAN is a transnational advocacy coalition consisting of more than 300 international and national ENGOs (CAN, 2004). Formed in 1989, CAN is a network for ENGOs 'who share a common concern for the problems of climate change and wish to cooperate in the development and implementation of short term and long term strategies to combat these problems' (CAN, 1997, p. vii). The network is divided into eight regions, each with its own coordinator. During international climate change negotiations, CAN seeks to serve as the voice of the environmental community and provides a forum for ENGOs to share ideas, debate issues and develop strategies for influencing the process and outcome of treaty negotiations.

At each negotiating session, CAN members meet daily to share information and coordinate their lobbying activities. They publish a daily newsletter, *ECO*, which is widely read by other participants and provides an opportunity for CAN to put its spin on issues under debate. Formally, transnational actors have limited access to state delegates during the negotiations. CAN members are typically permitted to make two formal statements during plenary sessions – one from the North and one from the South – and receive daily briefings from the Chair of the negotiations and their respective state delegations. Otherwise, transnational actors are not permitted on the floor during plenary sessions and are excluded from the closed-door, 'non-group' sessions in which most of the negotiations take place. Informally, however, CAN members capitalize on the relationships they have developed with delegates over the years, gathering information through corridor meetings and cellular phone conversations. Of course, this advantages those organizations and activists that have participated in the negotiations long enough to develop such relationships. In addition, some CAN members resort to more 'subversive' measures such as lurking in corridors, hotel lobbies and restrooms and searching for documents left in meeting rooms and on photocopiers. For CAN members, the problem of access has not been insurmountable but demands considerable time and resources.

CAN believes its leverage in international climate change negotiations derives from the specialized knowledge and expertise of its members. Many ENGOs arrive at the negotiations armed with studies about various issues under debate, and during the negotiations they work hard to secure the latest information about the status of the negotiations. Members try to ensure that all scientific information and 'intelligence' are verified before being made public. The network does not typically draw upon its position as representative of civil society as a source of leverage. While some CAN members organize demonstrations and protest activities to draw public and media attention to the negotiations and the issue of climate change, these are usually organized by a single organization rather than CAN as a whole and are not necessarily directly targeted at influencing the negotiations per se.

It is important to recognize the variety of transnational actors that participate in international climate change negotiations. In addition to CAN, there is a significant business/industry presence (Kolk, 2001; Newell, 2000). Their positions are extremely diverse, ranging from those organizations fundamentally opposed to any international regulation to those hoping to capitalize on increased demand for non-fossil-based energy sources. International climate change negotiations involve competition between transnational actors who often present conflicting versions of the 'truth' about the implications of issues under debate. It is in this context that I consider whether CAN has influenced the institutionalized global politics of climate change.

If CAN influences international climate change negotiations, then it should be possible to observe the effects of their activities independent of information about those activities (King et al., 1994). For example, we might expect to see CAN's position reflected in the final treaty text. During the Kyoto Protocol negotiations between 1995 and 1997, CAN advocated for an agreement that would (1) require industrialized countries to reduce their greenhouse gas (GHG) emissions 20 per cent below 1990 levels by 2005; (2) contain strong compliance and review mechanisms; and (3) not permit Parties to meet their commitments through emissions trading or credit for emissions absorbed by 'sinks'.[10] The Kyoto Protocol, however, requires industrialized countries to reduce their aggregate GHG emissions 5.2 per cent below 1990 levels by 2008–12, does not contain compliance and review mechanisms (as of 1997) and allows Parties to meet their commitments through trading and the use of sinks (United Nations, 1997). By this measure, it appears CAN had minimal influence on the negotiations.

However, a closer examination of the negotiations on targets and trading reveals a slightly different picture. Table 7.1 presents the various parties' positions on these issues. CAN's call for 20 per cent reductions was the most stringent position on the table and, given the domestic economic concerns in many industrialized countries, was never viewed as politically feasible. However, CAN's presence was instrumental in two ways. First, CAN, and European ENGOs in particular, pressured the EU to hold on to their 15 per cent reduction target in the face of US preference for a stabilization target. The European delegates only agreed to accept the use of trading (which was central to the US position) once the US agreed to accept a reduction (rather than stabilization) target. Second, American ENGOs appear to have encouraged then Vice President Al Gore to attend the Kyoto meeting and instruct the US delegation to be more 'flexible' in its negotiating position, opening the way for the US to accept a reduction rather than stabilization target. In the absence of CAN, the Europeans might have backed down sooner, the US might not have been pressured to accept reductions, and the Kyoto target might have been even lower.

Table 7.1 Positions on targets and trading, 1995–97

Actor	Position on targets	Position on trading
CAN, Alliance of Small Island States (AOSIS)	20% reductions below 1990 levels by 2005	No
EU, G77 (developing countries)	7.5% reductions below 1990 levels by 2005, 15% reductions by 2010	No
Japan	5% reductions by 2008–12 (with differentiated targets)	Yes
Business Council for a Sustainable Energy Future, International Climate Change Partnership (moderate business)	Moderate GHG emissions reduction targets	Yes
Australia, Canada, New Zealand	3% reductions below 1990 levels by 2010	Yes
United States	Stabilization at 1990 levels by 2008–12	Yes
OPEC, Global Climate Coalition (fossil-fuel industry)	No international regulations	Yes (if there are to be regulations)

Sources: Betsill (2000), Newell (2000), Oberthür and Ott (1999) and Raustiala (2001).

Many observers attribute CAN's influence on international climate change negotiations to its ability to coordinate ENGO activity . However, such coordination is becoming increasingly difficult due at least in part to the growing number of organizations participating in the negotiations. In the early 1990s, a relatively small number of ENGOs regularly attended the negotiations. As a result, CAN's organizational structure was fairly flat and decision-making was transparent. By the end of the 1990s, the number of ENGOs attending sessions grew considerably and older CAN members took on the task of educating these new groups about substantive and procedural issues. While recognizing that a larger membership could enhance the network's political reach, the core group of organizations and individuals who had been involved in the negotiations from the beginning began to meet separately to discuss the technical details of CAN's position. CAN's organizational structure has become much more hierarchical, leading to charges (from inside and outside) that CAN no longer speaks for the environmental community as a whole, but rather represents the views of a small group of organizations based primarily in the North (see Duwe, 2001).

conclusions

The subject of transnational actors in IEP has generated considerable scholarly attention since the early 1990s. As a result, we have a rich picture of the types of transnational actors in IEP and the ways in which they engage in world politics. However, some questions remain unanswered and new issues have emerged. In this section, I suggest that future research in the area of transnational environmental politics should address a number of methodological, normative, theoretical and substantive issues.

As discussed above, scholars must consider the methods used in the analysis of transnational actors in IEP and develop more sophisticated strategies in order to strengthen our theorizing about the significance of the differences between transnational actors and the variety of political arenas in which they engage in IEP. All transnational actors are not alike; they differ in terms of their interests, resources, strategies, and so on. Moreover, they participate in IEP in many different political arenas, each likely to involve different dynamics that in turn shape the ways that transnational actors participate, the goals they pursue, the strategies they use and the likelihood that they will achieve those goals (Betsill and Corell, 2001).

There is also a need for more comparative research, which will first of all require that scholars expand the range of cases they examine and include instances in which transnational actors have little impact on IEP and/or fail to participate altogether. Early research on transnational actors was driven by the observation that such actors were having unexpected effects on political processes. In other words, work focused on those instances in which transnational actors appeared to matter. As a result, we likely have an overrepresentation of positive cases in our overall sample giving the appearance that transnational actors matter all the time. Future research should examine negative cases, those instances in which transnational actors sought to affect political outcomes but failed to do so (Mahoney and Goertz, 2004). This would help isolate the conditions that shape the ability of transnational actors to shape world politics. Moreover, by studying cases in which transnational actors fail to participate, scholars may gain purchase on the question of whether the involvement of transnational actors strengthens environmental governance.

Future research should assess the impact of transnational actors in the broader context of IEP.[11] When we place transnational actors at the center of our analysis, it is not surprising that we conclude they matter. However, researchers should step back and evaluate the relative importance of transnational actors compared to other types of actors (for example, states) in particular issue areas rather than setting out to assess transnational actor impact. We may find that transnational actors are more likely to engage in and shape environmental politics in particular types of issue areas, at particular stages of the policy process and/or in distinct realms of activity. Scholars might also capitalize on the growing interest in multilevel governance in international relations and consider the role of transnational actors in the vertical and horizontal redistribution of authority in environmental politics.

There remains work to be done on the general question of who participates in transnational environmental politics. *Agenda 21* introduced the notion of 'major groups' including local authorities, religious organizations and youth (UNCED, 1992). To date, these groups have not been analysed in terms of transnational environmental politics to the same extent as other types of transnational actors (for example, environmental NGOs and business interests). What are the implications of expanding our conception of transnational actors to include such groups? Do these groups alter the nature of global environmental governance? For example, does their participation signal a more democratic system of governance? Who do these groups represent and to what extent are they accountable to their constituents and/or one another?

Finally, it would be interesting for scholars of transnational environmental politics to look beyond the environmental issue area. How do the findings about transnational actors in IEP compare to similar research in other areas of international relations? Are there insights from other areas of transnational relations research (in the field of human rights, for instance) that would enhance our study of transnational environmental politics?

notes

1. While studies of transnational environmental politics typically exclude intergovernmental organizations such as the UN Environment Programme (UNEP) (except as a target of transnational activity), several 'hybrid'

organizations whose members represent government as well as non-governmental interests, such as the International Union for the Conservation of Nature (IUCN) are frequently included (see Caldwell, 1990; Morphet, 1996; Willetts, 1996a). Intergovernmental organizations are covered in greater detail in Biermann's chapter in this volume.
2. This is by far the most common starting point for studies of scientists as transnational actors (see Andresen et al., 2000; Litfin, 1994; Miller and Edwards, 2001; Parson, 2003; Rowlands, 1995; Raustiala, 2001). Only Jasanoff (1997) and Newell (2000) use the term 'NGO' to identify scientific bodies.
3. For a critique of this distinction, see Sell and Prakash (2004).
4. I've borrowed this term from Smith et al. (1997, p. 73).
5. For many transnational actors, the significance of networking goes beyond its utility as a strategy to influence governmental decision-making processes. I return to this point in the section on global civil society.
6. On the significance of the distinction between negotiating process and outcome, see Betsill and Corell (2001).
7. A frame is 'an interpretive schemata that simplifies and condenses the 'world out there' by selectively punctuating and encoding objects, situations, events, experiences, and sequences of action within one's present or past environment' (Snow and Benford, 1992, p. 137).
8. The discussion in this section relies heavily on Betsill and Corell (2001) and subsequent work we have done in the context of our project 'NGO Influence in International Environmental Negotiations'.
9. This section draws on Betsill (2000). I gratefully acknowledge the support of the Institute for the Study of World Politics and the College of Liberal Arts at Colorado State University.
10. These points were articulated in the pages of *ECO* as well as in CAN interventions during this period and confirmed during subsequent interviews with CAN members.
11. This point was raised in a recent discussion on the gep-ed email list. See the 'NGOs and Climate Change Science and Policy Making' discussion 3–6 January 2005. Available at <www.mail-archive.com/gep-ed%40listserve1.allegheny.edu/index.html#00243>.

bibliography

Anderson, Kenneth (2001) 'The Limits of Pragmatism in American Foreign Policy: Unsolicited Advice to the Bush Administration on Relations with International Nongovernmental Organizations', *Chicago Journal of International Law* 2, pp. 371–88.

Andresen, Steinar, Tora Skodvin, Arild Underdal and Jørgen Wettestad (eds) (2000) *Science and Politics in International Environmental Regimes*, Manchester: Manchester University Press.

Arts, Bas (1998) *The Political Influence of Global NGOs: Case Studies on the Climate and Biodiversity Conventions*, Utrecht: International Books.

Arts, Bas (2001) 'Impact of ENGOs on International Conventions' in Bas Arts, Math Noortmann and Bob Reinalda (eds) *Non-State Actors in International Relations*, Aldershot: Ashgate, pp. 195–210.

Bernstein, Steven (2002) 'Liberal Environmentalism and Global Environmental Governance', *Global Environmental Politics* 2, 3, pp. 1–16.
Betsill, Michele M. (2000) 'Greens in the Greenhouse: Environmental NGOs, Norms and the Politics of Global Climate Change'. Unpublished PhD thesis, Department of Political Science, University of Colorado, Boulder.
Betsill, Michele M. (2002) 'Environmental NGOs Meet the Sovereign State: The Kyoto Protocol Negotiations on Global Climate Change', *Colorado Journal of International Environmental Law and Policy* 13, 1, pp. 49–64.
Betsill, Michele M., and Harriet Bulkeley (2004) 'Transnational Networks and Global Environmental Governance: The Cities for Climate Protection Program', *International Studies Quarterly* 48, pp. 471–93.
Betsill, Michele M., and Elisabeth Corell (2001) 'NGO Influence in International Environmental Negotiations: A Framework for Analysis', *Global Environmental Politics* 1, 4, pp. 65–85.
Biliouri, Daphne (1999) 'Environmental NGOs in Brussels: How Powerful are their Lobbying Activities?' *Environmental Politics* 8, pp. 173–82.
Bramble, Barbara J., and Gareth Porter (1992) 'Non-Governmental Organizations and the Making of US International Environmental Policy', in Andrew Hurrell and Benedict Kingsbury (eds) *The International Politics of the Environment: Actors, Interests and Institutions*, Oxford: Clarendon Press, pp. 313–53.
Broadhurst, Arlene I., and Grant Ledgerwood (1998) 'Environmental Diplomacy of States, Corporations and Non-Governmental Organizations: The Worldwide Web of Influence', *International Relations* XIV, 2, pp. 1–19.
Caldwell, Lynton Keith (1990) *International Environmental Policy: Emergence and Dimensions*, 2nd edn, Durham: Duke University Press.
CAN (1997) *Climate Action Network: International NGO Directory 1997*, Washington, DC: Climate Action Network.
CAN (2004) *About Climate Action Network*, Climate Action Network. available from <www.climatenetwork.org/>.
Carpenter, Chad (2001) 'Businesses, Green Groups and the Media: The Role of Non-Governmental Organizations in the Climate Change Debate', *International Affairs* 77, 2, pp. 313–28.
Charnovitz, Steve (1997) 'Two Centuries of Participation: NGOs and International Governance', *Michigan Journal of International Law* 18, 2, pp. 183–286.
Chartier, Denis, and Jean-Paul Deleage (1998) 'The International Environmental NGOs: From the Revolutionary Alternatives to the Pragmatism of Reform', *Environmental Politics* 7, pp. 26–41.
Chatterjee, Pratap, and Matthias Finger (1994) *The Earth Brokers: Power, Politics and World Development*, London: Routledge.
Clapp, Jennifer (1998a) 'Foreign Direct Investment in Hazardous Industries in Developing Countries: Rethinking the Debate', *Environmental Politics* 7, 4, pp. 92–113.
Clapp, Jennifer (1998b) 'The Privatization of Global Environmental Governance: ISO 14000 and the Developing World', *Global Governance* 4, 3, pp. 295–316.
Clark, Ann Marie (1995) 'Non-Governmental Organizations and the Influence on International Society', *Journal of International Affairs* 48, pp. 507–25.
Clark, Ann Marie, Elisabeth J. Friedman and Kathryn Hochstetler (1998) 'The Sovereign Limits of Global Civil Society: A Comparison of NGO Participation

in UN World Conferences on the Environment, Human Rights and Women', *World Politics* 51, pp. 1–35.

Close, David (1998) 'Environmental NGOs in Greece: The Acheloos Campaign as a Case Study', *Environmental Politics* 7, pp. 55–77.

Cohen, Jean L., and Andrew Arato (1992) *Civil Society and Political Theory*, Cambridge, Mass.: MIT Press.

Conca, Ken (1996) 'Greening the UN: Environmental Organisations and the UN System', in Thomas G. Weiss and Leon Gordenker (eds) *NGOs, the UN, and Global Governance*, Boulder: Lynne Rienner, pp. 103–19.

Corell, Elisabeth (1999) 'Non-state Actor Influence in the Negotiations of the Convention to Combat Desertification', *International Negotiation* 4, 2, pp. 197–223.

Corell, Elisabeth, and Michele M. Betsill (2001) 'A Comparative Look at NGO Influence in International Environmental Negotiations: Desertification and Climate Change', *Global Environmental Politics* 1, 4, pp. 86–107.

DeSombre, Elizabeth R. (2000) *Domestic Sources of International Environmental Policy: Industry, Environmentalists, and U.S. Power*, Cambridge, Mass.: MIT Press.

Deutsch, Karl W. (1957) *Political Community in the North Atlantic Area: International Organization in Light of Historical Experience*, Princeton: Princeton University Press.

Dodds, Felix (2001) 'From the Corridors of Power to the Global Negotiating Table: The NGO Steering Committee of the Commission on Sustainable Development', in Michael Edwards and John Gaventa (eds) *Global Citizen Action*, Boulder: Lynne Rienner, pp. 203–13.

Duwe, Matthias (2001) 'The Climate Action Network: Global Civil Society at Work?' *Reciel* 10, 2, pp. 1–14.

Edwards, Michael (2001) 'Introduction', in Michael Edwards and John Gaventa (eds) *Global Citizen Action*, Boulder: Lynne Rienner, pp. 1–14.

Edwards, Michael, and John Gaventa (eds) (2001) *Global Citizen Action*, Boulder: Lynne Rienner.

Falkner, Robert (2003) 'Private Environmental Governance and International Relations: Exploring the Links', *Global Environmental Politics* 3, 2, pp. 72–87.

Feraru, Anne T. (1974) 'Transnational Political Interests and the Global Environment', *International Organization* 28, 1, pp. 31–60.

Florini, Ann (ed.) (2000) *The Third Force: The Rise of Transnational Civil Society*, Washington, DC: Carnegie Institute for International Peace.

Fox, Jonathan A. and L. David Brown (eds) (1998) *The Struggle for Accountability: The World Bank, NGOs and Grassroots Movements*, Cambridge, Mass.: MIT Press.

Frank, David John, Ann Hironaka, John W. Meyer, Evan Schofer and Nancy Brandon Tuma (1999) 'The Rationalization and Organization of Nature in World Culture', in John Boli and George Thomas (eds) *Constructing World Culture: International Nongovernmental Organizations since 1875*, Stanford: Stanford University Press, pp. 81–99.

Friedman, Elisabeth Jay, Kathryn Hochstetler and Ann Marie Clark (2005) *Sovereignty, Democracy, and Global Civil Society: State–Society Relations at UN World Conferences*, Albany: SUNY Press.

Garcia-Johnson, Ronie (2000) *Exporting Environmentalism: U.S. Multinational Chemical Corporations in Mexico and Brazil*, Cambridge, Mass.: MIT Press.

Gemmill, Barbara, and Bamidele-Izu (2002) 'The Role of NGOs and Civil Society in Global Environmental Governance', in Daniel C. Esty and Maria. H. Ivanova (eds) *Global Environmental Governance: Options and Opportunities*, New Haven: Yale School of Forestry and Environmental Studies, pp. 77–100.

Gereffi, Gary, Ronie Garcia-Johnson and Erika Sasser (2001) 'The NGO-Industrial Complex', *Foreign Policy* (July/August), pp. 56–65.

Gough, Clair and Simon Shackley (2001) 'The Respectable Politics of Climate Change: The Epistemic Communities and NGOs', *International Affairs* 77, 2, pp. 329–45.

Gutman, Pablo (2003) 'What did the WSSD Accomplish?' *Environment* 45, 2, pp. 20–6.

Grundbrandsen, Lars H., and Steinar Andresen (2004) 'NGO Influence in the Implementation of the Kyoto Protocol: Compliance, Flexibility Mechanisms, and Sinks', *Global Environmental Politics* 4, 4, pp. 54–75.

Haas, Ernst B. (1958) *The Uniting of Europe: Political, Social and Economic Forces 1950–1957*, Stanford: Stanford University Press.

Haas, Peter M. (1992) 'Introduction: Epistemic Communities and International Policy Coordination', *International Organization* 46, 1, pp. 1–35.

Hall, John A. (ed.) (1995) *Civil Society: Theory, History, Comparison*, Cambridge, Mass.: Blackwell and Cambridge, UK: Polity Press.

Harper, Caroline (2001) 'Do the Facts Matter? NGOs, Research, and International Advocacy', in Michael Edwards and John Gaventa (eds.) *Global citizen action*, Boulder: Lynne Rienner, pp. 247–58.

Held, David (1999) 'The Transformation of Political Community: Rethinking Democracy in the Context of Globalization', in Ian Shapiro and Casiano Hacker-Cordón (eds) *Democracy's Edges*, Cambridge: Cambridge University Press, pp. 84–111.

Hochstetler, Kathryn (2002) 'After the Boomerang: Environmental Movements and Politics in the La Plata River Basin', *Global Environmental Politics* 2, 4, pp. 35–57.

Hogenboom, Barbara (2001) 'Co-operation and Discord: NGOs and the NAFTA', in Bas Arts, Math Noortmann and Bob Reinalda (eds) *Non-State Actors in International Relations*, Aldershot: Ashgate, pp. 177–93.

Humphreys, David (2004) 'Redefining the Issues: NGO Influence on International Forestry Negotiations', *Global Environmental Politics* 4, 2, pp. 51–74.

Jasanoff, Sheila (1997) 'NGOs and the Environment: From Knowledge to Action', *Third World Quarterly* 18, pp. 579–94.

Johnson, Brian (1972) 'The United Nations' Institutional Response to Stockholm: A Case in the International Politics of Institutional Change', *International Organization* 26, 2, pp. 255–301.

Jordan, Grant (2001) *Shell, Greenpeace and the Brent Spar*, Basingstoke: Palgrave.

Jordan, Lisa, and Peter Van Tuijl (2000) 'Political Responsibility in Transnational NGO Advocacy', *World Development* 28, pp. 2051–65.

Kakabadse, Yolanda N., with Sarah Burns (1994) *Movers and Shapers: NGOs in International Affairs*, Washington, DC: World Resources Institute.

Keck, Margaret E., and Kathryn Sikkink (1998) *Activists Beyond Borders: Advocacy Networks in International Politics*, Ithaca: Cornell University Press.

Kellow, Aynsley (2000) 'Norms, Interests and Environmental NGOs: The Limits of Cosmopolitanism', *Environmental Politics* 9, 3, pp. 1–22.

Keohane, Robert O., and Joseph S. Nye, Jr (1977) *Power and Interdependence*, Boston: Little, Brown.
Khagram, Sanjeev, James V. Riker and Kathryn Sikkink (2002) 'From Santiago to Seattle: Transnational Advocacy Groups Restructuring World Politics', in Sanjeev Khagram, James V. Riker and Kathryn Sikkink (eds) *Restructuring World Politics: Transnational Social Movements, Networks and Norms*, Minneapolis: University of Minnesota Press, pp. 3–23.
King, Gary, Robert O Keohane and Sidney Verba (1994) *Designing Social Inquiry: Scientific Inference in Qualitative Research*, Princeton: Princeton University Press.
Kolk, Ans (1996) *Forests in International Environmental Politics: International Organisations, NGOs and the Brazilian Amazon*, Utrecht, Netherlands: International Books.
Kolk, Ans (2001) 'Multinational Enterprises and International Climate Policy', in Bas Arts, Math Noortmann and Bob Reinalda (eds) *Non-State Actors in International Relations*, Aldershot: Ashgate, pp. 211–25.
Levy, David L., and Peter Newell (2000) 'Oceans Apart? Business Responses to Global Environmental Issues in Europe and the United States' *Environment* 42, 9, pp. 8–20.
Levy, David L., and Peter Newell (eds) (2005) *Business in International Environmental Governance: A Political Economy Approach*, Cambridge, MA: The MIT Press.
Lipschutz, Ronnie (1992) 'Reconstructing World Politics: The Emergence of Global Civil Society', *Millennium* 21, pp. 389–420.
Lipschutz, Ronnie D., with J. Mayer (1996) *Global Civil Society and Global Environmental Governance*, Albany: SUNY Press.
Litfin, Karen (1994) *Ozone Discourses: Science and Politics in Global Environmental Cooperation*, New York: Columbia University Press.
Mahoney, James, and Gary Goertz (2004) 'The Possibility Principle: Choosing Negative Cases in Comparative Research', *American Political Science Review* 98, 4, pp. 653–69.
Mathews, Jessica T. (1997) 'Power Shift', *Foreign Affairs* January, pp. 51–66.
McCormick, John (1993) 'International Nongovernmental Organizations: Prospects for a Global Environmental Movement' in Sheldon Kamieniecki (ed.) *Environmental Politics in the International Arena: Movements, Parties, Organizations, and Policy*, Albany: SUNY Press, pp. 131–43.
Miller, Clark, and Paul N. Edwards (eds) (2001) *Changing the Atmosphere: Expert Knowledge and Environmental Governance*, Cambridge, Mass.: MIT Press.
Mitrany, David (1966) *A Working Peace System*, Chicago: Quadrangle Books.
Mol, Arthur P. J. (2000) 'The Environmental Movement in an Era of Ecological Modernisation', *Geoforum* 31, pp. 45–56.
Mol, Arthur P. J. (2001) *Globalization and Environmental Reform: The Ecological Modernization of the Global Economy*, Cambridge, Mass.: MIT Press.
Morphet, Sally (1996) 'NGOs and the Environment', in Peter Willetts (ed.) *The Conscience of the World: The Influence of Non-governmental Organisations in the U.N.*, Washington, DC: Brookings Institute, pp. 116–46.
Newell, Peter (2000) *Climate for Change: Non-state Actors and the Global Politics of the Greenhouse*, Cambridge: Cambridge University Press.

Newell, Peter (2001a) 'Campaigning for Corporate Change: Global Citizen Action on the Environment', in Michael Edwards and John Gaventa (eds) *Global Citizen Action*, Boulder: Lynne Rienner Publishers, pp. 189–201.

Newell, Peter (2001b) 'Environmental NGOs, TNCs and the Question of Governance', in Dimitris Stevis and Valerie J. Assetto (eds) *The International Political Economy of the Environment: Critical Perspectives*, Boulder: Lynne Rienner Publishers, pp. 85–107.

Oberthür, Sebastian, Matthias Buck, Sebastian Müller, Stefanie Pfahl, Richard G. Tarasofsky, Jacob Werksman and Alice Palmer (2002) *Participation of Non-Governmental Organisations in International Environmental Governance: Legal Basis and Practical Experience*, Berlin: Ecologic-Institute for International and European Environmental Policy.

Oberthür, Sebastian, and Hermann E. Ott (1999) *The Kyoto Protocol: International Climate Policy for the 21st Century*, New York: Springer.

O'Brien, Robert, Anne Marie Goetz, Jan Aart Scholte and Marc Williams (2000) *Contesting Global Governance: Multilateral Economic Institutions and Global Social Movements*, Cambridge: Cambridge University Press.

Parson, Edward (2003) *Protecting the Ozone Layer: Science and Strategy*, Oxford: Oxford University Press.

Pasha, Mustapha Kamal, and David L. Blaney (1998) 'Elusive Paradise: The Promise and Peril of Global Civil Society', *Alternatives* 23, pp. 417–50.

Princen, Thomas (1994) 'NGOs: Creating a Niche in Environmental Diplomacy', in Thomas Princen and Matthias Finger (eds) *Environmental NGOs in World Politics: Linking the Local and the Global*, London: Routledge, pp. 29–47.

Princen, Thomas, and Matthias Finger (eds) (1994) *Environmental NGOs in World Politics: Linking the Local and the Global*, London: Routledge.

Raustiala, Kal (1997) 'States, NGOs and International Environmental Institutions', *International Studies Quarterly* 41, pp. 719–40.

Raustiala, Kal (2001) 'Nonstate Actors in the Global Climate Regime', in Urs Luterbacher and Detlef F. Sprinz (eds) *International Relations and Global Climate Change*, Cambridge, Mass.: MIT Press, pp. 95–118.

Reinalda, Bob (2001) 'Private in Form, Public in Purpose: NGOs in International Relations Theory', in Bas Arts, Math Noortmann and Bob Reinalda (eds) *Non-State Actors in International Relations*, Aldershot: Ashgate, pp. 11–40.

Risse, Thomas (2002) 'Transnational Actors and World Politics', in W. Carlsnaes, T. Risse and B. A. Simmons (eds) *Handbook of International Relations*, Thousand Oaks, Calif.: Sage, pp. 255–74.

Risse-Kappen, Thomas (1995) 'Bringing Transnational Relations Back In: Introduction', in Thomas Risse-Kappen (ed.) *Bringing Transnational Relations Back In: Non-State Actors, Domestic Structures and International Institutions*, Cambridge: Cambridge University Press, pp. 3–36.

Rootes, Christopher (1999) 'Acting Globally, Thinking Locally? Prospects for a Global Environmental Movement', in Christopher Rootes (ed.) *Environmental Movements: Local, National and Global*, London: Frank Cass, pp. 290–310.

Rosenau, James N. (2003) *Distant Proximities: Dynamics Beyond Globalization*, Princeton: Princeton University Press.

Rosenau, James N., and E. Czempiel (eds) (1992) *Governance without Government: Order and Change in World Politics*, Cambridge: Cambridge University Press.

Rowlands, Ian H. (1995) *The Politics of Global Atmospheric Change*, Manchester: Manchester University Press.
Rowlands, Ian H. (2000) 'Beauty and the Beast?: BP's and Exxon's Positions on Global Climate Change', *Environment and Planning C: Government and Policy* 19, pp. 339–54.
Rowlands, Ian H. (2001) 'Transnational Corporations and Global Environmental Politics', in Daphné Josselin and William Wallace (eds) *Non-State Actors in World Politics*, New York: Palgrave, pp. 133–49.
Seligman, Adam B. (1992) *The Idea of Civil Society*, New York: The Free Press.
Sell, Susan K., and Aseem Prakash (2004) 'Using Ideas Strategically: The Contest Between Business and NGO Networks in Intellectual Property Rights', *International Studies Quarterly* 48, pp. 43–175.
Skjelsbaek, Kjell (1971) 'The Growth of International Nongovernmental Organization in the Twentieth Century', *International Organization* 25, 3, pp. 420–42.
Skodvin, Tora, and Steinar Andresen (2003) 'Nonstate Influence in the International Whaling Commission, 1970–1990', *Global Environmental Politics* 3, 4, pp. 61–86.
Skodvin, Tora, and Jon Birger Skjærseth (2001) 'Shell Houston, We Have a Climate Problem!' *Global Environmental Change* 11, pp. 103–6.
Smith, J. Eric (1972) 'The Role of Special Purpose and Nongovernmental Organizations in the Environmental Crisis', *International Organization* 26, 2, pp. 302–26.
Smith, Jackie, Ron Pagnucco and Charles Chatfield (1997) 'Social Movements and World Politics: A Theoretical Framework', in Jackie G. Smith, Charles Chatfield and Ron Pagnucco (eds) *Transnational Social Movements and Global Politics: Solidarity Beyond the State*, Syracuse: Syracuse University Press, pp. 59–77.
Snow, David A., and Robert D. Benford (1992) 'Master Frames and Cycles of Protest', in Aldon D. Morris and Carol McClurg Mueller (eds) *Frontiers in Social Movement Theory*, New Haven: Yale University Press, pp. 133–55.
Speth, James Gustave (2003) 'Perspectives on the Johannesburg Summit', *Environment* 45, 1, pp. 24–9.
Tesh, Sylvia Noble (2000) *Uncertain Hazards: Environmental Activists and Scientific Proof*, Ithaca: Cornell University Press.
UIA (2003) 'Types of Organizations in the Yearbook', Union of International Associations, available from <www.uia.org/organizations/orgtypes/orgtyped.php>.
UNCED (1992) *Agenda 21*, New York: United Nations.
UNEP (2002) *Civil Society Consultations on International Environmental Governance, 22–23 May 2001* United Nations Environment Programme, available from <www.unep.org/IEG/Meetings.asp>.
United Nations (1997) *Kyoto Protocol to the United Nations Framework Convention on Climate Change*, UNFCCC Secretariat, available from <http://unfccc.int/resource/docs/convkp/kpeng.pdf>.
United Nations (2002) *Report of the World Summit on Sustainable Development*, New York: United Nations.
Wapner, Paul (1996) *Environmental Activism and World Civic Politics*, Albany: SUNY Press.

Wapner, Paul (1997) 'Governance in Global Civil Society', in Oran R. Young (ed.) *Global Governance: Drawing Insights from the Environmental Experience*, Cambridge, Mass.: MIT Press, pp. 66–84.

Wapner, Paul (1998) 'Reorienting State Sovereignty: Rights and Responsibilities in the Environmental Age', in Karen Litfin (ed.) *The Greening of Sovereignty in World Politics*, Cambridge, Mass.: MIT Press, pp. 275–97.

Wapner, Paul (2001) 'Horizontal Politics: Transnational Environmental Activism and Global Cultural Change'. Paper presented at the annual meeting of the International Studies Association, Chicago, 21–24 February.

Wapner, Paul (2002) 'Defending Accountability in NGOs'. Paper presented at the annual meeting of the International Studies Association, New Orleans, 24–27 March.

Wapner, Paul (2003) 'World Summit on Sustainable Development: Toward a Post-Jo'burg Environmentalism', *Global Environmental Politics* 3, 1, pp. 1–10.

Willetts, Peter (ed.) (1982) *Pressure Groups in the Global System: The Transnational Relations of Issue-Oriented Non-Governmental Organisations*, London: Pinter.

Willetts, Peter (1996a) *The Conscience of the World: The Influence of Non-Governmental Organisations in the UN System*, Washington, DC: Brookings Institute.

Willetts, Peter (1996b) 'Consultative Status for NGOs at the United Nations', in Peter Willetts (ed.) *The Conscience of the World: The Influence of Non-Governmental Organisations at the U.N.*, Washington, DC: Brookings Institute, pp. 31–62.

Willetts, Peter (1996c) 'From Stockholm to Rio and Beyond: The Impact of the Environmental Movement on the United Nations Consultative Arrangements for NGOs', *Review of International Studies* 22, pp. 57–80.

Williams, Marc, and Lucy Ford (1999) 'The World Trade Organisation, Social Movements and Global Environmental Management', *Environmental Politics* 8, pp. 268–89.

Wright, Brian G. (2000) 'Environmental NGOs and the Dolphin-Tuna Case', *Environmental Politics* 9, 4, pp. 82–103.

Yamin, Farhana (2001) 'NGOs and International Environmental Law: A Critical Evaluation of their Roles and Responsibilities', *Reciel* 10, 2, pp. 149–62.

Yearley, Steven (1994) 'Social Movements and Environmental Change', in Michael Redclift and Ted Benton (eds) *Social Theory and the Global Environment*, London: Routledge, pp. 150–68.

Young, Oran R. (ed.) (1997) *Global Governance: Drawing Insights from the Environmental Experience*, Cambridge, Mass.: MIT Press.

Young, Zoe (1999) 'NGOs and the Global Environmental Facility: Friendly Foes?' *Environmental Politics* 8, pp. 243–67.

Zürn, Michael (1998) 'The Rise of International Environmental Politics: A Review of Current Research', *World Politics* 50, pp. 617–49.

annotated bibliography

Keck, Margaret E., and Kathryn Sikkink (1998) *Activists Beyond Borders: Advocacy Networks in International Politics*, Ithaca: Cornell University Press. Keck and Sikkink's book carefully examines the work of activist networks in the areas of human rights, environment and violence against women and has prompted a great deal of work in the area of transnational networks to the field of IEP.

Moreover, their idea of the 'boomerang' pattern in which national NGOs reach out to international organizations in order to exert external pressure on their own governments has been important in highlighting the linkages between domestic and international politics.

Levy, David L., and Peter Newell (eds) (2005) *Business in International Environmental Governance: A Political Economy Approach*, Cambridge, Mass.: MIT Press. Levy and Newell's edited volume brings together several important works as well as some new material analysing the role of business interests in international environmental governance. Using a political economy approach, the contributions demonstrate how business and industry have both shaped and been shaped by international environmental policies and policy-making processes. Case studies examine environmental politics in the areas of climate change, ozone depletion, tropical logging and the development of environmental management standards.

Pasha, Mustapha Kamal, and David L. Blaney (1998) 'Elusive Paradise: The Promise and Peril of Global Civil Society', *Alternatives* 23, pp. 417–50. Though not focused on the environment per se, this article challenges scholars of transnational environmental politics to consider the relationship between global civil society and democratization. Pasha and Blaney argue that transnational activity has limited opportunity to transform global social and political space because it largely reflects existing relations within the broader capitalist structure of the global economy.

Princen, Thomas, and Matthias Finger (eds) (1994) *Environmental NGOs in world politics: Linking the Local and the Global*, London: Routledge. This edited volume initiated the recent wave of scholarship on transnational actors in IEP. Princen and Finger bring together a number of scholars to examine NGOs as agents of social change in world politics and argue that environmental NGOs have created a special niche in international environmental diplomacy by linking the local and the global. Half of the chapters address theoretical issues related to NGOs in IEP, while the remaining chapters are case studies of NGOs in several international environmental policy processes: the Great Lakes Water Quality Agreement, the ivory ban, the Antarctic Environmental Protocol and the Rio Earth Summit.

Wapner, Paul (1996) *Environmental activism and world civic politics*, Albany: SUNY Press. Wapner's detailed study of three environmental organizations (the World Wildlife Fund/World Wide Fund for Nature, Greenpeace and Friends of the Earth) offers excellent insight into how these groups operate internally. In addition, this book was important in moving the scholarly discussion of NGOs in international environmental politics beyond questions about how NGOs influence states. Wapner argues that NGOs shape world politics in the realm of 'global civil society' by promoting an environmental sensibility.

8
environmental security
larry a. swatuk

The study of environmental security revolves around a central idea that environmental problems – in particular, resource scarcity and environmental degradation – may lead to violent conflict between and among states and societies. Although these ideas are not new, they have gained momentum since environmental issues emerged on the international political agenda in the early 1970s (Gleditsch, 1998, p. 382). And while scholars such as Deudney (1990) lament the interlinking of environmental problems with security studies, for Dalby, since matters such as ozone depletion, pollution, and 'many situations with a vaguely environmental designation' are now 'part of international political discourse and policy initiatives, environment cannot be separated from matters of what is now called "global" security' (Dalby, 2002a, p. 95; Worldwatch Institute, 2005).

Proponents of environmental security argue that if environmental change is a potential source of social conflict, and if societies face dangers from environmental change, then security policies – indeed, the very concept itself – must be redefined to account for these threats (Conca and Dabelko, 1998). While new thinking about security had been ongoing throughout the 1980s (Mathews, 1989; Ullman, 1983; World Commission on Environment and Development, 1987), the end of the Cold War provided real intellectual space for considering how the environment might be accounted for in various approaches to security (Dalby, 2002b).

Thus the 1990s saw two interrelated discussions: one involving the redefinition of security (Baldwin, 1997; Booth, 1991; Buzan, 1991; Buzan et al., 1998; Krause and Williams, 1997; Lipschutz, 1995; Walt, 1991); the other involving questions about how environmental change threatens (individual, state, global) security (Deudney and Matthew, 1999; Myers,

1989, 1993; Ohlsson, 1999; Renner, 1989). As this chapter will show, there is little consensus on either of these issues. Yet because these debates are ongoing at the policy level, the practical application of various ideas regarding environmental security by a wide array of actors embodies all of the contradictions and controversies extant in the debates. This will be demonstrated in the case study below.

This chapter proceeds as follows. In the next section I outline the traditional, realist approach to 'security'. Following that I provide a number of critiques of this approach: institutionalist, neorealist, structuralist, and poststructural/postmodernist. I argue that the first three work within a problem-solving framework in an effort to 'rethink' security, whereas the last presents a 'critical' challenge to both the analysis and practice of security. I then turn to an extended discussion of the environment and security. I frame this section in terms of scholars working within the ambit of the dominant paradigm – what I call *environmental security studies* – and those who challenge this approach – *critical environmental security studies*. Lastly, I show how the analysis and practice of 'environmental security', contested though it is, manifests itself in the area of transboundary natural resource management (TBNRM) in Southern Africa.

rethinking security

The traditional approach to security studies, known as realism, focuses on the causes of war and the conditions of peace between and among states. Given the systemic condition of anarchy, wherein sovereign states pursue policies of self-help, states by their very existence threaten each other.[1] 'Security', then, may be defined as the protection of a state (and, by logical extension, its citizens) from the threat of other self-regarding states acting in their own interests. This protection is facilitated by military preparedness. Diplomacy is considered the practice of 'war by other means'. In this rendering, the 'state' is both means and end (or 'referent object') of 'security'. State security is achieved through the exercise of 'power', itself also both means and ends, to wit: a powerful state is made secure by the exercise of power. Power involves multiple factors, including centrally the threat and use of military force (Morgenthau, 1978).

Since the acquisition and projection of power is the main currency of the interstate system, self-regarding states usually find themselves caught up in global and regional security dilemmas. The end result may be an overall decrease in regional or global security, depending on the nature of the arms produced. For realists, stability of the interstate system depends on achieving a 'balance of power'.

To many, great power 'security' threatens the lives of everybody. Critiques of the realist approach to security proliferated during the 1970s and 1980s. There are many different ways to categorize these challenges. One is to group them according to general theoretical position. Liberal institutionalists, for example, work within the accepted framework of interstate relations but challenge the logic of states pursuing unilateral security policies (Hurrell, 1995; Ruggie, 1998). Liberal institutionalists argue in support of both a broadened understanding of 'security' (defined in military, economic and ecological terms) and a multilateral framework for addressing multidimensional challenges to global security (Commission on Global Governance, 1995).

Neorealists, working within the ambit of 'national security', have modified their position in acknowledgement of the multidimensional nature of threats to the state. For example, while retaining his focus on the state as the primary referent object and provider of security, Buzan (1991, p. 12) argues that rethinking security 'reflects the relentless pressure of interdependence on the older ways of thinking of both Realists and Idealists'. Alongside conventional military threats, Ullman (1983), for example, includes the inability to meet basic needs, environmental deterioration and natural disasters. So while security means freedom from these threats, the framework for analysis remains state-centric.

Structuralist analyses may or may not work within a statist ontology. For example, dependency and world-systems theorists generally retain a statist language while focusing attention on broad, structural processes like global capitalism. So states may be divided into core, semi-peripheral and peripheral (Wallerstein, 1974). In the context of late-twentieth-century and early-twenty-first-century globalization, whole regions are seen to merit core, intermediate, or peripheral zone status (Hettne, 1997). In terms of security,

> Regions in the core zone, North America, Europe and East Asia centred on Japan, are thus economically more advanced and normally growing, and they have stable – if not always democratic – regimes which manage to avoid interstate as well as intra-state conflicts ... Regions in the intermediate zone ... are under 'core guidance' ... Regions in the peripheral zone, in contrast, are politically turbulent and economically stagnant. War, domestic unrest, and underdevelopment constitute a vicious circle which make them sink to the bottom of the system, creating a zone of war and starvation. (Hettne, 2001, pp. 95–6)

Thus, for many structuralists, insecurity is a consequence of a state's or region's history of incorporation into the global capitalist system (Swatuk, 2001, pp. 278–83). Focusing attention on other structural processes and conditions – for example, race, class and gender – helps reveal a panoply of threats and insecurities individuals and groups suffer despite (or possibly because of) the existence of a 'strong', militarily-secure state (Tickner, 1995). Herein lies a more serious challenge to traditional renderings of security: the challenge is not merely to redefine the content of security (from military to military-plus), but to reconsider the very subject itself. If traditional approaches to security ignore or exacerbate those global social/ political/economic inequalities that constitute the basis for most forms of insecurity, whose interest does the traditional approach serve? When combined with the spectre of nuclear Armageddon, the answer would seem to be 'no one'. Yet Steven Walt (1991, p. 222), reflecting upon the 1980s' 'renaissance' in security studies, admonishes those who would ask such difficult questions:

> Security studies seeks cumulative knowledge about the role of military force. To obtain it, the field must follow the standard canons of scientific research: careful and consistent use of terms, unbiased measurement of criticial concepts, and public documentation of theoretical and empirical claims.

But is security studies the dispassionate search for objective truth? Post-structuralist and postmodernist perspectives suggest that it is nothing of the sort. To the contrary, they suggest that, as a discourse, it acts as the handmaiden of the powerful (Walker, 1997, p. 62). Traditional security studies have long been tied to the strategic interests of the world's most powerful states. So, answers given to questions regarding the security of (powerful) states lead directly to policies that serve the few and imperil the many. The postmodern/structural critique constitutes a broad but barely unified church, including, inter alia, feminism, environmental politics, political ecology, development studies and critical strands of international political economy (IPE). In Waever's (1997) terms, they are part of international relations (IR)'s 'fourth debate'. Most of these challengers may be thought of as constitutive of a nascent, open-ended 'critical theoretical' enterprise (George, 1994). Critical theory is to be contrasted with 'problem-solving theory' which, in Cox's terms, takes 'prevailing social and power relationships and the institutions into which they are organised ... as the given framework for action' (Cox, 1986, p.

206). For several key reasons, the environment stands at the epicentre of this fourth debate.

Environmental degradation, resource depletion, species loss – for critical theorists these conditions are emblematic of the crisis of contemporary society. Along with weapons of mass destruction and the Holocaust (Baumann, 1989; George, 1994), they are logical consequences of modernity whose overriding idea – progress – has for several hundred years been dependent upon a belief in the rightness of 'harnessing' nature, of rendering useful to 'rational man' all that is 'natural'. Given that the modern age arose out of a white, Western European Enlightenment, too often and for too long, this constructed category has included not just the physical environment, but animals, women, inferior European men, and all people of colour (Pettman, 1996).

There are two points of importance for the topic at hand. First, the national security state remains the primary institutional structure of the modern era. 'Security studies' as the handmaiden of the most powerful of these states is founded on key binaries: the possibilities of the political (inside the state where order is possible); the location of threats (outside the state where anarchy reigns); the means of discerning threats and proper responses (through the identification of objective facts by a rational 'knower'); the construction of a theory of security through practice so that one eventually arrives at an accurate representation of truth (that is, knowledge claims independent of subjectivity); the masculine exercise of 'power over' the 'irrational other' as proper form of response (in defence of the weak and feminine, but also, given the condition of anarchy, any other response would itself be weak and feminine).

For critical theorists, the practice of this form of 'social science' produces simultaneously wealth for the few, poverty for the many; creative invention and possibly irreversible environmental destruction; increased leisure and choice for some, increased toil and a lack of options for most. Any 'celebration of the age of rational science and modern technological society' therefore 'cannot simply be disconnected from' its myriad, historically specific negative consequences. They are, in fact, integrally related (George, 1994, p. 141; Peterson, 2003, pp. 22–9). In terms of 'the environment', the very forms and practices giving rise to its degradation cannot, therefore, be its salvation. As will be seen below, 'the state is not only unnecessary from a Green point of view, it is positively undesirable' (Paterson, 1995, p. 238).

The second, and related point is that for critical theorists, these outcomes are not the accidental result of policies based on poorly understood science. Rather, they are the direct outcome of a form of knowledge production dependent upon a positivist/empiricist methodology, a discipline whose ontological frameworks regarding both 'human nature' and social organization asks the wrong questions, or frames the right questions the wrong way, and whose epistemological claims 'to know' 'obstruct a relational, multi-dimensional understanding of the social' (Peterson, 2003, p. 39).

While constructivist methodologies take us some way toward a more nuanced understanding of the ways in which human language, thought and action create social reality – so, in my view, effectively contesting positivist claims of passive subjects studying objective reality – it is the poststructural, postmodernist perspective which lends most insight into this process.

In the face of dramatic change – for example, global warming, deforestation, the collapse of fisheries, the collapse of the Soviet empire – we desire explanations that will ensure stability or bring order. In other words, we seek to contain the 'threat'. But explanations regarding causality diverge markedly: for Realists, 'nature' remains somehow 'out there', as does a collapsed Soviet Union – an untamed other beyond the purview of ordered civilization within a (democratic) state. Security requires containment. For a majority of critical theorists, however, these events and conditions are linked. They are emblematic of the multidimensionality of and multiple paradoxes within the late modern era, wherein the state system, industrial capitalism, technological innovation (particularly its military dimension), individual liberty manifesting as consumerism, and the tyranny of bureaucracy are equally complicit (Hall et al., 1992). 'Stability' in this case requires a fundamental reordering of late modern society (Peterson, 2003). This is an ongoing academic debate whose impact on people, places and things is, as we will see below, very real. Having mapped the terrain of and debates within security studies, let us now turn to environmental security.

environment and security

Positing negative outcomes of environmental decay for humanity and/or the planet, particularly from an over-consumption perspective, has a proud lineage stretching back 75 years or more (Ehrlich, 1968; Leopold, 1949; Mumford, 1934). It has been particularly strong since the 1960s, however, with seminal texts being provided by Rachel Carson (1962) and the Sprouts (1965). As with realism, a variety of historical examples – for example, the 'lessons of Easter Island' – provide a foundational narrative of sorts. Since the end of the Cold War, however, the security focus has been dominated by state-centric approaches interrogating the implications of environmental degradation *in the global South* for security of states *in the global North*, an argument described by supporters and critics alike as 'Malthusian' or 'neo-Malthusian' (Dalby, 2002a; Gray, 2002, pp. 180–2).

The end of the Cold War meant that, for a time, Western state-makers were at a loss regarding a revised role for their militaries. If the West had arrived at 'the end of history', as hypothesized by Fukuyama (1992), from where would future threats emanate? Given 20 years of high-profile summitry regarding the human impact on the global environment, state-makers logically looked to the 'threat potential' of the environment.

Throughout the 1990s several extensive research programmes were undertaken regarding the hypothesized link between environmental degradation, resource scarcity and violent conflict, commonly called 'environmental security'. There have been a number of attempts to review and synthesize this vast literature (Barnett, 2001; Gleditsch, 1998), with the annual report of the Woodrow Wilson Center's Environmental Change and Security Project (ECSP) acting as a valuable 'rough guide' to the field (see <wwics.si.edu>). Matthew (2002) reviews the research in terms of studies explicitly concerned with the state and environmentally induced acute conflict probability and those that are more holistic, historical and critical in their approaches.

In essence, the first group are engaged in an intraparadigm debate regarding problem-solving theory. They are primarily concerned with methodology, data collection, appropriate scope and definitions in the hope that their research programmes will lead both to the accumulation of knowledge (Homer-Dixon, 2003, p. 90) and policy relevance (see also Hochstetler and Laituri in this volume).

The second group, in contrast, are centrally concerned with critical theory, though they come at it in a wide variety of ways. While no less concerned with scope and method, they are equally concerned with engaging those working within the problem-solving camp regarding ontological and epistemological questions. Their central concern is that problem-solving theory begins, rather like traditional security studies, from a false premise: that global society is primarily a world of self-regarding states, some of which have more (political, military, technological) power and are more economically developed than others. A state's place in the system is largely the result of its own efforts, and political order – indeed, 'political life' – is only possible within the boundaries of the sovereign state (Walker, 1997, p. 62).

Many scholars, mostly of the neo-institutionalist or political ecology variety, fall somewhere in between the two – at once holding post-structuralist positions regarding, for example, an ontology centred on individuals or ecosystems, and making policy advice for states and global institutions.

In the next section, I detail the findings of these two different groups. Following Krause and Williams (1997), I label these 'environmental security studies' and 'critical environmental security studies'.

environmental security studies

Gleditsch (1998) summarized historical concerns with the link between environmental degradation, resource scarcity and violence in terms of fights over territory, raw materials, continental shelves and islands, energy and food. In traditional Realist analysis, state power depends fundamentally on the natural resources contained within its territorially delimited space (Morgenthau, 1978). National power also depends on the ability to access key resources not contained within the state. In the late modern, industrial era, national power also depends on the ability of a state to transform these renewable and non-renewable resources into tradable consumables. Natural resources may be enhanced, depleted or transformed over time. According to Klare (2001), competition and control over critical natural resources will be the guiding principle behind the use of military force in the twenty-first century. The popular press throughout the world continues to play on these fears. The physical location of these events in the global South and its link to warfare and general human misery in the popular imagination has been facilitated by magazine and newspaper reporting focusing on the more bizarre and fantastical aspects of West African conflicts, with Robert Kaplan's 1994 article 'The Coming Anarchy' for the *Atlantic Monthly* being influential in policy circles (see Ellis, 1999, for a trenchant critique). In the 1990s, a number of systematic studies emerged to examine the hypothesized relationship between the environment and conflict.

renewable resource degradation and violent conflict

The Homer-Dixon-directed projects on 'Population, Environment and Security' and 'Environmental Change and Security' are undoubtedly the most well-known in North America of these systematic programmes of research (Homer-Dixon, 1991, 1994, 1999; Homer-Dixon and Blitt, 1998). Research focused on the causal link between violent conflict and the depletion of renewable resources, in particular agricultural land, water, forests and fisheries, argues that the social effects of large-scale environmental changes such as global warming and ozone depletion 'will not be seen until well into the [twenty-first] century' (in Conca and Dabelko, 1998, p. 289). 'Researchers sought to answer two questions. First, does environmental scarcity contribute to violence in developing countries? Second, if it does, how does it contribute?' (Homer-Dixon, 1998, p. 279).

Three hypotheses were used to link conflict with environmental change:

- decreasing supplies of physically-controllable resources (for example, water and agricultural land) would provoke 'simple scarcity' conflicts;
- 'group identity' conflicts would result from large population movements caused by environmental stress;
- severe environmental scarcity would simultaneously increase economic deprivation and disrupt key social institutions (most importantly, the state) and cause 'deprivation' conflicts.

Key to their analysis is the definition of scarcity. Scarcity can come about in one of three ways: (1) as a result of increased demand (demand-induced), for example, through population growth or increased per capita consumption; (2) as a result of decreased supply (supply-induced), for example, the erosion of cropland; and/or (3) as a result of unequal access to and distribution of a resource (structural) 'that concentrates it in the hands of relatively few people while the remaining population suffers from serious shortages' (Homer-Dixon, 1998, p. 280). These push-pull factors are said to often coexist and interact (Homer-Dixon and Blitt, 1998, p. 6).

In the event of scarcity, two processes are set under way: resource capture by those with the means to do so, ecological marginalization of those without. So what is important is not merely the absolute supply of a resource. 'What we should investigate, rather, is the resource's supply *relative to*, first, demand on the resource, and, second, the social distribution of the resource' (Schwarz et al., 2000, p. 79).

A mid-1990s pilot study from NATO advanced the analysis of 'environmental stress' – defined as a combination of resource scarcity and resource degradation – and violent conflict by combining aggregate data with the 'syndrome approach'[2] to add space and time dimensions to studies that had hitherto been overwhelmingly ahistorical and focused on Third World states. Nevertheless, it has proved difficult for Northern scholars to move away from both neo-Malthusian assumptions and a South-focused geography of conflict – even where evidence points in a different direction. For example, Tir and Diehl (1998), in their analysis of population patterns and conflict propensities among all states over 1930–89, show, among other things, that: (1) population growth rate varies inversely with escalation to war; (2) 'population density has no independent effect on the propensity for conflict involvement'; (3) 'there is little evidence that high population growth states are more likely to initiate militarised conflict'; and (4) while there is a weak positive correlation between population growth and conflict involvement, it is really only in combination with military capability that makes it significant. 'States with multiple borders and high military spending are more conflict-prone' (Tir and Diehl, 1998, pp. 332–6). Yet, in their conclusion the authors feel compelled to say, 'Generally, we have found population growth pressure to have a significant impact on the likelihood of a state becoming involved in military conflict' (Tir and Diehl, 1998, p. 336; dataset at <http://wsi.cso.uiuc.edu/polisci/faculty/diehl.html>).

A key finding of the Homer-Dixon project is that 'environmental scarcity does not inevitably or deterministically lead to social disruption and violent conflict'. The causal pathway from environmental degradation to social violence is neither direct nor unilinear. On the one hand, 'environmental scarcity ... increases society's demands on the state while decreasing its ability to meet those demands' (Homer-Dixon, 1998, p. 281). But on the other hand, 'environmental scarcity produces its effects within extremely complex ecological-political systems' (Homer-Dixon, 1999, p. 178). Therefore, a key finding is that 'environmental scarcity is not sufficient, by itself, to cause violence; when it does contribute to violence, research shows, it always interacts with other political, economic, and social factors. Environmental scarcity's causal role can never be separated from these contextual factors, which are often unique to the society in question' (Homer-Dixon, 1999, p. 178).

'maldevelopment' as contextual factor

The international research team under the direction of Günther Baechler examined these contextual factors, particularly the perceived links between maldevelopment, environmental transformation and conflict. While increasing population pressure on renewable resources remains important, population increase is but one of a number of consequences of maldevelopment in developing and transitional societies: '*Development and security dilemmas* are connected to a syndrome of problems which produces environmental conflicts of varying intensity and nature'

(Baechler, 1998, p. 24). Maldevelopment, in Baechler's terms, is most often the result of the ways in which developing and transitional societies have experienced modernization, the outcome typically being a weak state form imposed on a multiethnic society, but dominated by one ethnic group dependent on one or a few primary commodities for revenue. This, they argue, is the most likely setting in which conflict will occur.

In contrast to Homer-Dixon, Baechler suggests a somewhat longer causal chain, one that begins from the conditions that gave rise to environmental degradation. The important finding here is that states in both the North and South help constitute instability in the South. With regards to war-torn Sierra Leone, Richards (1996, p. xvii) states, '"we" and "they" have made this bungled world of Atlantic-edge rain-forest-cloaked violence together'. Maldevelopment is a consequence of historical processes and contemporary global links (also Duffield, 2001).

Based on 40 case studies, Baechler draws the following conclusions. First, 'while environmental discrimination plays different roles in the causation of a conflict, its intensity does not depend on the degree of the physical and chemical degradation of the landscape' (Baechler, 1998, p. 37). Rather, 'it is the political context that matters'. Second, 'there is no automatic spiral towards violence'. Indeed, 'environmental conflicts become a catalyst for cooperation if political compromises are seen as desirable and technical solutions are feasible' (Baechler, 1998, pp. 37–8). Third, there seems to be 'little empirical support for the first hypothesis that environmental scarcity causes violent conflicts or wars between states'. But, fourth, at the same time 'there is substantial evidence to support the second hypothesis that environmental scarcity causes large population movements, which in turn cause conflicts' (Baechler, 1998, p. 38). Migration-triggered conflicts vary depending on location and impacted groups (highland versus lowland producers; rural versus urban dwellers). These conflicts do not always manifest as intergroup identity conflicts; they may, in fact, involve 'one and the same ethnic group that may be divided by geographical or national boundaries' (Baechler, 1998, p. 38). This finding differs somewhat from Homer-Dixon's more general claim that migration triggers intergroup identity conflicts. Fifth, while Baechler acknowledges that there does seem to be some empirical support for the claim that environmental scarcity simultaneously increases economic deprivation and disrupts key social institutions (these are Homer-Dixon's terms), this is no guarantee that the problem will result in violent conflict.

Whereas Homer-Dixon relies heavily on the rather vague hope of 'ingenuity' as a way forward for societies suffering environmental degradation and resource depletion, Baechler's study, in my opinion, more readily leads toward practical policy choices linking environment to sustainable development, and by extension, Northern practices of production and consumption to Southern experiences of conflict and environmental transformation. They do this by identifying modernization, particularly its 'global socioecological side', as a primary causal factor in the occurrence of violent conflict in developing and transitional societies (Baechler, 1999, p. 227). This suggests, importantly, that solutions to environmental conflict are not dependent solely on the ingenuity available to particular societies suffering protracted conflict, but on the willingness of Northern states, individuals, corporations and non-governmental organizations to acknowledge the very real role they play in fomenting conflict in the South (Callaghy et al., 2001). Two relatively obvious pathways to peace present themselves: debt forgiveness and stemming the flow of Northern arms sales to Southern recipients. So political economic policies in the North can directly alleviate 'environmental stress' and human insecurity in the South (Swatuk, 2004).

While remaining primarily state- and region-based in analysis, the Baechler-led study points toward interconnections. Ingenuity may be important, but Baechler's world does not neatly divide into societies that have it in abundance and those that do not; nor does this study 'blame the victim' for the concentration of 'diffuse and persistent' violence in the South, as suggested so clearly by Kaplan's 'coming anarchy'.

governance, violent conflict and peace-building

Aspects of environmental security research also touched on democratic peace theory (Midlarsky, 1998) and the analysis of state failure (State Failure Task Force, 1999), sometimes yielding surprising results. Midlarsky (1998, p. 358), for example, concludes that 'instead of positive relationships between the extent of democracy and environmental protection, as much popular and recent scholarly writing have suggested, the associations found here are principally negative or non-existent'. The CIA-established 'State Failure Task Force', rather unsurprisingly, argued that 'partial democracies' are the most vulnerable to state failure (for example, revolutionary or ethnic warfare, adverse or disruptive regime transition, genocide or politicide) and that, based on available evidence, the environment plays no direct role here (State Failure Task Force, 1999).

Homer-Dixon's research suggests certain states are prone to environmentally induced conflict. 'The character of the state is particularly important: a representative state will receive [demands from civil society] and react quite differently to a non-representative state such as apartheid

South Africa' (Homer-Dixon, 1998, p. 281). Moreover, economically poor states lacking both financial and human capital, and being ethnically diverse seem less able to adapt to severe environmental challenges. They are said to lack adaptive capacity (Homer-Dixon and Blitt, 1998, p. 9). A cursory look at their case studies gives some idea as to the sorts of states Homer-Dixon and his colleagues consider short on adaptive capacity: Mexico, the Philippines, South Africa, Pakistan and Rwanda.

A clear trend across these studies is the distance they have travelled away from a fascination with conflict potential toward deliberately pursuing peace building opportunities through cooperation on environmental issues. For Lietzmann and Vest (1999, p. 40), 'violence is by no means the automatic outcome of conflict. Countless issues of conflict, particularly at the local or regional level are resolved cooperatively; only a limited number of conflicts reach a higher conflict intensity.' Environmental stress plays different roles along the 'conflict dynamic': as a structural source; a catalyst; or a trigger (Lietzmann and Vest, 1999, p. 41). To ensure that environmental stress does not reach levels at which violent conflict becomes likely, 'the development of early warning indicator systems, data bases and decision support systems are feasible and warranted'. The Executive Summary Report suggests a number of ways in which institutions (at local, national, regional, global levels) can be strengthened, adherence to agreements can be facilitated, and that 'existing prevention and dialogue mechanisms can be used to address the security impact of environmental issues, capitalise on the catalytic function of environmental cooperation for confidence building, and enhance dialogue and cooperation among themselves' (Lietzmann and Vest, 1999, pp. 46–8).

Environmental 'peace-making' is a very new trend in the environmental security studies literature. Conca and Dabelko (2002), for example, suggest two specific ways in which cooperation on environmental problems can help build peace: by altering the 'strategic climate' within which states operate; and by helping construct post-Westphalian forms of governance that may ultimately tie states into cooperative agreements and practices so facilitating 'learning'. Environmental peace-making may be understood as one way of bringing parties in conflict together 'to work on environmental issues in ways that build confidence and reduce political tensions'. However, certain resources may be more conducive to cooperative behaviour than others. For example, shared water resources seem to offer pathways to peaceful cooperation, whereas forests and minerals – due to the particular nature of the resource – appear more prone to induce conflict (Adelphi Research and Woodrow Wilson Center, 2004, pp. 2, 7–10; also Swatuk, 1996).

critical environmental security studies

Almost as soon as 'the environment' appeared on the policy map of state security apparatuses, dissenting and critical voices could be heard questioning the appropriateness of linking environmental issues to (national) security practices. Dalby (1997, 1998, 2002b), Peluso and Watts (2001), Barnett (2001), Gleditsch (1997, 1998) and Levy (1995) are among those who have made important critical interventions in the environment and security debate. Yet it was Deudney's 1990 article which remains, to my mind, the most compelling argument against this link, with all others taking their cue in one way or another from him. His discussion centres on three basic points. First, the structures developed to ensure *national* security are of little help as far as environmental problems – be they local or global – are concerned. National security is 'safeguarded' through a system of organized violence highly dependent on secrecy and technological expertise, whereas solutions to environmental problems require transnational cooperation, openness and creativity.

Second, given that national security discourses render all those people, places and things outside the state as a potential 'threat' – as an unknowable, untrustable 'other' – 'it seems doubtful that the environment can be wrapped in national flags without undercutting the "whole earth" sensibility at the core of environmental awareness' (Deudney in Conca and Dabelko, 1998, p. 309).

Third, while 'few ideas seem more intuitively sound than the notion that states will begin fighting each other as the world runs out of usable natural resources', global systems of trade, the substitutability of many raw materials, and 'the very multitude of interdependency in the contemporary world, particularly among the industrialised countries, makes it unlikely that intense cleavages of environmental harm will match interstate borders ... Resolving such conflicts will be a complex and messy affair, but the conflicts are unlikely to lead to war' (Deudney in Conca and Dabelko, 1998, pp. 310, 312).

Each of these three claims has been subject to, in Matthew's terms, 'a decade of environmental security research, debate and policy experimentation' (Matthew, 2002, p. 109). Military organizations have gone to great lengths to argue the value of their institutions and methods to environmental preservation. Others have revealed an abiding tendency, however, to 'securitize the environment' rather than 'green the military'

(Van Deveer and Dabelko, 1998; more generally, see Williams, 2003). Most importantly, in my estimation, however, is Deudney's claim that environmentalism calls into question the very idea of a world of self-regarding states seeking security through violent practices.

the problem of 'the state'

A common theme running through the critical environmental security studies literature is what to do about the state – as a constellation of power and practice and as an analytical concept. Many of those who first argued for an expanded definition of security, including environmental issues, did so from the vantage point of disciplines not wedded to the state – environmental studies, human geography, ecology, philosophy, anthropology, biology, feminist theory and gender studies – and/or from physical locations outside traditional networks of power and privilege. While many people working in these fields bring traditional methodologies to bear on their research, their starting points are generally conceptions of space and time at variance with state-based analyses. The world they 'see', therefore, is one quite different than those working within IR's dominant paradigm, be they (neo)realist or (neo-)institutionalist. And what they see, most often, is the negative consequences of modern industrial life dominated by a hierarchy of states.

What this disparate group shares, therefore, is a general desire to problematize practices and institutions that are considered 'natural' and/or immutable by, for example, most political scientists, security studies 'specialists' (be they experts or practitioners), or policy-making elites. For example, to many ecologists environmental security is about securing environmental health (within specific ecosystems; or at the level of the planetary biosphere) and, by extension, human well-being for humans are part of the biosphere, not separate from it. To ensure this 'security' requires a holistic understanding of the ways in which humans interact with 'nature'. It requires at minimum 'global environmental governance', not divisive national security policies (Peluso and Watts, 2003).

At the same time, however, they recognize the very real power wielded by state-makers and maintainers (Pettman, 1991). If 'anarchy is what you make of it', as Wendt suggests, so too is order. States, as powerful actors operating within a largely Realist logic, make a rather brutal reality. For this reason, most people concerned with the negative side of the modern era, in particular those activities undertaken in the name of 'order' and 'stability', are forced to work with states so that they might at minimum modify their practices. Just how difficult this is has been demonstrated in America's 'war on terror', its decision to reject the Kyoto Protocol and perhaps overturn its commitment to the Montreal Protocol, the latter long considered the shining example of global environmental governance.

In terms of dealing with the state as a constellation of power and practice, most academics, if called upon, are only too willing to give testimony in the halls of state power. The environment, being outside the traditional purview of security practitioners, requires a different kind of expertise so opening up the heavy (and heavily fortified) doors of the state to, among others, ecologists, geographers and historians (Matthew, 2002, pp. 118–19). In terms of dealing with the state as an analytical category, once again preferences differ. Matthew (2002), Lonergan (2000), and Barnett (2001) are among those who, as far as the environment goes, seek to decentre the state. The state is but one form of social organization that changes through time. More appropriate referents of security are the biosphere and the individual, together linked by the concept 'human security'.

a new language for new understanding?

A focus on human security located within a discourse of modernity takes us only part way toward a more inciteful understanding of the forces at work in creating what McKibbon (1998) labels 'Earth II'. They are limited by a constructivist epistemology which remains committed to 'better science for better policy-making'. What they are after is a framework which more accurately represents (socially constructed) reality. As such, they choose to work with and within a positivist understanding of the world that others feel is at the heart of not only environmental degradation but class, race and gender oppression (Escobar, 1996; Peet and Watts, 1996). It is left to scholars such as those included in the collections by Peluso and Watts (2001), and Peet and Watts (1996) to bring poststructural and postmodern (together we can call them 'interpretivist') understandings to bear in unearthing the roots of environmental degradation and resource depletion. The utility of an interpretive approach is stated more forcefully by George (1994, p. 166):

> A postmodernist politics of dissent ... is *post*modern in the sense that it seeks to confront, at every level, those aspects of modernity that undermine any potential people might have to produce, in their everyday lives, resistances to power relations that silence, demean and oppress them.

Language is key to this enterprise, for it is not neutral (Peterson, 2003, p. 41). Because 'development', 'security' and the 'environment' are dominated by 'expert' groups vested with specific technical/managerial knowledge, challenging their way of knowing and ordering the world requires, almost unavoidably, the use of their language. When a term like 'sustainable development' or 'security' is being used, its multiple meanings ensure (1) that different groups continue to talk past each other; and (2) reaching no consensus, or a consensus so vague as to allow participants in the discourse to walk away equally satisfied, the dominant group's approach remains ascendant.

Magnusson (1994) articulates world politics 'as a problem of urban politics'. Gadgil and Guha (1995) encourage us to think of resource use at global level in terms of biosphere and ecosystem people. McNeill et al. (1991) describe the 'ecological shadow' of a country as 'the environmental resources it draws from other countries and the global commons'. Wackernagel and Rees (1996; also Rees and Wackernagel, 1994) use the term 'ecological footprint' to help us better understand resource flows beyond 'the state' (see Dalby, 2002a, for details). These approaches all deploy a new and different language to that of state-centric security and environmental security studies.

Holistic, historically-rooted and -conscious approaches are also key to the critical environmental security studies enterprise. Traditional approaches to environmental security present, at best, a linear and 'bifurcated' understanding of history, with 'developed' and 'developing' states acting as simplified markers in this 'historical' process. States, societies and empires rise and fall through time and do so as a result of decisions taken within their spatial boundaries and/or as they impact upon one another through practices such as trade and war. Successful states persist due to their own 'ingenuity', to use Homer-Dixon's term.

For (human) geographers, (social) anthropologists, sociologists, environmental historians and ecologists, among others, this is only one part of a more complex story. Because political science and international relations focus primarily on the modern state, their frameworks are partial. As such, their understandings and answers to important questions are equally partial. Drawing on Grove's (1997) work on the environmental impacts of colonization, and on Mumford's (1934) notion of the emergence of 'carboniferous capitalism', Dalby articulates a history of environmental change that draws our attention to the way processes of modernization – of (industrial) imperial conquest leading to profound social and ecological transformation throughout the world – implicate 'developed' states in perceived 'resource scarcities' in 'developing' states.

Citing Alker and Haas (1993), who set their own argument about environmental security within the context of 'Vernadsky's ideas of a single biosphere and Braudel's historical formulations of the macropatterns of civilisations', Dalby argues in support of the importance of understanding nature in the long term (Dalby, 2002a, p. 72). This, he suggests, could lead toward thinking of political space in terms of ecopolitics, rather than through the conventional markers of states.

What makes Dalby and others like him a 'dissident voice' (Lowi and Shaw, 2000) among a majority of environmental security scholars and policy makers, is that a holistic, historical perspective that emphasizes interconnections, dynamism and complexity directly challenges each and every assumption underlying mainstream scholarship: inter alia the disconnectedness of an interstate system where order is possible within states and a mature anarchy is the preferable order without. Critical scholars argue that Western 'wealth' and Southern 'poverty' are mutually constituted, and that accelerated and extended forms of ecological disruption are due to 'European expansion, carboniferous industrialisation, and contemporary globalisation' (Dalby, 2002a, p. 81; Peet and Watts, 1996; Peluso and Watts, 2001). To be sure, this 'offers a very different history and a more comprehensive causal sequence for understanding environmental insecurity' (Dalby, 2002a, p. 81). More importantly, however, such an analysis exposes to critical scrutiny all those who benefit from the current world 'disorder'. State-makers, more concerned with containment and continuity rather than fundamental change, are unlikely to yield to critical insights arguing in favour of (radical) transformation. Hence, Dalby's (2002a) hope for 'reconceptualisation and synthesis' is, in my view, rather idealistic.

Nevertheless, what the critical school makes clear is that, among other things, 'environmental security has been written by the rich omnivores in their comfortable offices and libraries' (Dalby, 2002a, p. 184). This must change, for it suggests that traditional approaches to environmental security are being folded into dominant security routines designed in the main to ensure the stability of the state system, and US primacy therein. 'Stability' versus 'transformation' ensures that 'environmental security' will remain an essentially contested concept for the foreseeable future.

environmental security in practice: transboundary natural resource management

Southern Africa provides an excellent geographical location to observe the myriad, often contradictory and sometimes complementary ways in which environmental security is practised. The region figures centrally in the global discourse of environmental security. In the mid-1990s, the World Bank identified the region as one in which water scarcity could lead to violent conflict. Recently, the Bank has turned around arguing that water scarcity provides opportunity for regional peace-building. Each of these perspectives has generated a considerable literature. Beyond water resources, Southern Africa has featured in each of the major studies on environmental conflict discussed above.

Importantly, Southern Africa has been at the centre of traditional discourses and practices of global security and development at least since the end of World War II (Vale, 2003). Most recently, this manifests in US foreign policy and development discussions regarding pivotal states, failed states, transitions to democracy and theories of democratic peace, neoliberal macroeconomic policy, popular participation in rural development, and regional integration, including regional approaches to natural resources management. Duffield (2001) shows how these once separate discourses are now inextricably intertwined.

The environmental security discourse in Southern Africa is dominated by states and non-governmental organizations (NGOs), both local and global. State-makers in the region generally regard environmental issues as a 'Northern', and therefore imposed, concern. The tendency in state houses around the region is to slot 'the environment' within a specific Ministry and to treat it as tangential to the main concerns of developing states, that is, economic growth. Local environmental actors, for example, the Group for Environmental Monitoring in South Africa, are often regarded as special interest groups who sometimes stand in the way of the state-led and -determined development process. In other cases, they may be regarded as mouthpieces for 'foreign' interests, be they states through donor agencies, or international non-governmental organizations (INGOs). This is not to argue that a healthy environment is not a (security) concern among states and citizens in the region. Rather, it is to suggest that policy-makers regard 'biodiversity preservation' and 'global warming', for example, as deeply political issues, the consensual language of 'global goods' notwithstanding (Vale and Swatuk, 2002).

The regional turn toward transboundary natural resource management should be understood within this context. TBNRM is as much discursive political site as policy programme. Its strongest supporters are international actors (for example, donor states, the World Conservation Union, and NGOs such as Conservation International) in league with the privately funded South African Peace Parks Foundation. These actors link to national nodes (generally parks and wildlife departments) who coordinate their actions via the Southern African Development Community (SADC) – an interstate body devoted to integrated regional development. This shared belief in 'peace parks' does not extend much beyond actors directly involved. Indeed, given South Africa's historical dominance of the region, many people in Southern Africa are sceptical of the peace-building motives said to be driving TBNRM (Wirbelauer et al., 2003).

TBNRM has been defined as 'any process of cooperation across boundaries that facilitates or improves the management of natural resources (to the benefit of all parties in the area concerned)' (Griffin, 1999, in Jones and Chonguiça, 2001, p. 1). Boundaries, in this instance, are those between states, although many other boundaries – for example, those demarcating different forms of land tenure, land use and administrative jurisdictions – come into play.

Ingram et al. (1994) state that resource management in border areas requires special attention because borders are areas where 'inequities surface and conflicts erupt'. For Katerere et al. (2001, p. 9), 'in response to the problem of resource management in border areas, arrangements and initiatives focused on TBNRM have emerged with the following objectives: (1) to improve conservation of shared resources that are being depleted or degraded at unsustainable rates; (2) to ensure that communities and other stakeholders benefit from sustainable use of resources (in particular, to counter inequitable resource distribution associated with land and resource appropriation by local elite and foreign investors); and (3) to optimize regional distribution of benefits from resource use. In other words, TBNRM is hypothesized to simultaneously provide biosphere security, national security in a regional context, and human security.

TBNRM has taken a number of different forms in the region. Jones and Chonguiça (2001, pp. 2–5) delineate these as transfrontier conservation areas (TFCAs), transboundary natural resource management areas (TBNRMAs), informal networks of resource use across boundaries, spatial development initiatives (SDIs), and development corridors. These different forms of TBNRM often intersect and overlap.

TFCAs are initiatives undertaken by state conservation agencies in support of biodiversity conservation and focus mainly on expanding protected areas within one country by linking them to a protected area or areas in one or more neighbouring countries (Jones and Chonguiça, 2001, p. 2). Another feature distinguishing TFCAs from TBNRMAs is the proposed object of security: in the former, it is primarily the environment itself with biodiversity conservation as the driving force; in the latter it is sustainable use for sustainable livelihoods, with people – particularly rural people and those living in remote areas – being the main object of security. There are at present more than half a dozen TFCAs under way in Southern Africa. Some of the more advanced TFCAs in the region are:

- Kgalagadi Transfrontier Park between Botswana and South Africa. The total area involved is 37,991 km^2 – three-quarters of which is in Botswana – involving state-owned protected areas in the two countries and communal land within the South African protected area.
- Gaza/Kruger/Gonarezhou (GKG) TFCA between Mozambique, South Africa and Zimbabwe. The total land area is 99,800 km^2 (with two-thirds being in Mozambique), involving state-owned protected areas, private ranches, private game reserves and communal areas.

TBNRMAs, in contrast, are more complex, so involving inter alia a number of government departments, communities, companies and local and international NGOs. According to USAID (Chengeta et al., 2003, p. 1), TBNRMA is 'a relatively large area, which straddles a frontier between two or more countries and covers a large-scale natural system (ecosystem)'. The definition is flexible enough to include either a portion of a river basin (for example, the 'Four Corners' project, see below) or, where integrated environmental management is the driving principle, an entire river basin (for example, Zambezi River Basin which includes portions of eight states, tens of millions of people, and all economic activities, including conservation). There are numerous TBNRMAs currently at the planning stage. Some of the more advanced TBNRMAs in the region are:

- The Four Corners initiative involving 200,500 km^2 in Botswana, Namibia, Zambia and Zimbabwe. Land use/tenure in the area is state owned protected areas, communal land and community wildlife areas.
- Every River Has Its People initiative involving 160,000 km^2 in the Okavango River Basin states of Angola, Namibia and Botswana. Communal land, state-owned hunting areas, protected areas and community wildlife areas are involved.

TFCAs and TBNRMAs are directly linked to the language of 'peace'. The fascination with, indeed great hope for, TBNRM initiatives in the region derives in part from SADC states' recent transition from conflict to peace and from colonial/authoritarian rule to unconsolidated democracies (Chengeta et al., 2003; Ramutsendela and Tsheola, 2002). The twin catalysts were the serial endings of the global Cold War and of apartheid rule in Namibia (1990) and South Africa (1994). South Africa's move to majority-rule put an end to a decade of military and economic aggression against its neighbours. A pressing question became 'What do we do with the large militaries developed by SADC states?' Aside from demobilization, one anticipated 'peace dividend' was the possible shifting of military personnel and technology to environmental protection. The IUCN was an early proponent of such activity (Steiner, 1993; Swatuk and Omari, 1997). Given the place of mountains, forests, and national parks as shelters and headquarters for rebel movements, redeployment and reconstruction became key elements in the discourse of 'peace parks'. In the words of one long-time observer of environmental issues in Southern Africa, 'nature has the power to heal old wounds' (Koch, 1998).

Because TBNRM in general and 'peace parks' in particular appear to be all things to all people, practice is rife with contradiction. The most 'successful' TFCA thus far has been the relatively non-controversial establishment of the Kgalagadi transfrontier park between the governments of South Africa and Botswana. The park is jointly managed by Botswana's Department of Wildlife and National Parks and South Africa's Department of National Parks (De Villiers, 1999, pp. 127–43). Yet even here where 5,712 tourist entries were recorded on the Botswana side in 2000, generating roughly US$110,000, the parties to the agreement deemed it necessary to begin the preamble with 'recognising the principle of sovereign equality and territorial integrity of their states' (De Villiers, 1999, p. 136; data from Government of Botswana, 2003, p. 240): joint management is to pose no challenge to the state's ultimate authority. For international donors interested primarily in conservation, this seems a reasonable concession.

The picture is much more complicated in other, more densely populated and economically active, parts of the region. The GKG TFCA, for example, brings together the region's strongest (South Africa) and weakest (Mozambique) states in a border area historically characterized by its apartheid-era electrified fence. Many Mozambicans crossed this fence, then attempted to cross South Africa's Kruger National Park, in a desperate attempt to flee Mozambique's long civil war – a civil war whose main antagonist, the rebel group RENAMO, was supported by South Africa (Hanlon, 1986). Occasionally, newspapers would carry reports of refugees having been devoured by lions in the park.

Beyond Mozambique's civil war, many displaced peasants are seeking to return to their homes in the border lands. An estimated 160,000 people live in and around Zinhave and Banhine National Parks and the hunting concession known as Coutada 16 (Koch, 1998, pp. 65–6). The land they now occupy is contiguous with Kruger and integral to the GKG 'peace park'. Little forethought has been given to questions concerning the place of rural peoples within the borders of immanent peace parks. To remove them is to reproduce the worst of colonial and apartheid era state conservation practices. To leave them without a clear understanding of the resource use and personal safety issues for rural peoples living within a TFCA is to subject these citizens to a number of forms of insecurity at individual, household and village levels. Whereas Van der Linde et al. (2000, p. 73) argue that 'the complexity of TBNRM makes it imperative for stakeholders to undertake a very clear appraisal of the opportunities and risks of embarking on such a program', Katerere et al. (2001, p. 12) state: 'In practice, however, TFCAs have been pushed forward at a rapid pace without much time for consultation with communities and other stakeholders. While there has been little implementation yet, individual countries have signed agreements committing themselves to TFCAs with very little understanding of the consequences.' States have gone ahead with high-level deals in the name of peace-building, economic benefit-sharing, ecosystem protection and the like. Entrepreneurs interested in exploiting tourism-related business opportunities have already begun 'resource raiding' through, for example, the acquisition of freehold land for the establishment of private conservancies bordering the TFCAs (Katerere et al., 2001). A devastated state such as Mozambique, classified as a Highly-Indebted Poor Country, has virtually no capacity at local government level to implement decisions taken at higher levels, or regulate activities in their political jurisdictions. In the case of TFCAs, the working assumption is that the stronger partner, South Africa, will bring its human, technical and capital resources to bear in the sustainable management of the shared resource.

Focusing on the real and potential insecurities of rural people arising from the establishment of TBNRM projects facilitates a critical interrogation of the statist discourses and practices of development/environment/security. It reveals a constructed landscape – the 'peace park' – the establishment of which mirrors vast power asymmetries within states, among states in the region, and between Africa and actors external to the continent. The discourse is driven by 'experts' working within an epistemic community whose self-perception is one based on virtue and justice: peace, sustainable development, the empowerment of Africans at every level of society. The practice, however, reveals how already empowered actors – state-makers, entrepreneurs, conservation organizations – seek to maintain their dominant positions in local/regional/global society. Where interests conflict, the flexibility of the language facilitates compromise: so conservation organizations see the establishment of their TFCA; entrepreneurs receive various concessions to develop; and states ensure sovereignty through co-management rather than common property relations. In almost every case, rural peoples benefit least. In some cases, rural livelihood strategies are seriously threatened by 'peace parks'. At best, rural people are passive participants in multi-stakeholder practices.

Various international NGOs have undertaken to preserve biodiversity and empower communities through the establishment of community-based natural resources management (CBNRM) projects (Jeffery and Vira, 2001; Vira and Jeffery, 2001). Many have deliberately attempted to begin from village level and only involve the state when necessary and/or unavoidable. Where the resource being used – for example, a veld product such as thatching grass or a type of fruit – improves (personal/ household/community) incomes but the amount is relatively small, projects have been sustainable. However, where projects (potentially) involve state-owned resources, such as wildlife, challenge existing forms of land tenure, or possibly truly empower local people such that they are no longer dependent on central government for survival, the state invariably gets involved, often in an obstructive way (Swatuk, 2005). Katerere et al. (2001) suggest that this is one reason why USAID shifted support from country-specific CBNRM projects to region-wide TBNRM projects: CBNRM, through the empowerment of local people, challenges state power; TBNRM, in contrast, needs the state to succeed. So, while not losing sight of the desire to empower local people through the sustainable management of natural resources, donors acknowledge that state interests must be respected.

In summary, environmental security is a highly contested concept. For state-makers it means, first and foremost, threats posed to sovereign states by environmental change. This has led state-makers in Southern Africa to consider new ways of sustainably managing natural resources – specifically, through TBNRM – and the potential role of the military therein – for example, through anti-poaching units or technical expertise in drought monitoring. However, implementation of these projects raises a number of questions regarding beneficiaries of 'environmental security'. This is particularly the case when contrasting the language of 'peace parks' with the reality of impacted rural communities. Donors, as the primary drivers of these activities, are forced to make compromises and often put regional cooperation, biodiversity preservation and peace-building at state level ahead of the interests of rural people living in borderlands. This Realist decision is masked by a discourse of development that presumes failure in rural settings (Ferguson, 1990). A small corps of critical thinkers regularly exposes contradictions in theory and practice, but they remain marginal forces in decision-making (Swatuk and Vale, 1995). As such, environmental security in practice reflects the continuing hegemony of the statist paradigm as articulated by Homer-Dixon and Klare, and others. This is not because the theoretical framework is more persuasive, but because it better serves dominant interests.

directions for future research

Perhaps reflecting my own preference for critical analytical approaches, and in contrast to Deudney, my feeling is that a great deal of interdisciplinary research needs to be conducted linking the environment to security – especially that which privileges non-state-centric spatial and organizational frameworks. First and foremost, the ways in which Western patterns of consumption – at the levels of the region, the city, the state, the company and in broad aggregate terms – contribute to resource degradation and conflict in the global South must be investigated. Critical political ecology provides a useful framework for drawing together scholars from different disciplines – for example, ecology, geography, anthropology, politics – to investigate specific examples of 'who gets what, when and where?' (which is, after all, a question of resource use) and what the (in)security implications are – for individuals, households, communities, ecosystems and the biosphere. Clearly, there will be no shortage of funds available for individual and coordinated research within the dominant paradigm. Western states will continue to undertake their own programmes of research designed to answer the question of how to maintain their own dominant positions in the current world order. Scholars who feel that the policy resulting from such research will continue to compromise the sustainability of the biosphere and/or exacerbate social instability in the global South need to gather systematic evidence in order to speak truth to power.

There are many studies devoted to specific resources (for example, water, forests) and physical areas (such as watersheds and ecosystems in Africa, Latin America, Asia) that begin with the assumption that resource degradation leads to competition and (violent) conflict. There are only a few studies that take the possibility for cooperation as well as conflict as a point of departure. It is important that scholars begin to investigate the ways in which (renewable) resource scarcities and/or particular resource uses in specific locations create avenues for peace-making and political cooperation. UNEP has begun to use environmental clean-up in post-conflict settings – the Balkans, for instance – as a confidence-building tool. How fungible are these activities? Beyond the clean-up, do they have positive political spillover effects?

Similarly, case studies have overwhelmingly focused on the population–resource degradation–violent conflict triad found in specific (usually historically unstable) Third World states and regions. There is therefore a need for case studies chosen to challenge these abiding assumptions. For example, Percival and Homer-Dixon (in Homer-Dixon and Blitt, 1998) present the case of KwaZulu-Natal in South Africa as a region where resource degradation is likely to have contributed to violent conflict in the pre-1994 period. But what is so different about that region today? The same authors in the same volume also look at Rwanda, concluding that environmental scarcity had a real but 'surprisingly limited' role (1998: p. 217). Yet similar resource scarcities and ethnic cleavages (between Hutu and Tutsi) existed (and continue to exist) in northern Tanzania: why has there been no violent conflict there? Scholars must follow on from Homer-Dixon's summary findings in his 1999 study and investigate the political, social and economic frameworks that give rise to resource capture and ecological marginalization, and resist the temptation to say that weak states with too many people are a recipe for environmental degradation and violent social conflict.

In analysing links between resource degradation and violent conflict, there is also a tendency to lump together resources as if they are all equal – for example, forests, minerals, agricultural land and water (despite Homer-Dixon's early assessment that freshwater was an unlikely contributor to violent conflict). A promising area of future research is in comparative resource studies, either a single resource in different physical settings or several resources in similar or different settings. For example, diamonds are exploited in a variety of ways: can it be argued that alluvial diamonds as opposed to those found and exploited in kimberlite deposits more easily contribute to state breakdown and civil/regional war? Answering this question may lead directly to realistic policy interventions.

It is unfortunate but understandable that environmental security studies have been dominated by investigations concerned with the impact on specific human communities (generally, states) of resource degradation, scarcity and competition. In my opinion, there needs to be more study of the way in which resource use contributes to biosphere insecurity and to particular human insecurities across space and time. As stated earlier, there is a strong current of research concerned with these questions but which generally fails to make convincing recommendations for policy. Granted, getting state-makers to change the way resources are accessed and allocated particularly where such decisions may harm short term (personal, party political) interests is not easy. All the more reason for careful empirical analyses into the consequences of particular resource decisions across time and space.

Some years ago, scholars hypothesized possible post-conflict environmental 'peace dividends' where, for example, formerly antagonistic militaries might work together for transboundary conservation ends. Some scholars dubbed this 'greening' the military. Over the last 15 years, militaries themselves have established environmental security units, held seminars concerning environmental security, and conducted their own research into environmental change scenarios. Militaries have also been heavily involved in natural disaster relief exercises – offering support when dramatic environmental change threatens human security. It seems an appropriate time to investigate the record of the military in environmental security.

Lastly, given empirical trends toward regionalization – from the EU to APEC and the African Union – there is need for studies into the ways in which resource access, allocation and use across state boundaries create new sites of conflict and cooperation. The assumptions of enhanced transboundary resource cooperation leading to interstate political cooperation inform efforts toward TBNRM around the world. Does the reality confirm the theory? Mapping existing networks of (formal and informal) political power over particular resource geographies (for example, the watershed; the forest) will help reveal possible (sub- and trans-)state sites of cooperation and conflict. Such a research programme will add nuance to a field of study dominated by unfounded and often unfortunate assumptions regarding population–resource degradation–violent conflict.

notes

1. See also Paterson's discussion in Chapter 2 of this volume regarding those studies taking international anarchy as their points of departure.
2. Sixteen different 'syndromes' were identified and grouped into three categories: utilization syndromes (for example, 'sahel syndrome', 'overexploitation syndrome'); development syndromes (for example, 'aral sea syndrome', 'urban sprawl syndrome'); and sink syndromes (for example, 'smokestack syndrome').

bibliography

Adelphi Research and Woodrow Wilson Center (Environmental Change and Security Project) (2004) *Environment, Development and Sustainable Peace: Finding Paths to Environmental Peacemaking.* Report of a conference held at Winston House, Wilton Park, UK, 16–19 September.

Alker, Hayward, and Peter M. Haas (1993) 'The Rise of Global Ecopolitics', in Nazli Choucri (ed.) *Global Accord: Environmental Challenges and International Responses*, Cambridge, Mass.: MIT Press, pp. 133–71.

Baechler, Günther (1998) 'Why Environmental Transformation Causes Violence: A Synthesis', *Environmental Change and Security Project Report* 4 (Spring), pp. 24–44.

Baechler, Günther (1999) *Violence Through Environmental Discrimination: Causes, Rwanda Arena, and Conflict Model*, Dordrecht: Kluwer.

Baldwin, David (1997) 'The Concept of Security', *Review of International Studies* 23, 1, pp. 5–26.

Barnett, Jon (2001) *The Meaning of Environmental Security: Ecological Politics and Policy in the New Security Era*, London: Zed Books.

Baumann, Zygmunt (1989) *Modernity and the Holocaust*, Cambridge: Polity Press.

Booth, Ken (1991) 'Security and Emancipation', *Review of International Studies* 17, 4, pp. 313–26.

Buzan, Barry (1991) *People, States and Fear*, 2nd edn, Boulder: Lynne Rienner.

Buzan, Barry, Ole Waever and Jaap de Wilde (1998) *Security: A New Framework for Analysis*, Boulder: Lynne Rienner.

Callaghy, Thomas M., Ronald Kassimir and Robert Latham (eds) (2001) *Intervention and Transnationalism in Africa*, Cambridge: Cambridge University Press.

Carson, Rachel (1962) *Silent Spring*, Boston: Houghton Mifflin.

Chengeta, Zuma, Jamare Jamare and Nyasha Chishakwe (2003) *Assessment of the Status of Transboundary Natural Resources Management Activities in Botswana*, Gaborone: Printing and Publishing Company Botswana.

Commission on Global Governance (1995) *Our Global Neighbourhood*, Oxford: Oxford University Press.

Conca, Ken, and Geoff D. Dabelko (eds) (1998) *Green Planet Blues: Environmental Politics from Stockholm to Kyoto*, 2nd edn, Boulder: Westview Press.

Conca, Ken, and Geoff D. Dabelko (eds) (2002) *Environmental Peacemaking*, Washington, DC: Johns Hopkins University Press.

Cox, Robert W. (1986) *Production, Power and World Order*, New York: Columbia University Press.

Dalby, Simon (1997) 'Contesting an Essential Concept: Reading the Dilemmas in Contemporary Security Discourse', in Keith Krause and Michael C. Williams (eds) *Critical Security Studies: Concepts and Cases*, Minneapolis: University of Minnesota Press.

Dalby, Simon (1998) 'Ecological Metaphors of Security: World Politics in the Biosphere', *Alternatives* 25, 3, pp. 291–320.

Dalby, Simon (2002a) *Environmental Security*, Minneapolis: University of Minnesota Press.

Dalby, Simon (2002b) 'Security and Ecology in the Age of Globalization', *Environmental Change and Security Project Report* 8 (Summer), pp. 95–108.

Deudney, Daniel (1990) 'The Case against Linking Environmental Degradation and National Security', *Millennium* 19, 3, pp. 461–76.

Deudney, Daniel, and Richard A. Matthew (eds) (1999) *Contested Grounds: Security in the New Environmental Politics*, New York: SUNY Press.

De Villiers, Bertus (1999) *Peace Parks: The Way Ahead*, Pretoria: HSRC.

Duffield, Mark (2001) *Global Governance and the New Wars*, London: Zed Books.

Ellis, Stephen (1999) *The Mask of Anarchy*, New York: New York University Press.

Erhlich, Paul (1968) *The Population Bomb*, New York: Ballantine.
Escobar, Arturo (1996) 'Constructing Nature: Elements for a Poststructural Political Ecology', in Richard Peet and Michael Watts (eds) *Liberation Ecologies*, London: Routledge, pp. 46–68.
Ferguson, James (1990) *The Anti-Politics Machine: 'Development', Depoliticization and Bureaucratic Power in Lesotho*, Cambridge: Cambridge University Press.
Fukuyama, Francis (1992) *The End of History and the Last Man*, New York: Free Press.
Gadgil, Madhav, and Ramachandra Guha (1995) *Ecology and Equity: The Use and Abuse of Nature in Contemporary India*, London: Routledge.
George, Jim (1994) *Discourses of Global Politics: A Critical (Re)Introduction to International Relations*, Boulder: Lynne Rienner.
Gleditsch, Nils Petter (1997) *Conflict and the Environment*, Dordrecht: Kluwer.
Gleditsch, Nils Petter (1998) 'Armed Conflict and the Environment: A Critique of the Literature', *Journal of Peace Research* 35, 3, pp. 381–400.
Government of Botswana (2003) *National Development Plan 9*, Gaborone: Government Printer.
Gray, John (2002) *Straw Dogs: Thoughts on Humans and Other Animals*, London: Granta.
Griffin, John (1999) *Transboundary Natural Resource Management in Southern Africa: Main Report*, Washington, DC: Biodiversity Support Program.
Grove, Richard (1997) *Ecology, Climate and Empire: Colonialism and Global Environmental History, 1400–1940*, Cambridge: White Horse.
Hall, Stuart, David Held and Tony McGrew (eds) (1992) *Modernity and Its Futures*, Cambridge: Polity Press.
Hanlon, Joseph (1986) *Beggar Your Neighbours: Apartheid Power in Southern Africa*, London: James Currey.
Hettne, Bjorn (1997) 'Development, Security and World Order: A Regionalist Approach', *European Journal of Development Research* 9, 1, pp. 83–106.
Hettne, Bjorn (2001) 'Regional Cooperation for Security and Development in Africa', in P. Vale, L. A. Swatuk and B. Oden (eds) *Theory, Change and Southern Africa's Future*, Basingstoke: Palgrave.
Homer-Dixon, Thomas (1991) 'On the Threshold: Environmental Changes as Causes of Acute Conflict', *International Security* 16, 2, pp. 76–116.
Homer-Dixon, Thomas (1994) 'Environmental Scarcities and Violent Conflict: Evidence from Cases', *International Security* 19, 1, pp. 5–40.
Homer-Dixon, Thomas (1998) 'Environmental Scarcities and Violent Conflict: Evidence from Cases' (abridged version), in Ken Conca and Geoff Dabelko (eds) *Green Planet Blues: Environmental Politics from Stockholm to Kyoto*, Boulder: Westview Press, pp. 287–97.
Homer-Dixon, Thomas (1999) *The Environment, Scarcity and Violence*, Princeton: Princeton University Press.
Homer-Dixon, Thomas (2003) 'Debating Violent Environments', *Environmental Change and Security Project Report* 9, pp. 89–96.
Homer-Dixon, Thomas, and Jessica Blitt (eds) (1998) *Ecoviolence: Links Among Environment, Population and Security*, Lanham, Md: Rowman and Littlefield.
Hurrell, Andrew (1995) 'International Political Theory and the Global Environment', in Ken Booth and Steve Smith (eds), *International Relations Theory Today*. Oxford: Clarendon Press, pp. 129–53.

Ingram, Helen, Leonard Milich and R. G. Varady (1994) 'Managing Transboundary Resources: lessons from Ambos Nogales', *Environment* 36, 4, pp. 6–9; 28–38.
Jeffery, Roger, and Bhaskar Vira (eds) (2001) *Analytical Issues in Participatory Natural Resources Management*, Basingstoke: Palgrave.
Jones, Brian T. B., and Ebenezario Chonguiça (2001) *Review and Analysis of Specific Transboundary Natural Resource Management Initiatives in the Southern African Region*, IUCN-ROSA Series on Transboundary Natural Resource Management, No. 2, Harare: IUCN-ROSA.
Kaplan, Robert (1994) 'The Coming Anarchy', *Atlantic Monthly* (February), pp. 44–77.
Katerere, Yemi, Sam Moyo and Ryan Hill (2001) *A Critique of Transboundary Natural Resource Management in Southern Africa*, IUCN-ROSA series on Transboundary Natural Resource Management, No. 1, Harare: IUCN-ROSA.
Klare, Michael (2001) *Resource Wars: The New Landscape of Global Conflict*, New York: Henry Holt and Company.
Koch, Eddie (1998) 'Nature Has the Power to Heal Old Wounds: War, Peace and Changing Patterns of Conservation in Southern Africa', in D. Simon (ed.) *South Africa in Southern Africa: Reconfiguring the Region*, London: James Currey.
Krause, Ken, and Michael C. Williams (eds) (1997) *Critical Security Studies*, Minneapolis: University of Minnesota Press.
Leopold, Aldo (1949) *A Sand County Almanac: And Sketches Here and There*, Oxford: Oxford University Press.
Levy, Marc A. (1995) 'Is the Environment a National Security Issue?' *International Security* 20, 2, pp. 35–62.
Lietzmann, Kurt M. and Gary Vest (1999) 'Environment and Security in an International Context: Executive Summary Report, NATO/Committee on the Challenges of Modern Society Pilot Study', *Environmental Change and Security Project Report* 5 (Summer), pp. 34–48.
Lipschutz, Ronnie (1995) *On Security*, New York: Columbia University Press.
Lonergan, Steve (2000) 'Human Security, Environmental Security and Sustainable Development', in Miriam R. Lowi and Brian R. Shaw (eds) *Environment and Security: Discourses and Practices*, New York: St Martin's Press, pp. 66–83.
Lowi, Miriam R., and Brian R. Shaw (eds) (2000) *Environment and Security: Discourses and Practices*, New York, St Martin's Press.
Magnusson, Warren (1994) 'Social Movements and the Global City', *Millennium* 23, 3, pp. 621–45.
Mathews, Jessica T. (1989) 'Redefining Security', *Foreign Affairs* 68, 2, pp. 162–77.
Matthew, Richard A. (2002) 'In Defense of Environment and Security Research', *Environmental Change and Security Project Report* 8 (Summer), pp. 109–24.
McKibbon, Bill (1998) 'A Special Moment in History', *Atlantic Monthly* (May), pp. 55–78.
McNeill, James, Peter Winsemius and Taizo Yakaushiji (1991) *Beyond Interdependence*, New York: Oxford University Press.
Midlarsky, Manus (1998) 'Democracy and the Environment: An Empirical Assessment', *Journal of Peace Research* 35, 3, pp. 341–62.
Morgenthau, Hans (1978) *Politics Among Nations*, 5th edn, New York: Knopf.
Myers, Norman (1989) 'Environment and Security', *Foreign Policy* 47, pp. 23–41.
Myers, Norman (1993) *Ultimate Security: The Environmental Basis of Political Stability*, New York: W. W. Norton.

Mumford, Lewis (1934) *Technics and Civilization*, New York: Harcourt Brace and World.
Ohlsson, Leif (1999) *Environment, Scarcity and Conflict: A Study of Malthusian Concerns*, Goteborg: PADRIGU.
Paterson, Matthew (1995) 'Green Politics', in Scott Burchill, Richard Devetak, Andrew Linklater, Matthew Paterson, Christian Reus-Smit and Jacqui True (eds) *Theories of International Relations*, New York: St Martin's Press, pp. 277–307.
Peet, Richard, and Michael Watts (eds) (1996) *Liberation Ecologies: Environment, Development, Social Movements*, London and New York: Routledge.
Peluso, Nancy, and Michael Watts (eds) (2001) *Violent Environments*, Cornell: Cornell University Press.
Peluso, Nancy, and Michael Watts (2003) 'Violent Environments: Responses', *Environmental Change and Security Project Report* 9, pp. 89–96.
Peterson, V. Spike (2003) *A Critical Rewriting of Global Political Economy: Integrating Reproductive, Productive and Virtual Economies*, London and New York: Routledge.
Pettman, Jan Jindy (1996) *Worlding Women: A Feminist International Politics*, London and New York: Routledge.
Pettman, Ralph (1991) *International Politics*, Boulder: Lynne Rienner.
Ramutsendela, Maano, and Johannes Tsheola (2002) 'Transfrontier Conservation Areas: A Framework for Managing Peace and Nature in Southern Africa?', in Tor Arve Benjaminsen, Ben Cousins and Lisa Thompson (eds) *Contested Resources: Challenges to the Governance of Natural Resources in Southern Africa: Papers from the International Symposium on 'Contested Resources: Challenges to Governance of Natural Resources in Southern Africa: Emerging Perspectives from Norwegian–Southern African Collaborative Research' Held at the University of the Western Cape, Cape Town, 18–20 October 2000*, Cape Town: Programme for Land and Agrarian Studies, School of Government, University of the Western Cape.
Rees, William E., and Mathis Wachernagel (1994) 'Ecological Footprints and Appropriated Carrying Capacity: Measuring the Natural Capital Requirements of the Human Ecology', in A. M. Jannson, M. Hammer, C. Folke and R. Constanza (eds) *Investing in Natural Capital*, Washington, DC: Island Press.
Renner, Michael (1989) *National Security: The Economic and Environmental Dimensions*, Worldwatch Paper No. 89. Washington, DC: Worldwatch Institute.
Richards, Paul (1996) *Fighting for the Rainforest: War, Youth and Resources in Sierra Leone*, Oxford: James Currey.
Ruggie, John G. (1998) *Constructing the World Polity: Essays on International Institutionalism*, London: Routledge.
Schwarz, Daniel M., Tom Deligiannis and Thomas Homer-Dixon (2000) 'The Environment and Violent Conflict: A Response to Gleditsch's Critique and Some Suggestions for Future Research', *Environmental Change and Security Project Report* 6 (Summer), pp. 77–94.
Sprout, Harold, and Margaret Sprout (1965) *The Ecological Perspective on Human Affairs with Special Reference to International Politics*, Princeton: Princeton University Press.
State Failure Task Force (1999) 'State Failure Task Force Report: Phase II Findings', *Environmental Change and Security Project Report* 5 (Summer), pp. 49–72.
Steiner, Achim M. (1993) 'The Peace Dividend in Southern Africa: Prospects and Potentials for Redirecting Military Resources Towards Natural Resources

Management'. Paper presented at the UNDP conference on Military and the Environment: Past Mistakes and Future Options, New York, 22–23 February.

Swatuk, Larry A. (1996) *Power and Water: The Coming Order in Southern Africa*, Bellville: Centre for Southern African Studies.

Swatuk, Larry A. (2001) 'Southern Africa Through Green Lenses', in Peter Vale, Larry A. Swatuk and Bertil Oden (eds) *Theory, Change and Southern Africa's Future*, New York: Palgrave.

Swatuk, Larry A. (2004) 'The United States and Africa: Cybernetic Foreign Policy, Continental Decline', *Journal of Military and Strategic Studies* 6, 4, online journal available at <www.jmss.org/2004/summer/articles/swatuk.pdf>.

Swatuk, Larry A. (2005) 'From "Project" to "Context": Community Based Natural Resource Management in Botswana', *Global Environmental Politics* (special issue entitled *The Global South and the Environment: Essays in Memory of Marian Miller*).

Swatuk, Larry A. and Abillah H. Omari (1997) 'Regional Security: Southern Africa's Mobile "Frontline"', in Larry A. Swatuk and David R. Black (eds) *Bridging the Rift: The New South Africa in Africa*, Boulder: Westview Press.

Swatuk, Larry A., and Peter Vale (1999) 'Why Democracy is Not Enough: Southern Africa and Human Security in the Twenty-first Century', *Alternatives* 24, 3, pp. 361–89.

Tickner, J. Anne (1995) 'Re-visioning Security', in Ken Booth and Steve Smith (eds) *International Relations Theory Today*, Cambridge: Polity Press.

Tir, Jaroslav, and Paul F. Diehl (1998) 'Demographic Pressure and Interstate Conflict: Linking Population Growth and Density to Militarized Disputes and Wars, 1930–89', *Journal of Peace Research* 35, 3, pp. 319–39.

Ullman, Richard (1983) 'Redefining Security', *International Security* 8, pp. 129–53.

USAID/Regional Center for Southern Africa (1997) *Regional Integration Through Partnership and Participation: RCSA Strategic Plan 1997–2003*, Gaborone: USAID/RCSA (August).

Vale, Peter (2003) *Security and Politics in South Africa: The Regional Dimension*, Boulder: Lynne Rienner.

Vale, Peter, and Larry A. Swatuk (2002) 'Sunset or Sunrise? Regime Change and Institutional Adjustment in South Africa – A Critical Analysis', in Helge Hveen and Kristen Nordhaug (eds) *Public Policy in the Age of Globalization: Responses to Environmental and Economic Crises*, Basingstoke: Palgrave.

Van der Linde, Harry, Judy Oglethorpe, Trevor Sandwith, Deborah Snelson, Yemeserach Tessema, Anada Tiega and Thomas Price (2000) *Beyond Boundaries: A Framework for Transboundary Natural Resource Management in Sub-Saharan Africa*, Washington, DC: Biodiversity Support Program.

Van Deveer, Stacy, and Geoff D. Dabelko (1998) 'Redefining Security Around the Baltic: Environmental Issues in Regional Context', *Global Governance* 5, 2, pp. 221–49.

Vira, Bhaskar, and Roger Jeffery (eds) (2001) *Conflict and Cooperation in Participatory Natural Resource Management*, Basingstoke: Palgrave.

Wackernagel, Mathis and William Rees (1996) *Our Ecological Footprint: Reducing Human Impact on the Earth*, Philadelphia: New Society.

Waever, Ole (1997) *Concepts of Security*, Copenhagen: Institute of Political Science.

Walker, R. B. J. (1997) 'The Subject of Security', in Ken Krause and Michael C. Williams (eds) (1997) *Critical Security Studies*, Minneapolis: University of Minnesota Press.

Wallerstein, Immanuel (1974) *The Capitalist World Economy*, Cambridge: Cambridge University Press.

Walt, Steve (1991) 'The Renaissance of Security Studies', *International Security Studies Quarterly* 35, pp. 211–39.

Williams, Michael C (2003) 'Words, Images, Enemies: Securitization and International Politics', *International Studies Quarterly* 47, 4, pp. 511–31.

Wirbelauer, Cathrine, Phemo Kgomotso and L. Innocent Magole (2003) *Proceedings of Stakeholder Consultation Workshops and a National TBNRM Stakeholder Meeting, Botswana, October 2002–August 2003*, Gaborone: Printing and Publishing Company Botswana.

World Commission on Environment and Development (1987) *Our Common Future*, New York: Oxford University Press.

Worldwatch Institute (2005) *State of the World 2005: Global Security*, London: Earthscan.

annotated bibliography

Dalby, Simon (2002a) *Environmental Security*, Minneapolis: University of Minnesota Press. In Chapter 2 of this volume, Paterson points out that most studies of international environmental politics take as their starting point the effect of environment change on human community; few begin by asking what kind of politics is necessary for biosphere sustainability (that is, planetary security)? In my view, this is the perspective taken by most scholars of critical environmental security studies discussed above. Here, Dalby's book is a useful starting point. *Environmental Security* espouses the 'whole earth sensibility' that Deudney found to depart so markedly from the state-specific particularities of traditional 'security studies'. A geographer by training, Dalby takes the sustainability of the biosphere (in all its complexity – so also interested in preserving local processes) as his point of departure. The book is an excellent summary of current debates within the field of environmental security, but falters at the end where the question is asked, 'what is to be done?' As with most non-state-centred, critical studies, finding a way forward after having cogently critiqued planetary-wide processes of late-modern industrial capitalism is not so easy. Like many of his contemporaries, Dalby generally combines a critical analytical approach with a pragmatic policy approach, so linking with neo-institutional approaches to environmental security: flawed and problematic though the state may be, it cannot be ignored where questions of environmental security are concerned.

Homer-Dixon, Thomas (1999) *The Environment, Scarcity and Violence*, Princeton: Princeton University Press. This book marks the high point of the Toronto school's multiyear research programme into environmental change and 'acute conflict probability'. This study synthesises the empirical findings of the project and reflects carefully both on their content and on claims made by those associated with the project in earlier publications. Homer-Dixon modifies some and reiterates more forcefully other earlier claims made regarding the connections between scarce renewable resources and violent conflict, bringing his findings more clearly in line with those of Baechler. But given Homer-Dixon's high media profile, particularly in association with Robert Kaplan's article 'The Coming Anarchy' for the *Atlantic Monthly* (1994), this text is essential reading. Despite the nuance of his conclusions, what remains contestable in my view are (1) his abiding state-centredness; and (2) his belief in the role 'ingenuity' should and can play in overcoming state-specific problems.

Klare, Michael (2001) *Resource Wars: The New Landscape of Global Conflict*, New York: Henry Holt and Company. In my view, Klare articulates quite clearly how leaders such as George W. Bush, Vladimir Putin and Tony Blair regard natural resources. Guided by the question 'How will resource scarcity impact on interstate relations and Great Power security?', Klare presents the reader with a classical Realist interpretation of environmental politics. Klare argues that control and competition over key natural resources – timber, minerals, gemstones, oil, water – will be the guiding principle behind the use of military force in the twenty-first century. He highlights the way competition for these resources will be both inter- and intrastate in character. Lacking nuance, the policy options for the most powerful states are obvious: containment of weak states engaged in intrastate (or intraregional) resource wars, such as West Africa; direct control of those states containing resources key to Western stability (for example, Iraq). There are many shortcomings of this book, but its importance lies in the way it so clearly mirrors the way Realists in Western state houses think about 'the environment' and security. After having read Klare, one should be able to explain why the Montreal Protocol is regarded as a potential threat to US national sovereignty by the Bush Jr administration.

9
global governance and the environment
frank biermann

'Global governance' has become a key term of the discourse on world politics at the dawn of the twenty-first century. While an internet search conducted in 1997 revealed only 3418 references to 'global governance' and in January 2004 less than 90,000, in August 2004 the World Wide Web listed 184,000 pages that mentioned the term. Global governance became a rallying call for policy advocates who hail it as panacea for the evils of globalization; a global menace for opponents who fear it as the universal hegemony of the many by the powerful few; and an analytical concept that has given rise to much discussion among scholars of international relations, including the successful launch of the journal *Global Governance* in 1995.

Global governance is also part and parcel of many of the chapters in this volume on international environmental policy. Global governance is largely a response to economic, social and ecological globalization (Kütting and Rose in this volume); it is characterized by the increased participation of civil society in world politics (Betsill in this volume), including the growing relevance of knowledge and science for global decision-making. Likewise, the concept of global environmental governance is inseparable from questions of sustainability (Bruyninckx in this volume) institutional effectiveness (Wettestad in this volume), and environmental justice (Parks and Roberts in this volume).

This chapter shall explore the concept of global governance in some detail. I proceed in three steps. First, I sketch the historic trajectory and the different current uses of the term 'global governance' in the literature. In the second section, I highlight three key characteristics of global governance that make it different from traditional international relations. The last section offers the reader two examples from recent policy debates on the reform of the existing system of global environmental governance.

different notions of global governance

As is common with many chapters of this book, most of what is conceptualized today as 'global environmental governance' is not without predecessors. The concept of global governance builds on a substantial pedigree of studies that have analysed international environmental cooperation long before, starting with the 1972 Stockholm Conference on the Human Environment, which led to a first wave of academic studies on intergovernmental environmental cooperation and organization (for example, Caldwell, 1984; Johnson, 1972; Kennan, 1970). The most relevant precursor of the concept of global governance is the debate on international environmental regimes of the 1980s (Krasner, 1983; Young, 1980, 1986, 1989) and 1990s (see Wettestad in this volume), including the discussions on the creation of environmental regimes, on their maintenance, and on their eventual effectiveness (for example, Bernauer, 1995; Brown Weiss and Jacobson, 1998; Haas et al., 1993; Keohane and Levy, 1996; Mitchell and Bernauer, 1998; Young, 1994, 1997, 1999a, 1999b; Zürn, 1998). Important earlier research also addressed intergovernmental environmental organizations (Bartlett et al., 1995; Kay and Jacobson, 1983) and non-state environmental organizations (Conca, 1995; Princen et al., 1995; Raustiala, 1997; Wapner, 1996), both of which have received new attention in the global governance discourse. However, while this earlier research has provided important groundwork for the current debate on global environmental governance, there is also much that is different.

The modern discourse on global environmental governance is hence the focus of this chapter. Because the term itself remains vaguely defined despite the recent prolific debate, I will start with a brief discussion of its meanings. Some time ago, 'governance' became a widely discussed concept within the field of domestic politics (van Kersbergen and van Waarden, 2004), often used for new forms of regulation that differed from traditional hierarchical state activity ('government'). Generally, the term implies notions of self-regulation by societal actors, of private–public cooperation in the solving of societal problems, and of new forms of multilevel policy, especially in the European Union. In the discourse on development policy, the term has also received some relevance in the 1990s, frequently with the contested qualifier 'good governance' (de Alcántara, 1998, p. 105).

The more recent notion of global governance builds on these earlier debates among political scientists working on domestic issues, and tries to capture similar developments at the international level. Clear definitions of 'global governance', however, have not yet been agreed upon: global governance means different things to different authors (for example, Dingwerth and Pattberg, forthcoming; Gupta, 2005). There are essentially two broad categories of meanings for 'global governance' – one phenomenological, one normative: global governance as an emerging new phenomenon of world politics that can be described and analysed, or global governance as a political programme or project that is needed to cope with various problems of modernity (the affirmative-normative perspective) or that is to be criticized for its flaws and attempts at global domination of weak states through the powerful few (the critical-normative perspective). Other differentiations seem to be less relevant, for example between governance as a system of rules, an activity, or a process (for example, Finkelstein, 1995; Smouts, 1998). In the three perspectives described below, governance can simultaneously be seen as regulative system, regulative activity, and regulatory process.

phenomenological notions of global governance

Within the group of writers who employ a phenomenological definition of global governance, various subcategories can be identified, differing in the breadth of their definitional scope. First, some writers restrict the term to problems of foreign policy and more traditional forms of world politics. Oran Young, for example, sees global governance as 'the combined efforts of international and transnational regimes' (Young, 1999b, p. 11). Lawrence S. Finkelstein defines the concept in his conceptual essay 'What is Global Governance?' as 'doing internationally what governments do at home' and as 'governing, without sovereign authority, relationships that transcend national frontiers' (Finkelstein, 1995, p. 369). The problem with these narrow phenomenological understandings of global governance is the need to distinguish the term from traditional international relations, because it is often not clear what is gained by using the term 'global governance' instead of 'international relations' or 'world politics'.

Other writers try to address this problem by broadening the term to encompass an increasing number of social and political interactions. James Rosenau, for example, writes that 'the sum of the world's formal and informal rules systems at all levels of community amount to what can properly be called global governance' (Rosenau, 2002, p. 4). In an earlier paper, Rosenau had defined global governance equally broadly as 'systems of rules at all levels of human activity – from the family to the international organization – in which the pursuit of goals through the exercise of control has transnational repercussions' (Rosenau, 1995, p. 13). The UN Commission on Global Governance (1995, pp. 2–3) described governance similarly vaguely as 'the sum of the many ways individuals and institutions, public and private, manage their common affairs. It is a continuing process through which conflicting or diverse interests may be accommodated and cooperative action taken. It includes formal institutions and regimes empowered to enforce compliance, as well as informal arrangements that people and institutions either have agreed to or perceive to be in their interest.' When transferred to the global level, such all-encompassing definitions hardly leave room for anything that is not global governance. Given the increasing international interdependence at all levels, few political rules will have no repercussions beyond the borders of the nation state. In this broad usage the term threatens to become synonymous with politics. In the second section of this chapter, I will try to sketch a middle ground between these two extremes.

normative notions of global governance

A different strand of literature views global governance as a political programme or 'project', mainly in an affirmative sense that demands the construction of a 'global governance architecture' as a counterweight to the negative consequences of economic and ecological globalization. Typically, this involves the call for the creation of new institutions, such as multilateral treaties and conventions, of new and more effective international organizations, and of new forms of financial mechanisms to account for the dependence of current international regimes on the goodwill of national governments. The UN Commission on Global Governance (1995) adhered also to this understanding of the term and elaborated a plethora of more or less far-reaching reform proposals to deal with problems of modernization: global governance is seen here as a solution, as a tool that politicians need to develop and employ to solve the problems that globalization has brought about.

This use of the term is popular especially in continental Europe. A group of writers based at the University of Duisburg, Germany, for example, view global governance as 'a guiding programme to re-gain political governing capacity in an interdependent world' (Messner and Nuscheler, 1998, p. 31, own translation). Quite similar to the Duisburg school are recent papers by a commission of inquiry on globalization of the German Bundestag that define global governance as the 'problem-adequate re-organization of the international institutional environment' (Deutscher Bundestag, 2002, pp. 415, 450). French analyst Marie-Claude Smouts (1998, p. 88) similarly views global governance not as an 'analytical reflection on the present international system [but as a] standard-setting reflection for building a better world'. This understanding of global governance as a political programme on a worldwide scale, however, is not restricted to recent European discourses on global governance. The US academics Leon Gordenker and Thomas G. Weiss (1996, p. 17), for example, also see global governance as 'efforts to bring more orderly and reliable responses to social and political issues that go beyond capacities of states to address individually' (see also, for example, de Alcántara, 1998, p. 111).

Several authors have adopted the programmatic definition of global governance, yet without its affirmative connotation. These authors can be divided into three broad camps, which all share the same concern: that increasing global governance is subduing national sovereignty through some form of supranational hierarchy. First, some neoconservative writers see global governance as the attempt of the United Nations and others to limit the unilateral freedom of action of powerful states (typically with reference to US power). Second, writers in the tradition of postfordism and neomarxism view global governance, in the words of Ulrich Brand (2003), as 'a means to deal more effectively with the crisis-prone consequences caused by [post-Fordist-neoliberal social transformations]'. A third group of writers view global governance through the lens of North–South power conflicts. The Geneva-based South Centre, for example, cautioned in 1996 that in 'an international community ridden with inequalities and injustice, institutionalizing "global governance" without paying careful attention to the question of who wields power, and without adequate safeguards, is tantamount to sanctioning governance of the many weak by the powerful few' (South Centre, 1996, p. 32).

an empirical definition of global governance

Which definition or conceptualization is then preferable? All definitions offered in the current debate have pros and cons depending on the specific context in which they are used. Given the increasing complexity and interdependence of world society in the face of economic and ecological globalization, more effective global regimes and organizations are needed, and there is nothing wrong to call this political reform programme 'global governance'. Also, today's international relations differ from the 1950s and 1960s in many respects, and it seems appropriate to denote these new forms of international regulation as 'global governance'. The term should be restricted, however, to qualitatively new phenomena of world politics. Not much analytical insight can be expected if all forms of human interaction, or all forms of interstate relations, are relabelled as 'global governance'. Instead, I argue that empirically, 'global governance' is defined by a number of new phenomena of world politics that make the world of today different from what it used to be in the 1950s.

First, global governance describes world politics that is no longer confined to nation-states, but characterized by increased participation of actors that have so far been largely active at the subnational level. This multiactor governance includes private actors such as networks of experts, environmentalists, human rights lobby groups and multinational corporations, but also new agencies set up by governments, including intergovernmental organizations and international courts. Second, this increased participation has given rise to new forms of institutions in addition to the traditional system of legally binding documents negotiated by states. Politics are now often organized in networks and in new forms of public–private and private–private cooperation, and they are negotiated between states and private entities. Third, the emerging global governance system is characterized by an increasing segmentation of different layers and clusters of rule-making and rule-implementing, fragmented both vertically between supranational, international, national and subnational layers of authority and horizontally between different parallel rule-making systems maintained by different groups of actors.

None of this is entirely new. Some non-state actors such as the Catholic Church have been influential and engaged in treaty-making with governments for centuries. Politics among nations has always been a multilevel process, with governmental delegations being forced to seek support from domestic constituencies. Also, not all areas of politics follow the new paradigm of global governance, and the term may not aptly describe quite a few real-world conflicts especially in the area of war and peace. On the other hand, global governance is there. It is more frequent, and it is on the rise. It is a reaction to the complexities of modern societies and to the increasing economic, cultural, social and ecological globalization (Kütting and Rose in this volume). Whereas globalization denotes the harmonization and mutual dependence of once separate, territorially defined spheres of human activity and authority, global governance catches the political reaction to these processes. New degrees of global interdependence beget the increasing institutionalization of decision-making beyond the confines of the nation state, with a resulting transformation of the ways and means of global politics. Quantity – the increasing number of functional areas that require global regulation and of international regulatory regimes – creates shifts in quality: new types of actors have entered the stage; new types of institutions have emerged; with new types of interlinkage problems as a result.

Trade integration, for example, required international regulation of more and more 'trade-related' issue areas beyond the key concerns of custom liberalization; the impacts of this drive for institutionalization then brought the world trade regime on the radar screen of a variety of new actors beyond the traditional world of interstate politics: unions, business associations or environmentalists pay close attention to the emergence of the world trade regime and become actors of global governance in their own right. The globalization of environmental problems, from global climate change to the loss of biodiversity, creates new interdependencies between nation states that require new regulatory institutions at the global level. These institutions, however, do not remain isolated from the continuing debates within nation states, a situation which results in governance systems that stretch from local environmental politics to global negotiations and back. I will now elaborate on the key characteristics of global environmental governance.

characteristics of global environmental governance

increased participation: diversity through inclusion

The new system of global governance departs from international politics, first, because of the degree of participation by different actors that were earlier confined to the national sphere. The Westphalian system of international politics was characterized as politics among states. Non-state actors were either non-existent, or lacked sufficient power to influence affairs beyond territorial borders. There have been exceptions – such as the Catholic Church with its highly centralized system of authority or the transnational antislavery movement in the nineteenth century – yet those remained rare and confined to specific historic circumstances. The notion of global governance departs from traditional state-centred politics in accepting a host of non-state entities as new influential actors in transnational relations. The field of environmental policy provides ample illustrations for this evolution of a 'multi-actor governance system'.

The new role of non-governmental lobbying organizations in world politics, for example, has been acknowledged and analysed for decades. Activist groups, business associations and policy research institutes now provide research and policy advice, monitor the commitments of states, inform governments and the public about the actions of their own diplomats and those of negotiation partners, and give diplomats at international meetings direct feedback (Conca, 1995; Betsill and Corell, 2001; Princen et al., 1995; Raustiala, 1997; Wapner, 1996; see Betsill in this volume). Carefully orchestrated campaigns of environmentalists have proved to be able to change foreign policy of powerful nation-states – markedly in the campaign against the dumping of the Brent Spar – or to initiate new global rules, such as the global campaign on banning anti-personnel landmines (see Betsill in this volume).

Second, networks of scientists have assumed a new role in providing complex technical information that is indispensable for policy-making on issues marked by both analytic and normative uncertainty. While the new role of experts in world politics is evident in many policy areas, it is particularly prevalent in the field of global environmental policy (Hisschemöller et al., 2001). New international networks of scientists and experts have emerged, in a mix of self-organization and state-sponsorship, to provide scientific information on both the kind of environmental problem at stake and the options for decision-makers to cope with it. Such scientific advice for political decision-making is not new in world politics; negotiations on fishing quotas for example have long been assisted by the International Council for the Exploration of the Sea. These early examples, however, have been significantly increased in both number and impact, which is mirrored in the substantial academic interest in global scientific networks in recent years (Biermann, 2002; Farrell and Jäger, 2005; Haas, 1990; Jäger, 1998; Jasanoff and Long Martello, 2004; Mitchell, 1998; Mitchell et al., forthcoming).

Third, business has taken a more prominent direct role in international decision-making. Again, the influence of major companies on international affairs is not new, and in some social theories, such as Marxism, business actors have been granted centre stage in global affairs. However, this 'old' influence by the corporate sector was mainly indirect through its influence on national governments. Today, many corporations take a more visible, direct role in international negotiations as immediate partners of governments, for example in the framework of the United Nations and of the Global Compact that major corporations have concluded with the world organization (Cutler et al., 1999; Hall and Biersteker, 2002; Higgot et al., 1999).

Fourth, global governance is marked by an increasing influence of intergovernmental organizations (Biermann and Bauer, 2004). In the field of environmental policy, more than two hundred international organizations have been set-up in the form of secretariats to the many international environmental treaties concluded in the last two decades. Whether the creation of a new 'world environment organization' would help or harm global environmental governance, has been debated for more than 30 years, with no conclusive answer (see the related discussion below).

Fifth, global governance is characterized by new, more powerful forms of supranational jurisdiction. While the International Court of Justice in The Hague has been available for the settlement of interstate disputes for almost eighty years without ever being involved in major conflicts, new tribunals have been established recently, with a considerable and unprecedented degree of compliance by state governments. These include the international criminal tribunals in The Hague, the dispute settlement body under the World Trade Organization, the International Tribunal for the Law of the Sea, as well as the new International Criminal Court. While states remain the eventual sources of authority through their power to alter the legal standards that international courts may apply, and through their remaining option to reject a court's judgment or jurisdiction, current evidence suggests that even powerful nations accept international jurisdiction on sensitive issues, notably in the area of trade. The role of international courts in environmental policy remains yet to be seen.

increased privatization: negotiation through partnerships

Global governance is also defined by new forms of cooperation beyond the traditional intergovernmental negotiation of international law. The influence of non-state actors is not confined to lobbying in such negotiations: more and more, private actors become formally part of norm-setting and norm-implementing institutions and mechanisms in global governance, which denotes the shift from intergovernmental regimes to public–private and increasingly private–private cooperation and policy-making at the global level (Cutler et al., 1999; Hall and Biersteker, 2002; Higgot et al., 1999; Pattberg, 2004). Private actors became partners of governments in the implementation of international standards, for example as quasi-implementing agencies for many programmes of development assistance administered through the World Bank or bilateral agencies. At times, private actors venture to negotiate their own standards, such as in the Forest Stewardship Council or the Marine Stewardship Council, two standard-setting bodies created by major corporations and environmental advocacy groups without direct involvement of governments. The new institutions set up by scientists and experts to advise policy, while formally often under governmental control, also enjoy a large degree of private autonomy from state control.

At times it seems that traditional intergovernmental policy-making through diplomatic conferences is being replaced by such networks, which some see as being more efficient and transparent. Yet the distribution of global public policy networks is often linked to the particular interests of private actors that have to respond to their particular constituencies, and serious questions of the legitimacy of private standard-setting remain. For example, the World Commission on Dams has been hailed as a new and effective mechanism that has quickly generated widely accepted standards, which had earlier been difficult to negotiate due to the persistent resistance of affected countries. Yet this success of private standard-setting gives rise to other voices pointing to the inherent problems of legitimacy that are part and parcel of private policy-making, which cannot relate back to democratic elections or other forms of formal representation (Dingwerth, 2003).

increased segmentation: complexity through fragmentation

Finally, global governance is marked by a new segmentation of policy-making, both vertically (multilevel governance) and horizontally (multipolar governance). First, the increasing institutionalization of world politics at the global level does not occur, and is indeed not conceivable, without continuing policy-making at national and subnational levels. Global standards need to be implemented and put into practice at the local level, and global norm-setting requires local decision-making to set the frames for global decisions. This results in the coexistence of policy-making at the subnational, national, regional and global levels in more and more issue areas, with the potential of both conflicts and synergies between different levels of regulatory activity. The international regulation of trade in genetically modified organisms serves as a prime example for such multilevel governance (Gupta, 2000, 2004).

Likewise, the increasing institutionalization of world politics at the global level does not occur in a uniform manner that covers all parts of the international community to the same extent. In the case of the 1987 Montreal Protocol on Substances that Deplete the Ozone Layer, for example, various recent amendments have provided for new standards and timetables that are not accepted by all parties to the original agreement from 1987. This leads to a substantial multiplicity of sub-regimes within the overall normative framework. The most prominent example of such horizontal fragmentation of policies is humankind's response to the global warming problem. Here, we observe the emergence of parallel policy approaches that include equally important segments of international society and may develop into divergent regulatory regimes in global climate governance.

Divergent policy approaches within a horizontally and vertically segmented policy arena pose significant challenges. Lack of uniform policies may jeopardize the success of the segmented approaches adopted by individual groups of countries or at different levels of decision-making. Regarding climate policy, for instance, the global emissions trading regime as envisaged by the 1997 Kyoto Protocol may create perverse incentives if the United States is not party to the mechanism. The possibly strong economic implications of a stringent climate policy adopted by one group of states may have severe ramifications for other policy arenas such as the world trade regime (Biermann and Brohm, 2005). On the other hand, a segmented policy arena may also have advantages. Distinct policy arenas allow for the testing of innovative policy instruments in some nations or at some levels of decision-making, with subsequent diffusion to other regions or levels (see, for example, Jänicke and Jörgens, 2000; Kern et al., 2001; Tews and Busch, 2002; Vogel, 1995). Also, sensible international policies could mitigate the negative political consequences of a horizontally and vertically segmented governance architecture, and innovative policies may assist in the step-by-step convergence of parallel approaches.

These challenges of interlinkages within a segmented governance system, however, have only poorly been addressed by students of global governance. Most scholars have focused on the emergence of international regimes and on their effectiveness in particular issue areas (see Wettestad in this volume). The interlinkages of regimes in different environmental policy areas have been addressed but only recently (for example, Chambers, 2001; IISD, 2001a, 2001b; Rosendal, 2001a, 2001b; Stokke, 2000; Velasquez, 2000). Yet interlinkages of parallel policies and regimes within a horizontally and vertically segmented governance system in the same issue area have hardly been studied; there is a need to explore the consequences of divergent policies in global environmental governance and to analyse what sets of compatible or diverging norms and rules exist, how they predetermine the political opportunities for coordination, and what response strategies policy-makers could avail themselves of. This research will also require better collaboration between distinct communities of researchers, especially those focusing on the international level and on international relations, and those concentrating on the national level and on comparative environmental politics (Biermann and Dingwerth, 2004; on the problem of interplay, see also IDGEC Science Plan, IHDP 1998; Young, 2002).

current reform debates

Global governance is a political response to economic, cultural, social and ecological globalization. It is not initiated and developed by some centralized decision-making body, but by an amalgam of centres of authority at various levels. The efficacy of the current system of global governance has been the subject of intense debate. It is not only a normative discussion on 'more global governance', but likewise a debate on 'better global governance'. I will sketch two of these reform debates in this section; both are related to environmental policy, and each attends to a particular aspect of global governance that has been highlighted above.

participation and privatization: institutionalizing civil society involvement

The first example of a reform debate deals with the increased participation of non-state actors in global environmental governance. This participation has not been without friction. Developing countries, in particular, often object to increases in the influence of non-governmental organizations in international forums because they view these groups as being more favourable to Northern agendas, perspectives and interests. Developing countries argue that most associations are headquartered in industrialized countries, that most funds donated to their cause stem from Northern organizations, both public and private, and that this situation influences the agenda of these groups to be more accountable to Northern audiences (South Centre, 1996). However, these suspected biases in the work of non-governmental actors should not lead to a decrease in the participation of civil society, but rather to the establishment of mechanisms that ensure a balance of opinions and perspectives.

I offer as an example the recent institutionalization and formalization of the advice of scientists and other experts on climate change. The key institution here is the Intergovernmental Panel on Climate Change (IPCC). The evolution of the IPCC is typical for the functioning of global governance: it has been initiated not by governments but by international organizations – the World Meteorological Organization (WMO) and UNEP. It is comprised of private actors – experts, scientists and their autonomous professional organizations – who are nonetheless engaged in a constant dialogue with representatives from governments. The final summary conclusions of IPCC reports are drafted by scientists, but are submitted to line-by-line review by governmental delegates. The reports from the IPCC are partially commissioned by public institutions – the UN climate convention – but are structured and organized by the expert community itself.

Typical for global environmental governance has been the continuous struggle for influence in this body, especially between industrialized and developing countries (Agrawala, 1998a, 1998b; Siebenhüner, 2002a, 2002b, 2003; Biermann, 2002, forthcoming). When the IPCC was set up in 1988, only a few experts and scientists from developing countries were actively involved. This has led, as many observers from developing countries argued, to a notable lack of credibility, legitimacy and saliency of these reports in the South. Continuous complaints from delegates from developing countries led to a number of reforms since 1989, which resulted in an increasing institutionalization of the involvement of private actors in this subsystem of global governance (Agrawala, 1998b; IPCC, 1997). For example, current IPCC rules of procedure now require each working group of scientists to be chaired by one developed and one developing country scientist. Each chapter of assessment reports must have at least one lead author from a developing country. Participation of developing country scientists in the IPCC thus appears much more visible than previously. The IPCC's governance structure now has a quota system that rather resembles public political bodies such as the meetings of parties to the Montreal Protocol, the executive committee of the ozone fund or the Global Environment Facility, all of which are governed by North–South parity procedures.

These changes have ameliorated, yet not abolished existing inequalities between North and South in global governance. Regarding the second IPCC assessment report from 1996, the percentage of Southern 'contributing lead authors', 'lead authors' and 'contributing authors' in IPCC working groups still ranged from only 5.1 per cent to 25.0 per cent. Likewise, the percentage of Southern peer reviewers in the working groups was small, reaching from only 8.5 per cent to 11.1 per cent and 14.9 per cent (Dingwerth, 2001). Financing, in particular, remains a problem. Most research institutions in developing countries lack funds to send their scientists to professional conferences abroad. This has been addressed for direct participation in IPCC working groups. Still, general communication between Southern and Northern scientists is scarce compared to transatlantic or intra-European cooperation (Agrawala, 1998b, p. 632; Kandlikar and Sagar, 1999). Nonetheless, the institutionalization of the involvement of scientists in the IPCC has helped to increase the legitimacy of the panel in the South.

This form of institutionalization of private participation within the IPCC could even evolve into a pattern for other areas of global governance. An interesting model for achieving the balance between private actors from North and South is the decision-making procedure of the International Labour Organization (ILO). Each member state is represented with four votes, two of which are assigned to governments and one each to business associations and labour unions. The ILO procedure – if adopted for environmental institutions – would attend to the basic problem of a private participation in global environmental governance, namely that environmental groups can often not adequately compete with the financial clout of business associations, and that non-governmental organizations of developing countries lack standing vis-à-vis the financially well-endowed non-governmental organizations of industrialized countries. An ILO-type structure would thus grant business interests and environmental interests at least formally equal rights, and it would guarantee that the Southern non-governmental associations would have a clout in accordance with the population represented by them. The ILO formula is far from perfect, in particular given the higher degree of complexity in environmental policy compared to ILO's more clear-cut 'business versus labour'-type of conflicts. And yet, the ILO experience provides a conceptual model along which ideas for an equitable participation of civil society in global environmental governance could be developed.

segmentation: the debate on a world environment organization

Another current reform debate in the field of global environmental governance concerns the organizational and institutional fragmentation of global environmental policy. Many observers have pointed to the paradoxical situation that strong and powerful international bodies oriented towards economic growth – such as the World Trade Organization, the World Bank or the International Monetary Fund – are hardly matched by UNEP, the modest UN programme for environmental issues. The same imbalance is revealed when UNEP is compared to the plethora of influential UN specialized agencies in the fields of labour, shipping, agriculture, communication or culture. As a mere programme, UNEP has no right to adopt treaties or any regulations upon its own initiative, it cannot avail itself of any regular and predictable funding, and it is subordinated to the UN Economic and Social Council. UNEP's staff hardly exceeds 300 professionals – a trifle compared to its national counterparts such as the German Federal Environment Agency with 1,043 employees and the United States Environmental Protection Agency with a staff of 18,807.

This situation has led to a variety of proposals to grant the environment what other policy areas long had: a strong international agency with a sizeable mandate, significant resources and sufficient autonomy. The debate on such a 'world environment organization' (WEO) – or a global environmental organization, as it is sometimes being referred to (for example, Runge, 2001) – has been going on for some time. Magnus Lodewalk and John Whalley (2002) have reviewed no less than 17 recent proposals for a new organization, and they have not even covered all proposals that can be found in the literature, which dates back 34 years to George Kennan (1970; see Bauer and Biermann, 2005; Charnovitz, 2002, 2005, for an overview). In recent years, many opponents of a new agency have also taken the floor (for example, Juma, 2000; Oberthür and Gehring, 2005; von Moltke, 2001, 2005).

Proponents of a world environment organization can be divided into more pragmatic and more radical approaches. The more radical strand in the literature demands the abolition of major agencies such as the World Meteorological Organization, the creation of a new agency with enforcement power – for example, through trade sanctions – or for the creation of a new agency in addition to UNEP, which would have to transfer many of its functions to the new organization (Esty, 1994, 1996; Kanie and Haas, 2004). Most of these radical designs are both unrealistic and undesirable. Abolishing UN agencies has been rare in post-1945 history and seems politically unfeasible or unnecessary for most agencies today. Trade sanctions to enforce environmental treaties would unfairly focus on less powerful developing countries while leaving the big industrialized countries sacrosanct (Biermann, 2001). Establishing a new agency in addition to UNEP would create new coordinating problems while attempting to solve them and would likely result in an imbalance between supposedly global issues – to be addressed by a new global environmental organization – and local issues, which would then be addressed by the remaining UNEP.

Pragmatists, instead, propose to maintain the current system of decentralized, issue-specific international environmental regimes along with existing specialized organizations active in the environmental field while strengthening the interests of environmental protection by upgrading UNEP from a mere UN programme to a full-fledged international organization. This organization would have its own budget and legal personality, increased financial and staff resources, and enhanced legal powers. In this model, a world environment organization would function among the other international institutions and organizations, whose member states might then be inclined to shift some competencies related to the environment to the new agency. Additional financial and staff resources could be devoted to the fields of awareness raising, technology transfer and the provision of environmental expertise to international, national and subnational levels. The elevation of UNEP to a world environment organization of this type could be modelled on the World Health Organization and the International Labour Organization, that is, independent international organizations with their own membership.

There are three chief arguments brought forward in favour of a new agency (Biermann, 2005). First, upgrading UNEP to a WEO as a UN specialized agency could ameliorate the coordination deficit in the global governance architecture that results in substantial costs and suboptimal policy outcomes. When UNEP was set up in 1972, it was still a comparatively independent player with a clearly defined work area. Since then, however, the increase in international environmental regimes has led to a considerable fragmentation of the system. Norms and standards in each area of environmental governance are set up by distinct legislative bodies – the conferences of the parties – with little respect for repercussions and for links with other fields. While the decentralized negotiation of rules and standards in separate functional bodies may be defensible, this is less so regarding the organizational fragmentation of the various convention secretariats, which have evolved into medium-sized bureaucracies with strong centrifugal tendencies. In addition, most specialized international organizations and bodies have initiated their own environmental programmes independently from each other and with little policy coordination among themselves and with UNEP. The situation on the international level might come close, if compared to the national level, to the abolishment of national environment ministries and the transfer of their programmes and policies to the ministries of agriculture, industry, energy, economics or trade: a policy proposal that would not find many supporters in most countries.

Streamlining environmental secretariats and negotiations into one body would especially increase the voice of the South in global environmental negotiations. The current system of organizational fragmentation and inadequate coordination causes special problems for developing countries. Individual environmental agreements are negotiated in a variety of places, ranging – for example, in ozone policy – from Vienna to Montreal, Helsinki, London, Nairobi, Copenhagen, Bangkok, Nairobi, Vienna, San José, Montreal, Cairo, Beijing and Ouagadougou. This nomadic nature of a 'travelling diplomatic circus' also characterizes most subcommittees of environmental conventions. Developing countries lack the resources to attend all these meetings with a sufficient number of well-qualified diplomats and experts (Rajan, 1997). The creation of a world environment organization could help developing countries to build up specialized 'environmental embassies' at the seat of the new organization, which would reduce their costs and increase their negotiation skills and respective influence.

Second, if UNEP were upgraded to a WEO as a UN-specialized agency, the body would be better poised to support regime-building processes, especially by initiating and preparing new treaties. The ILO could serve as a model. The ILO has developed a comprehensive body of 'ILO conventions' that come close to a global labour code. In comparison, global environmental policy is far more disparate and cumbersome in its norm-setting processes. It is also riddled with various disputes among the UN specialized organizations regarding their competencies, with UNEP in its current setting being unable to adequately protect environmental interests. A specialized UN organization could also approve – by qualified majority vote – certain regulations, which are then binding on all members, comparable to Articles 21 and 22 of the WHO statute. The WEO Assembly could also adopt draft treaties which have been negotiated by subcommittees under its auspices and which would then be opened for signature within WEO headquarters. The ILO Constitution, for example, requires its parties in Article 19(5) to process, within one year, all treaties adopted by the ILO General Conference to the respective national authorities and to report back to the organization on progress in the ratification process. Although governments remain free to not ratify an ILO treaty adopted by the ILO assembly, the ILO mandate still goes much beyond the powers of the UNEP Governing Council, which cannot pressure governments in the same way as ILO can.

Third, upgrading UNEP to a WEO as a UN specialized agency could assist in the build-up of environmental capacities in developing countries. Strengthening the capacity of developing countries to deal with global and domestic environmental problems has become one of the most essential functions of global environmental regimes (for example, Keohane and Levy, 1996). The demand for financial and technological North–South transfers is certain to grow when global climate, biodiversity and other policies are more intensively implemented in the South. Yet the current organizational setting for financial North–South transfers suffers from an adhocism and fragmentation that does not fully meet the requirements of transparency, efficiency and participation of the parties involved. At present, most industrialized countries strive for a strengthening of the World Bank and its recent affiliate, the Global Environment Facility (GEF), to which they will likely wish to assign most financial transfers. Many developing countries, on the other hand, view this development with concern, given their perspective of the Bank as a Northern-dominated institution ruled by decision-making procedures based on contributions. Though the GEF has been substantially reformed in 1994, it still meets with opposition from the South. A way out would be to move the tasks of overseeing capacity building and financial and technological assistance for global environmental policies to an independent body that is specially designed to account for the distinct character of North–South relations in global environmental policy, that could link the normative and technical aspects of financial and technological assistance, and that is strong enough to overcome the fragmentation of the current multitude of inefficient single funds. Such a body could be a world environment organization. This does not need to imply the set-up of large new bureaucracies. Instead, a world environment organization could make use of the extensive expertise of the World Bank or UNDP, including their national representations in developing countries. However, by designating a world environment organization as the central authoritative body for the various financial mechanisms and funds, the rights of developing countries over implementation could be strengthened without necessarily giving away advantages of the technical expertise and knowledge of existing organizations.

An organization, as opposed to a programme, could allow for a system of regular, predictable and assessed contributions of members, instead of voluntary contributions, as is the case with UNEP. A more comprehensive reform that leads to the creation of a new agency could also involve the reassembling and streamlining of the current system of independent (trust) funds, including the ozone fund under the Montreal Protocol and the GEF of the World Bank (jointly administered with UNEP and UNDP). The norm-setting functions of the GEF, for example regarding the criteria for financial disbursement, could be transferred to the WEO Assembly, in a system that would leave GEF the role of a 'finance ministry' under the overall supervision and normative guidance of the WEO Assembly. This would unite the economic and administrative expertise of GEF's staff with the 'legislative' role of a world environment organization.

In sum, creating a world environment organization would pave the way for the elevation of environmental policies on the agenda of governments, international organizations and private organizations; it could assist in developing the capacities for environmental policy in African, Asian and Latin American countries; and it would improve the institutional environment for the negotiation of new conventions and action programmes as well as for the implementation and coordination of existing ones.

conclusion

The current global governance discourse reveals that more theoretical debate as well as empirical research is needed. I will emphasize three needs for further discussion.

First, the debate on the very term 'global governance' and its conceptualization is not yet sufficiently concluded. There are a number of conceptual approaches, which in part have been reviewed in this chapter. Yet none of these has mustered sufficient support within the community. The second main section of this chapter has argued for an empirical understanding of global governance as a concept to denote essentially new phenomena in world politics that cannot be analysed adequately in the framework of traditional concepts such as international relations. This does not deny that global governance is also an important political programme. Yet it remains crucial to clearly demarcate the use of the term and to state whether any given analysis employs the phenomenological or the normative notion of the term 'global governance'.

Both uses of the term also suggest the need for further research. The phenomenological conceptualization directly defines a research programme. First, multi-actor governance requires us to better understand the behaviour and the influence of the new actors of world politics. While environmentalist lobbyist groups and scientists have been studied in some detail as actors of global environmental governance, significantly less knowledge is available regarding the increasing role of intergovernmental organizations and of business actors. This is one of the exciting new research frontiers in this field.

Second, the new mechanisms of global governance, such as private–public partnerships, also point to a new research programme that helps us to better understand the emergence, maintenance, effectiveness and finally legitimacy of these new regulatory mechanisms. Some work on private–public and private–private cooperation in the field of global environmental governance has already been done (see above), yet what is needed is a larger research effort that equals the substantial series of comparative studies on international environmental regimes in the 1980s and 1990s.

Third, the increasing segmentation of world politics is, again, also an empirical development in need of more research. We need to better understand in what ways governance between different levels occurs. This, in particular, requires new approaches of linking academic sub-disciplines that have been apart for long: international relations and comparative politics. Research programmes on the international climate regime, for example, must be better integrated with comparative work on national or local energy politics. This requires a number of essentially new research programmes on 'interlinkages' and on the 'interplay' within global environmental governance.

All this eventually needs to feed back into the actual reform debates, which have been exemplified in this chapter by the institutionalization of expert advice and the strengthening of the existing system of global environmental governance through creating a new world environment organization. However, these reform efforts toward a more effective and more legitimate system of environmental institutions and environmental organizations require, first and foremost, a better basic understanding of the set of phenomena that have been conceptualized in this chapter as global environmental governance.

bibliography

Agrawala, Shardul (1998a) 'Context and Early Origins of the Intergovernmental Panel on Climate Change', *Climatic Change* 39, pp. 605–20.
Agrawala, Shardul (1998b) 'Structural and Process History of the Intergovernmental Panel on Climate Change', *Climatic Change* 39, pp. 621–42.
Bartlett, Robert V., Priya A. Kurian and Madhu Malik (eds) (1995) *International Organizations and Environmental Policy*, Westport: Greenwood Press.
Bauer, Steffen, and Frank Biermann (2005) 'The Debate on a World Environment Organization: An Introduction', in Frank Biermann and Steffen Bauer (eds) *A World Environment Organization. Solution or Threat for Effective International Environmental Governance?*, Aldershot: Ashgate, pp. 1–23.
Bernauer, Thomas (1995) 'The Effect of International Environmental Institutions', *International Organization* 49, 2, 351–77.
Betsill, Michele, and Elisabeth Corell (2001) 'NGO Influence in International Environmental Negotiations: A Framework for Analysis', *Global Environmental Politics* 1, 4, pp. 65–85.
Biermann, Frank (2001) 'The Rising Tide of Green Unilateralism in World Trade Law: Options for Reconciling the Emerging North–South Conflict', *Journal of World Trade* 35, 3, pp. 421–48.
Biermann, Frank (2002) 'Institutions for Scientific Advice: Global Environmental Assessments and their Influence in Developing Countries', *Global Governance* 8, 2, pp. 195–219.
Biermann, Frank (2005) 'The Rationale for a World Environment Organization', in Frank Biermann and Steffen Bauer (eds) *A World Environment Organization. Solution or Threat for Effective International Environmental Governance?*, Aldershot: Ashgate, pp. 117–44.
Biermann, Frank (forthcoming) 'Whose Experts? The Role of Geographic Representation in Assessment Institutions', in Ronald B. Mitchell, William C. Clark and David W. Cash (eds), *Global Environmental Assessments: Information, Institutions and Influence*, Cambridge, Mass.: MIT Press.
Biermann, Frank, and Steffen Bauer (2004) 'Assessing the Effectiveness of Intergovernmental Organizations in International Environmental Politics', *Global Environmental Change*, 14, 2, pp. 189–93.
Biermann, Frank, and Steffen Bauer (eds) (2005) *A World Environment Organization: Solution or Threat for Effective International Environmental Governance?*, Aldershot: Ashgate.
Biermann, Frank, and Rainer Brohm (2005) 'Implementing the Kyoto Protocol Without the United States: The Strategic Role of Energy Tax Adjustments at the Border', *Climate Policy*, 4, 3, pp. 289–302.

Biermann, Frank, and Klaus Dingwerth (2004) 'Global Environmental Change and the Nation State', *Global Environmental Politics*, 4, 1, pp. 1–22.
Brand, Ulrich (2003) 'Nach dem Fordismus: Global Governance als der Neue Hegemoniale Diskurs des Internationalen Politikverständnisses' [After Fordism: Global Governance as the New Hegemonial Discourse of the International Understanding of Politics], *Zeitschrift für Internationale Beziehungen* 10, 1, pp. 143–65.
Brown Weiss, Edith, and Harold K. Jacobson (eds.) (1998) *Engaging Countries. Strengthening Compliance with International Environmental Accords*, Cambridge, Mass.: MIT Press.
Caldwell, Lynton Keith (1984) *International Environmental Policy: Emergence and Dimensions*, Durham: Duke University Press.
Chambers, W. B. (ed.) (2001) *Inter-Linkages: The Kyoto Protocol and the International Trade and Investment Regimes*, Tokyo: United Nations University Press.
Charnovitz, Steve (2002) 'A World Environment Organization', *Columbia Journal of Environmental Law* 27, 2, pp. 321–57.
Charnovitz, Steve (2005) 'Toward a World Environment Organization: Reflections upon a Vital Debate', in Frank Biermann and Steffen Bauer (eds) *A World Environment Organization. Solution or Threat for Effective International Environmental Governance?*, Aldershot: Ashgate, pp. 87–115.
Commission on Global Governance (1995) *Our Global Neighbourhood. The Report of the Commission on Global Governance*, Oxford: Oxford University Press.
Conca, Ken (1995) 'Greening the United Nations: Environmental Organizations and the UN System', *Third World Quarterly* 16, 3, pp. 441–57.
Cutler, Claire, Virginia Haufler and Tony Porter (eds) (1999) *Private Authority and International Affairs*, Albany: SUNY Press.
de Alcántara, Cynthia Hewitt (1998) 'Uses and Abuses of the Concept of Governance', *International Social Science Journal* 155, pp. 105–13.
Deutscher Bundestag (2002) *Schlussbericht der Enquete-Kommission 'Globalisierung der Weltwirtschaft'* [Final report of the Enquete Commission 'Globalization of the World Economy'], Opladen, Germany: Leske and Budrich.
Dingwerth, Klaus (2001) *Die Wissenschaft in der internationalen Klimapolitik: Partizipation, Kommunikation und Effektivität* [The science in international climate politics: Participation, communication and effectiveness], Berlin: Freie Universität Berlin, mimeo (on file with author).
Dingwerth, Klaus (2003) *The Democratic Legitimacy of Global Public Policy Networks: Analysing the World Commission on Dams*. Global Governance Working Paper No. 6 (English version). Potsdam, Berlin, Oldenburg: Global Governance Project. Available at <www.glogov.org>.
Dingwerth, Klaus, and Philipp Pattberg (forthcoming) *The Meaning of Global Governance*. Global Governance Working Paper Series. Amsterdam, Berlin, Potsdam, Oldenburg: Global Governance Project. Available at <www.glogov.org>.
Esty, Daniel C. (1994) 'The Case for a Global Environmental Organization', in Peter B. Kenen (ed.) *Managing the World Economy: Fifty Years After Bretton Woods*, Washington, DC: Institute for International Economics, pp. 287–309.
Esty, Daniel C. (1996) 'Stepping Up to the Global Environmental Challenge', *Fordham Environmental Law Journal* 8, 1, pp. 103–13.

Farrell, Alex and Jill Jäger (eds) (2005) *Assessments of Regional and Global Environmental Risks: Designing Processes for Effective Use of Science in Decisionmaking*, Washington, DC: Resources for the Future Press.
Finkelstein, Lawrence S. (1995) 'What is Global Governance?' *Global Governance* 1, 3, pp. 367–72.
Gordenker, Leon, and Thomas G. Weiss (1996) 'Pluralizing Global Governance: Analytical Approaches and Dimensions', in Thomas G. Weiss and Leon Gordenker (eds) *NGOs, the UN, and Global Governance*, Boulder: Lynne Rienner, pp. 17–47.
Gupta, Aarti (2000) 'Governing Trade in Genetically Modified Organisms: The Cartagena Protocol on Biosafety', *Environment* 42, 4, pp. 23–33.
Gupta, Aarti (2004) 'When Global is Local: Negotiating Safe Use of Biotechnology', in Sheila Jasanoff and Marybeth Long-Martello (eds), *Earthly Politics: Local and Global in Environmental Governance*, Cambridge, Mass.: MIT Press, pp. 127–48.
Gupta, Joyeeta (2005) 'Global Environmental Governance: Challenges for the South from a Theoretical Perspective', in Frank Biermann and Steffen Bauer (eds) *A World Environment Organization. Solution or Threat for Effective International Environmental Governance?*, Aldershot: Ashgate, pp. 57–83.
Haas, Peter M. (1990) *Saving the Mediterranean: The Politics of International Environmental Cooperation*, New York: Columbia University Press.
Haas, Peter M., Robert Keohane and Marc A. Levy (eds) (1993), *Institutions for the Earth: Sources of Effective International Environmental Protection*, Cambridge, Mass.: MIT Press.
Higgot, Richard A., Geoffrey D. Underhill and Andreas Bieler (eds) (1999) *Non-State Actors and Authority in the Global System*, London: Routledge.
Hisschemöller, Matthijs, Rob Hoppe, William N. Dunn and Jerry R. Ravetz (eds) (2001) *Knowledge, Power and Participation in Environmental Policy Analysis*, New Brunswick and London: Transaction Publishers.
IHDP (International Human Dimensions Programme on Global Environmental Change) (1998) *Science Plan of the Institutional Dimensions of Global Environmental Change core project*, IHDP report no. 9, Bonn: IHDP.
IISD (2001a) 'Summary of the WSSD International Eminent Persons Meeting on Inter-Linkages: Bridging Problems and Solutions to Work Towards Sustainable Development, 3–4 September 2001', *Sustainable Developments* 57, 1, pp. 1–9.
IISD (2001b) 'Summary Report of the Informal Regional Consultation on Inter-linkages: Synergies and Coordination among Multilateral Environmental Agreements', *Sustainable Developments* 48, 1, pp. 1–11.
IPCC (Intergovernmental Panel on Climate Change) (1997) *The IPCC Third Assessment Report Decision Paper*. Approved at the XIIIth session of the IPCC, 21–8 September 1997, Republic of the Maldives, available at <www.ipcc.ch>).
Jäger, Jill (1998) 'Current Thinking on Using Scientific Findings in Environmental Policy Making', *Environmental Modeling and Assessment* 3, pp. 143–53.
Jänicke, Martin, and Helge Jörgens (2000) 'Strategic Environmental Planning and Uncertainty: A Cross-national Comparison of Green Plans in Industrialized Countries', *Policy Studies Journal* 28, 3, pp. 612–32.
Jasanoff, Sheila, and Marybeth Long-Martello (eds) (2004) *Earthly Politics: Local and Global in Environmental Governance*, Cambridge, Mass.: MIT Press.
Johnson, Brian (1972) 'The United Nations Institutional Response to Stockholm: A Case Study in the International Politics of Institutional Change', *International Organization* 26, 2, pp. 255–301.

Juma, Calestous (2000) 'Stunting Green Progress', *Financial Times*, 6 July.
Kandlikar, Milind, and Ambuj Sagar (1999) 'Climate Change Research and Analysis in India: An Integrated Assessment of a South–North Divide', *Global Environmental Change* 9, 2, pp. 119–38.
Kanie, Norichika, and Peter M. Haas (eds) (2004) *Emerging Forces in Environmental Governance*, Tokyo: United Nations University Press.
Kay, David A., and Harold K. Jacobson (eds) (1983) *Environmental Protection: The International Dimension*, Totowa, NJ: Allanheld, Osmun and Co.
Kennan, George F. (1970) 'To Prevent a World Wasteland: A Proposal', *Foreign Affairs* 48, 3, pp. 401–13.
Keohane, Robert O., and Marc A. Levy (eds) (1996) *Institutions for Environmental Aid: Pitfalls and Promise*, Cambridge, Mass.: MIT Press.
Kern, Kristine, Helge Jörgens and Martin Jänicke (2001) *The Diffusion of Environmental Policy Innovations: A Contribution to the Globalisation of Environmental Policy*. Discussion Paper FS II 01–302 of the Social Science Research Centre Berlin, Berlin: Social Science Research Centre Berlin.
Krasner, Stephen D. (ed.) (1983) *International Regimes*, Ithaca: Cornell University Press.
Lodewalk, Magnus, and John Whalley (2002) 'Reviewing Proposals for a World Environmental Organisation', *World Economy* 25, 5, pp. 601–17.
Messner, Dirk, and Franz Nuscheler (1998) 'Globale Trends, Globalisierung und Global Governance' [Global trends, globalization and global governance]. In Stiftung Entwicklung und Frieden (ed.), *Globale Trends 1998*, Frankfurt am Main: Fischer, pp. 27–37.
Mitchell, Ronald B. (1998) 'Sources of Transparency: Information Systems in International Regimes,' *International Studies Quarterly* 42, pp. 109–30.
Mitchell, Ronald B., and Thomas Bernauer (1998) 'Empirical Research on International Environmental Policy: Designing Qualitative Case Studies', *Journal of Environment and Development* 7, 1, pp. 4–31.
Mitchell, Ronald B., William C. Clark and David W. Cash (eds) (forthcoming) *Global Environmental Assessments: Information, Institutions and Influence*, Cambridge, Mass.: MIT Press.
Oberthür, Sebastian, and Thomas Gehring (2005) 'Reforming International Environmental Governance: An Institutional Perspective on Proposals for a World Environment Organization', in Frank Biermann and Steffen Bauer (eds) *A World Environment Organization. Solution or Threat for Effective International Environmental Governance?*, Aldershot: Ashgate, pp. 205–34.
Pattberg, Philipp (2004) *The Institutionalisation of Private Governance: Conceptualising an Emerging Trend in Global Environmental Politics*. Global Governance Working Paper No. 9, Amsterdam, Berlin, Oldenburg, Potsdam: Global Governance Project.
Princen, Thomas, Matthias Finger and Jack Manno (1995) 'Nongovernmental Organizations in World Environmental Politics', *International Environmental Affairs* 7, 1, pp. 42–58.
Rajan, Mukund Govind (1997) *Global Environmental Politics: India and the North–South Politics of Global Environmental Issues*, Delhi: Oxford University Press.
Raustiala, Kal (1997) 'States, NGOs, and International Environmental Institutions', *International Studies Quarterly* 42, 4, pp. 719–40.

Rosenau, James N. (1995) 'Governance in the Twenty-First Century', *Global Governance* 1, 1, pp. 13–43.

Rosenau, James N. (2002) 'Globalization and Governance: Sustainability Between Fragmentation and Integration'. Paper presented at the Conference on Governance and Sustainability: New Challenges for the State, Business, and Civil Society, Berlin, 30 September.

Rosendal, G. K. (2001a) 'Impacts of Overlapping International Regimes: The Case of Biodiversity', *Global Governance* 7, 1, pp. 95–117.

Rosendal, G. K. (2001b) 'Overlapping International Regimes: The Case of the Intergovernmental Forum on Forests (IFF) between Climate Change and Biodiversity', *International Environmental Agreements: Politics, Law and Economics* 1, 4, pp. 447–68.

Runge, C. Ford (2001) 'A Global Environment Organization (GEO) and the World Trading System', *Journal of World Trade* 35, 4, pp. 399–426.

Siebenhüner, Bernd (2002a) 'How Do Scientific Assessments Learn? Part 1. Conceptual Framework and Case Study of the IPCC', *Environmental Science and Policy* 5, pp. 411–20.

Siebenhüner, Bernd (2002b) 'How Do Scientific Assessments Learn? Part 2. Case Study of the LRTAP Assessment and Comparative Conclusions', *Environmental Science and Policy* 5, pp. 421–7.

Siebenhüner, Bernd (2003) 'The Changing Role of Nation States in International Environmental Assessments: The Case of the IPCC', *Global Environmental Change* 13, 2, pp. 113–23.

Smouts, Marie-Claude (1998) 'The Proper Use of Governance in International Relations', *International Social Science Journal* 155, pp. 81–9.

South Centre (1996) *For a Strong and Democratic United Nations: A South Perspective on UN Reform*, Geneva: South Centre.

Stokke, Olaf Schram (2000) 'Managing Straddling Stocks: The Interplay of Global and Regional Regimes', *Ocean and Coastal Management* 43, 2–3, pp. 205–34.

Tews, Kerstin, and Per-Olof Busch (2002) 'Governance by Diffusion? Potentials and Restrictions of Environmental Policy Diffusion', in Frank Biermann, Rainer Brohm and Klaus Dingwerth (eds) *Global Environmental Change and the Nation State: Proceedings of the 2001 Berlin Conference on the Human Dimensions of Global Environmental Change*, Potsdam: Potsdam Institute for Climate Impact Research, pp. 168–182, available at <www.glogov.de/publications/bc2001/-tews.pdf>.

van Kersbergen, Kees, and Frans van Waarden (2004) '"Governance" as a Bridge Between Disciplines: Cross-disciplinary Inspiration Regarding Shifts in Governance and Problems of Governability, Accountability and Legitimacy', *European Journal of Political Research*, 43, 2, pp. 143–71.

Velasquez, Jerry (2000) 'Prospects for Rio+10: The Need for an Inter-linkages Approach to Global Environmental Governance', *Global Environmental Change* 10, 4, pp. 307–12.

Vogel, David (1995) *Trading Up: Consumer and Environmental Regulation in a Global Economy*, Cambridge, Mass.: Harvard University Press.

von Moltke, Konrad (2001) 'The Organization of the Impossible', *Global Environmental Politics* 1, 1, pp. 23–8.

von Moltke, Konrad (2005) 'Clustering International Environmental Agreements as an Alternative to a World Environment Organization', in Frank Biermann and Steffen Bauer (eds) *A World Environment Organization. Solution or Threat*

for *Effective International Environmental Governance?*, Aldershot: Ashgate, pp. 175–204.
Wapner, Paul (1996) *Environmental Activism and World Civic Politics*, Albany: SUNY Press.
Young, Oran R. (1980) 'International Regimes: Problems of Concept Formation', *World Politics* 32, 4, pp. 331–56.
Young, Oran R. (1986) 'International Regimes: Toward a New Theory of Institutions', *World Politics* 39, 104.
Young, Oran R. (1989) *International Cooperation: Building Regimes for Natural Resources and the Environment*, Ithaca: Cornell University Press.
Young, Oran R. (1994) *International Governance. Protecting the Environment in a Stateless Society*, Ithaca: Cornell University Press.
Young, Oran R. (1996) 'Institutional Linkages in International Society: Polar Perspectives', *Global Governance* 2, 1, pp. 1–24.
Young, Oran R. (ed.) (1997) *Global Governance. Drawing Insights from the Environmental Experience*, Cambridge, Mass.: MIT Press.
Young, Oran R. (ed.) (1999a) *Effectiveness of International Environmental Regimes: Causal Connections and Behavioral Mechanisms*, Cambridge, Mass.: MIT Press.
Young, Oran R. (1999b) *Governance in World Affairs*, Ithaca: Cornell University Press.
Young, Oran R. (2002) *The Institutional Dimension of Environmental Change: Fit, Interplay and Scale*, Cambridge, Mass.: MIT Press.
Zürn, Michael (1998) 'The Rise of International Environmental Politics', *World Politics* 50, 4, pp. 617–49.

annotated bibliography

Biermann, Frank, and Steffen Bauer (eds.) (2005) *A World Environment Organization: Solution or Threat for Effective International Environmental Governance?* Aldershot: Ashgate. This edited volume is a collection of the key contributions to the debate on a world environment organization, comprising three articles in favour and three articles against such a new agency, along with two articles that provide the general background for the debate.

Rosenau, James N. (1995) 'Governance in the Twenty-First Century', *Global Governance* 1, 1, pp. 13–43. This is one of the key articles in the debate on global governance by one of the most influential scholars in this area.

Young, Oran R. (ed.) (1997) *Global Governance: Drawing Insights from the Environmental Experience*, Cambridge, Mass.: MIT Press. This book was one of the first to introduce the global governance concept into research on environmental politics. It seeks to both draw on the general global governance discourse to better understand environmental politics, and to contribute research findings from environmental politics to the global governance debate.

Young, Oran R. (2002) *The Institutional Dimension of Environmental Change: Fit, Interplay and Scale*, Cambridge, Mass.: MIT Press. This book represents the key outline of the research strategy of the Institutional Dimensions of Global Environmental Change (IDGEC) core project of the International Human Dimensions Program on Global Environmental Change (IHDP), written by IDGEC's initiator and long-time chair.

part iii
normative frameworks for evaluating international environmental politics

10
sustainable development: the institutionalization of a contested policy concept

hans bruyninckx

Very few concepts have made such a fast and pervasive career in policy discourses as sustainable development. Since its introduction as a guiding policy principle during the period spanning from the publication of the Brundtland Report in 1987 to the Rio Conference in 1992, it has been accepted as a framework for policy agendas as widely different as macroeconomic development and the provision of basic health care services. Less than ten years after its introduction, it was the central concept in areas such as environmental policy, economic planning, spatial planning, development policy and foreign aid policy, at all levels of policy-making (although especially at the national and international levels). Outside of government and policy making, it has also been a defining concept for non-governmental organizations (NGOs) of different types, of business associations, labour unions and even churches. Yet, at the same time, the concept remains contested at different levels. Critics point to the vagueness of the concept, the level of aggregation that is not adapted for pragmatic policy-making, the Western or Northern bias, and its voluntaristic and unrealistic view of the role of economic dynamics.

This chapter will first provide an overview of the conceptual history of sustainable development and its basic content. Next, the main elements of debate that have crystallized in the 15 years that the concept has been used by policy-makers and other actors will be discussed. Because of the rather complex content, multiple examples of the way in which sustainable development is being implemented in actual policy processes will be given by looking at the institutionalization of sustainable development at different levels of government and governance and by different actors. Finally, an example of what is increasingly labelled as sustainable development regimes, namely the United Nations Convention to Combat Desertification (UNCCD), will be elaborated.

the conceptual history of sustainable development

Although sustainable development as a concept has become important only since the publication of the Brundtland report, *Our Common Future*, in 1987 (World Commission on Environment and Development (WCED), 1987) and the United Nations Conference on Environment and Development (UNCED), the so-called Rio Conference in 1992, it is clearly embedded in a number of currents that have existed much longer.

The early scenario builders, of which the Club of Rome was undoubtedly the most important or influential example, have clearly shaped our thinking about the interaction between human systems of production and consumption, population dynamics, and the fundamental environmental and natural resource basis on which our society is dependent (see Stevis, this volume). In their *Limits to Growth* report (Club of Rome, 1971) predictions about resource scarcity and pollution were projected into the 21st century. The conclusion was as simple as it was sobering: current trends are not sustainable and will lead to serious problems.[1] A similar exercise was undertaken about ten years later in the *Global 2000 Report to the President* (Barney, 1982), which was prepared for US President Jimmy Carter. Given the important improvements in computer technology and data availability, it can be seen as an elaborated and better-illustrated version of the *Limits to Growth*.[2] The message was very similar: our current path of use, or even usurpation, of natural resources and the negative side effects, such as pollution, are not tenable – or sustainable – in the long run. The emphasis in these reports was clearly on the environmental and natural resource aspects of our current system of production and consumption.[3]

Another origin of the sustainable development concept can be found in the developmentalist literature and a number of critical international reports on the enormous differences between dynamics in rich and in poor countries. Early reports include the Tinbergen Report (1970) and the Brandt Commission Report (1977), both of which were rather structuralist in their analysis of global inequalities (see Paterson in this volume). They explained the differences between North and South primarily through the fundamental imbalances in the global economic system of trade and production. What these two and other reports also share is that they describe the unacceptable and dangerous continuation of these differences. They explicitly call for strong international policies to close the wealth gap and to come to a New International Economic Order (NIEO), based on a more equal distribution of costs and benefits (Lozoya, 1980).

An important step forward in the international policy debate occurred in the early 1970s. During the preparation of the first global convention on the environment, the United Nations Convention on the Human Environment, held in Stockholm in 1972, the first internationally recognized and carefully developed link was formulated between environmental problems and poverty (Caldwell, 1990). From that point onwards, it was almost unthinkable that environmental problems would be delinked from their developmental aspects at the international level. In the aftermath of Stockholm, we see the emergence of a growing literature on international environment and development issues, which not only further elaborates these ideas, but increasingly puts the emphasis on the connections between the economic dimensions of North–South relations and their impact on the environment (see Clapp, Kütting and Rose, and Stevis in this volume). To underline this line of reasoning, during the second half of the 1980s and the 1990s, a number of global environmental issues were discovered and placed high on the international agenda. It is also increasingly recognized that solutions to these problems can only be formulated at the global level (Caldwell, 1990; Haas et al., 1993; Hurrell and Kingsbury, 1992).

This global dimension is strongly present in the Brundtland Report of the WCED, which was commissioned by the United Nations in the middle of the 1980s after it became clear that international environmental and developmental policies were not leading to satisfactory results (WCED, 1987). Although the obvious differentiations between North and South are made by the Commission in terms of impacts, capacities, responsibilities, and so on, the underlying message is that we have entered a period of global problems that require global solutions; hence the title of the report *Our Common Future*. The Brundtland Commission broadly introduced and defined sustainable development as development that 'meets the needs of the present without compromising the ability of future generations to meet their own needs' (WCED, 1987, p. 8). From that moment on, the term was increasingly used in international literature, negotiations and policy-making.

Yet the first really prominent use of the term sustainable development came from the International Union for the Conservation of Nature and Natural Resources (IUCN) in 1980, in the World Conservation Strategy.

This important policy document promoted 'the overall aim of achieving *sustainable development* through the conservation of living resources' (IUCN, 1980). We mention this also because it underlines the rather strong environmental emphasis that has been placed on the concept from the beginning.

During the United Nations Conference on Environment and Development, held in Rio de Janeiro (Brazil) in 1992, sustainable development was the central concept around which the debates were organized. One of the political advantages of the concept was that regardless of strong differences on a number of issues, hardly anybody could be against the basic ideas behind the concept as such. The main documents of the Rio Conference, the *Rio Declaration* and *Agenda 21*, further defined sustainable development and gave it a more policy-oriented content. The emphasis became the 'balancing' of environmental, economic and social goals: a stable economy should be able to produce enough welfare for everybody, and to distribute the benefits and the cost in a much more equitable way, without endangering the environment on which the whole system is based (Fisher, 1993).

Indeed, the changes towards a sustainable society will involve rather far-reaching transitions in core issues of our system of production and consumption, in the distribution of wealth on a global scale, in economic pricing mechanisms, and in the functioning of government institutions. This is where the rather broad consensus on the concept has started to become a real debate. Change involves actors that have to give up certain views and positions, institutions that have to change and – definitely in the short run – costs to be incurred by some.

Another way to approach sustainable development is, in fact, to evaluate some of the widely accepted policy principles that characterize sustainable development policy processes. Both *Our Common Future* and *Agenda 21* mention several of these principles (Bruyninckx, 2002). Although there are some nuances and differences across policy texts on these principles, a number of them seem to have reached (at least in their theoretical dimension) consensual status. These include integration, equity, intergenerational solidarity, internalization and participatory policy-making. The application of these principles is expected to lead to more sustainable policy-making, and at the same time they form a set of preconditions to come to an operationalization of the more holistic or meta-goals of sustainable development (Bartelmus, 1994).

An absolute core principle is the necessity to integrate different policy domains. Horizontal *policy integration* is defined as recognition of the linkages between different policy domains and the need to approach them together. Vertical integration refers to the need to come to better policy coherence between different levels of policy-making and implementation, for example, the local, regional and national (Wilkinson, 1997). *Equity* forms the strong normative foundation for the social dimension of sustainable development (Ikeme, 2003). Both production and consumption have to be based on a more equitable distribution of the costs and benefits associated with them (Cohen and Murphy, 2001; Wilk, 2002). This is true both within Northern and Southern countries and between them (Agyeman et al., 2003). *Intergenerational solidarity* refers to the – until now – often absent long-term planning that is needed to create fundamental changes in our society. It will be increasingly necessary to take the next generations into account when we make decisions. The *internalization of social and environmental costs* is another key principle (Bartelmus, 1994). Until now, we have functioned on the assumption that the market price of goods reflected the costs of production. It has become increasingly clear, however, that this is not always the case, which leads to price distortions. Finally, *participatory policy-making* involves both normative and instrumental hypotheses (Hemmati, 2002). More participation by stakeholders is believed to result in better policy-making and especially policy implementation, as actors have been involved, and will accept the proposed solutions more easily and support them through the required behavioural changes. Participation in this case, is an instrument to improve policy-making. But there is also a more fundamental element of participation. A participatory society is believed to be a better society, as it fundamentally recognizes the role of citizens and social groups for the legitimacy of policy-making processes.

Since the Rio Conference, these principles have been accepted as guidelines for international policy debates. Actors have committed to them and have adopted myriad programmes and changes in order to further the sustainable development agenda. In 2002, they met again in Johannesburg (South Africa) for the Rio+10 Conference known as the World Summit on Sustainable Development (WSSD). The main issue on the agenda was the lack of strong implementation in the decade since UNCED. States and other actors discussed better strategies to push forward the common agenda. It had become clear that implementing the multifaceted concept was far more difficult, and required far more political commitment than was generally admitted (Nierynck et al., 2003; Tritten et al., 2001). Debates had become tenser as it had been difficult to come to functioning global strategies on key issues such as biodiversity and climate change. In addition to these environmental issues, the countries of the South were not planning to have a second Summit with an explicit environmental agenda. They absolutely wanted development issues to play a much more central role. After ten years, it was clear that sustainable development was still central in the debate, but the debate was becoming more political and difficult or even conflictual.

academic debates about sustainable development as a concept

The Brundtland and Rio definitions can be considered as meta-concepts, which capture a broad, integrated, not to say holistic vision of the future (Bruyninckx, 2002; Glasby, 2002; Lumley and Armstrong, 2004). The translation of sustainable development as a policy concept has proven to be very difficult, both conceptually and in its implementation, and much less consensual than was expected by some (Pearce, 1999). From the start of the international policy process in preparation of the UNCED meeting several critiques have developed.

the vagueness of the concept

The broad use of sustainable development has led some to claim that *sustainable* has become an adjective that can be placed in front of nearly anything nowadays. One of the criticisms about the concept is that it is vague and means something different to all actors in the debate (Lee and Kirkpatrick, 2000; O'Riordan, 1985; Redclift, 1987). This is probably correct, although it is, to a certain extent, not surprising (Paehlke, 2001; Spangenberg, 2004). To make an illustrative comparison: the concepts that were central in the social conflicts and changes of the nineteenth and twentieth century were very similar in that respect. Think of freedom, liberty, and social equality: these are *essentially contested concepts*. Different social actors (for example, unions and employers) have played a central role not only in translating these theoretical concepts on the political agenda, but also in bringing them to life (Bruyninckx, 2002). The realization of the welfare state and of workers' rights, both on the job and in a broader social context, which now may seem self-evident to many in Western Europe, have been a long, fierce battle and a fundamental social transition. One could argue that sustainable development is (becoming) a similar concept (Bruyninckx, 2001; Lee and Kirkpatrick, 2000; Spangenberg, 2004). It holds the promise of a fundamental transition, toward a different kind of society, based on different principles and distinct types of social interaction (Zaccai, 2002). In that sense the debate surrounding the concept is useful, necessary and expected. Because different actors are using it to frame concrete planning and actions it is realizing itself in social and political reality.

environmental sustainability or broader interpretations?

Although most actors defend a broad definition, based on the integration of social, economic and environmental goals, the policy translation of sustainable development often has a strong environmental bias. A good example can be found in the national sustainable development plans of some of the Scandinavian countries, such as Sweden and Finland. Critics of this approach on the other hand emphasize the equilibrium between the three constitutive elements as the central and defining difference between the traditional external integration of environmental policies and sustainable development (Zaccai, 2002). This debate about the ecologicalization of sustainable development is important, as it largely determines the locus of institutionalization, the stakeholders and the stakes. The difference is also of great importance in the North–South debate.

There are, indeed, several explanations for the important position of environmental interpretations of sustainable development. First, there clearly exists an ecological essentialism in its foundation. The ecosystem is seen as an essential precondition for human functioning in its social and economic dimensions. In that sense, there is no real 'balance' in the three elements of sustainable development. It is obvious to many authors that the environmental dimension forms the fundament for the other two (McLaren, 2003). A second explanation is that environmental groups have from the start been the strongest proponents of the sustainability concept, and have hence had a very significant impact on the debates. This has been the case in the North, but also to a certain extent in the South, where regardless of the social and economic dimensions, environmental interpretations have been very influential in the discourse. Third, the policy-oriented translations of sustainable development have had a rather obvious environmental bias. This has been very visible in the dominance of environmental elements in the planning and actual policy implementation. In addition even the application of sustainable development to other policy fields is often based on some form of environmental policy integration.

However, it is important to point out that a number of countries and actors have chosen the more holistic interpretation of sustainable development. They emphasize the balance between the three basic elements and have developed policies accordingly. Examples include countries in Europe (for example, France and the Netherlands), which have included very strong social and economic goals in their national policy plans.

Probably the strongest social and economic proponents of sustainable development can be found in some developing nations, which are putting *human development* central in the whole enterprise (Mestrum 2003). This means that basic economic welfare, social development in terms of education, healthcare, and access to services, are the central elements and the real basis for the programme of social progress that serves itself of the concept of sustainable development. Obviously, there are strong links with food production and hence with issues such as soil degradation and deforestation or with environmental degradation and health risks, so there is emphasis on elements such as sanitation, drinking water and waste management.

the impact of the concept on the north–south debate

Although the Brundtland Report and also the preparation of the UNCED placed a heavy emphasis on the North–South dimension of sustainable development, the concept is increasingly becoming an element of debate between North and South. This debate is simplistically narrowed down, by many, to a development versus environment debate (Mestrum, 2003). Yet it is much more complex and refers to both development and environmental dynamics in industrialized and in developing countries, and in addition to the connection between those two (Faber and McCarthy, 2003; McLaren, 2003). In that sense, sustainable development can be interpreted as an essential concept in the globalization debate (see Kütting and Rose in this volume).

Indeed, some of the harshest criticisms on sustainable development are based on either the fact that it is the nth concept coming from Northern intellectuals trying to capture global inequalities. Or even more fundamental, that the concept is reaffirming precisely those power structures that underlie the issues for which it claims to be a cure (Faber and McCarthy, 2003; Lélé, 1991; Lohmann, 1990). This rather essentialist or structuralist critique questions both the fundamental analysis that is behind the use of the concept and the sincerity of the real agenda behind its use.

Conceptualizations of sustainable development by authors from the South usually go in one of several directions. Some approach the issue as closely connected to structural elements of the global economy and examine the impacts it has on socioeconomic conditions and subsequently also the environment (Kütting, 2000; Stevis and Assetto,

2001). Sustainable development then refers to fundamental changes in international economic parameters. The other approach is much more linked to poverty as a pervasive phenomenon in the South (Mestrum, 2003). This leads to recommendations in the sphere of basic needs and hence also on comparative analyses of the 'needs' concept which is part of the Brundtland definition. One other aspect often much more emphasized by developmentalist approaches to sustainability is the bottom-up or communitarian approach. It is defended either by the weakness of state institutions in developing countries, or because it would be culturally more fitting for the circumstances (Fisher, 1993; Hemmati, 2002; Velasquez, 2000).

is there really much beyond the discourse?

The Johannesburg Summit in 2002 increased attention to the implementation of *Agenda 21* and sustainable development in general (Nierynck et al., 2003; Pallemaerts, 2003). Although the Johannesburg Plan of Implementation indeed covers some of the pressing issues, a growing number of critics are claiming that sustainable development is staying at the discourse level and little effective implementation is being stimulated. In addition, whatever implementation planning and action there is at the international or global level is so limited in comparison to the challenges that are recognized in official documents, that the implementation gap is truly enormous (Lightfoot and Burchell, 2004).

One part of the critique is that the capacity to really implement changes is not made available (Velasquez, 2000). For example, the Global Environment Facility (GEF), which is so often mentioned (and with certain pride) by policy-makers as the global financial mechanism that drives financial transfers from North to South, has a budget that is ridiculously small and is in fact nothing more than a drop in a bucket. The level of funds for the 2002–06 period was increased to US$3 billion after negotiations. Although this might be a significant improvement over the previous period (tripling the budget), it is almost scandalous that the complete group of countries from the North announce this as a serious victory and commitment on their part, when in fact this amount equals, for example, two days' military spending in the US, or is only a fraction of the annual profits of large energy companies such as Shell or British Petroleum. Much more fundamental mechanisms of global distribution will be needed, according to critics, to really provide the needed capacity to countries in the South, to implement necessary policies. Fundamental trade mechanisms, debt relief schemes, and financial transactions will need to be discussed from the perspective of sustainable development (Petrella, 2003).

Another fundamental criticism is that the real political will for social change towards sustainable development is largely absent (Lightfoot and Burchell, 2004; Van Ypersele, 2003). Most political leaders pay the necessary lip service to the ideas of sustainable development without translating this into further results-oriented commitment. During the preparations of the Rio Summit, for example, George H. W. Bush left the world in doubt about his coming to the final meeting. Only after Maurice Strong struck a deal that there would be no discussion on the 'American way of life' (meaning unsustainable consumption and production patterns that are spreading quickly around the world) did President Bush attend. He did give a speech in support of sustainable development, however. Ten years later, President George W. Bush was one of the only leaders who openly decided not to come to Johannesburg. Some claim that at least this was an honest position on his part, whereas several other leaders with very bleak policy records on sustainable development gladly used the Johannesburg forum to give their verbal support to sustainable development and other issues such as global problem-solving, solidarity and global warming policies.

The most fundamental critique is that the current direction of sustainable development policies is far from providing the fundamental changes that will be necessary to turn things around (Lafferty and Meadowcraft, 1996). The term 'fundamental' is to be taken literally according to this reasoning. Some of the fundamental threats to human society will have to move in a radically different direction, including our energy system, which needs to change from largely carbon-based to other sources of energy. Also, the gap between rich and poor in the world absolutely needs to decrease. The more than 1 billion people who live on less than one or two dollars a day have been the subject of international promises for several decades by now. Such promises include the endless number of times that the North has said it will spend 0.7 per cent of its GDP on official development aid (a promise kept by only four countries), the technology transfer mantra in numerous international regimes, the debt restructuring promises, and the fair trade talks. Although things have changed in certain parts of the world in terms of economic development and the emergence of a middle class of (global) consumers, the fact remains that the gap between North and South is not closing and the number of extreme and absolute poor has not diminished. This is important because the unacceptably uneven distribution of wealth on a global level has extremely bad consequences for the local and global environment.

In short, although sustainable development has widely been accepted as a relevant and useful concept there is much debate on its meaning and its applicability. Some of this debate is about definitional issues and interpretations, but it is clear that there is also a more fundamental debate both in academia and among other actors about some of the key elements of this appealing concept. Both the much longer debate about the global distribution of wealth, and the debate about some basic features of our system of production and consumption are being reframed by the sustainable development concept. And lastly, the truly huge implementation gap is ammunition for those who claim that there is not much beyond the discourse of primarily northern political and economic policy-makers.

academic debates and research on the institutionalization and practices of sustainable development

Much of the academic literature on sustainable development is embedded in different interpretations of the concept and how these are reflected in its implementation. In the broadest sense one can make a distinction between institutionalist approaches and more structuralist approaches to the issues associated with sustainable development. Structuralists are not necessarily questioning the sort of meta-ideas embedded in the sustainable development concept. Rather, they are critical of the belief that incremental changes, largely within the boundaries of the existing world order and global policy dynamics, will be able to bring about the social transformations required to come to a sustainable society. Institutionalists, on the other hand, either strongly believe that incremental institutional adaptation and its problem-solving capacity will eventually lead to, or at least holds the best promise for social change in the direction of sustainable development (Gupta, 2002). Strong proponents of this current are to be found among regime theorists, neo-institutionalists and idealists.

These two basic views on the functioning of global politics are reflected in further academic debates about governance for sustainable development and about the role of different actors and agency. In the following pages I will look at several dimensions of the institutionalization of sustainable development through academic debates on levels of governance. These include the global governance idea and more regional approaches to governance as well as the role of the state. Another way to approach sustainable development is through debates on the actors involved in changes towards sustainability. Civil society and stakeholder approaches contrast with the latter. In conclusion the type of knowledge claims that have been made regarding sustainable development will be discussed.

sustainable development and global governance

One of the central debates in the sustainable development literature is about the necessity for, and the feasibility of, a functioning system of global governance for sustainable development (see Biermann in this volume). The necessity of such a system is defended because reconciling global sustainability with the current economic forces of globalization will require some sort of governance regime. Joyeeta Gupta clearly states that 'sustainable development governance refers to the interactive network of regimes at the international level that try and integrate the various elements of sustainable development' (2002b, p. 363). This issue can be further placed in the context of the expressed need for multilateral governance if sustainable development has any meaning as a competing vision against purely market-driven globalization (Faber and McCarthy, 2003; Pallemaerts, 2003, p. 275).

Yet in light of the distinction between institutionalist approaches and structuralist world visions, the answer to the global governance for sustainability question can be approached in different ways. Institutionalists tend to start from the UN as an institutional anchoring point (Pallemaerts, 2003). A number of organizational steps, which form a sort of skeleton of a global sustainable governance regime, have been taken. For example, after the Rio Conference, the UN Commission for Sustainable Development (UNCSD) was founded. It plays a key role in the international dynamics as an inter- or transnational forum for discussion. It has been the catalyst in central elements of institutionalization such as national planning and reporting, the development of sustainability indicators, as well as a place for meetings between the North and South. Countries are expected to report to the UNCSD at regular times about the state of sustainable development policies and planning efforts. This has been an impetus for many countries to set up working groups or commissions to develop national policy plans or reporting structures. Another part of the global efforts includes the numerous global conferences on partial themes of sustainable development (themes occurring in *Agenda 21*). Conferences on the role of women (Fourth World Conference on Women, 1995, Beijing), on sustainable housing (United Nations Conference on Human Settlements (Habitat II), 1996, Istanbul), on issues related to indigenous people (Indigenous Peoples International Summit on Sustainable Development, 2002, Kimberley, South Africa), and on numerous other issues, are used to demonstrate the institutional dynamics of the system (Pallemaerts, 2003). Global conventions such as the Convention on Biodiversity (1992) and the Convention to Combat Desertification (1994) illustrate that the sustainability paradigm has entered global policy-making dynamics and are labelled as post-Rio regimes (Bruyninckx, 2004).

The typical governance characteristics of these institutional steps are further emphasized by discussing the involvement of different stakeholders in the design and implementation of elements of a global governance system for sustainability (Gupta, 2002; von Moltke, 2002). The Johannesburg Declaration clearly refers to partnerships with the so-called major groups in this respect (Pallemaerts, 2003). Idealists within this institutional tradition take the global governance efforts one step further and have supported the idea of a global environmental and developmental body at the level of the UN that could coordinate the activities and have a sort of hierarchical precedence over other bodies and a position of moral superiority (Biermann in this volume; Gupta, 2002).

Structuralists or historical materialists, on the other hand, base their analysis on the essential unsustainable character of the global political economy. The structurally embedded imbalances between economic performance and social and environmental consequences, and the unequal distribution of wealth between and within North and South are essential characteristics of the current system of production and consumption and are reflected in the institutional outcomes at the level of global governance (Wilk, 2002; Zaccai, 2002). These social and material foundations of the current system prevent social change in the direction of sustainable development (Petrella, 2003). Incremental institutional steps are not able to overcome these essential elements (Campos Mello, 1999; Cox, 1987; Saurin, 1996). It is obvious that from a structuralist perspective, the current functioning of the global governance arrangements on sustainable development is inadequate. Power relationships in the global political economy define the context and the content of the debate, and lead to institutional outcomes reflecting those relationships (Ikeme, 2003).

The debate on fundamental changes in the UN system can be placed in this context. As long as the South is in the marginal position, and as long as strong states, which are catering to specific interests, are dominant, little fundamental change is to be expected. In addition to these critiques of the UN system, structuralists point at other major international institutions which fall outside of the idealist and institutionalist vision of governance for sustainable development. In their opinion, the Group of 8 (G8), the World Economic Forum (WEF), the World Trade Organization (WTO), and the Bretton Woods institutions (World Bank system and International Monetary Fund (IMF)) are more important to study in a systematic and historically and socially embedded way as promoters of unsustainable development. They also put rather strong emphasis on the continued role of the state as an important agent in global sustainable development dynamics (Frickel and Davidson, 2004). This is clearly against more institutional and idealist views on the demise of the state, or governance as a stateless social reality (see below).

Another intellectual current, which one could describe as literally 'materialist' is based in the factual analysis of current environmental and developmental trends. Glasby clearly argues that 'at present, the term sustainable development is misleading, because we actually live in a markedly unsustainable world and conditions will become even more unsustainable in the 21st century' (2002, p. 333). Some authors in this tradition use the neo-Malthusian argument (Adger and O'Riordan, 2000; Brown et al., 1999; Ehrlich, 1968). Others are based more in the scenario-building tradition (for example, the Global Environmental Outlook Report). The fundamentally structuralist element in their often empirical argumentation lies in the fact that mass consumption and its negative environmental externalities are the flipside of the more traditional historical materialist accounts which focus on the political economy on the production side. However, one could easily argue that consumption, as a social practice, is at this moment just as globalized as production, and equally embedded in power relationships and state–society arrangements (Barber, 2003; Spaargaren and Van Vliet, 2000; Wilk, 2002).

regional organizations and governance for sustainable development

In light of the previous section on governance, regional institutions have been put forward as one of the elements in multi-level arrangements on sustainable development. During the Johannesburg Summit, clear reference was made to regional arrangements such as the Economic Commission for Latin America and the Caribbean (ECLAC) or the New Partnership for Africa's Development (NEPAD). Yet, when it comes to institutionalizing sustainable development, the EU can be considered a special case (Bruyninckx, 2003). It is certainly the only international organization that has competencies in all relevant policy areas that are important to sustainable development policy-making, and has received much attention in the sustainable development literature because of this. The EU has, for example, a particularly strong impact on the environmental policies of member states, on agricultural matters, on transport policy, on trade relations with countries in the South through the Africa–Caribbean–Pacific liaison, and so on. Sustainable development has been added as a central objective of the EU. This has a strong impact on all policy domains and plans. The EU also emphasizes the integration idea more strongly than any other organization of that scale. Old functionalists (Haas, 1964) would probably see a further spillover of economic integration in this reality. Neo-institutionalists tend to approach the EU as an institution sui generis, and generally are less optimistic about the possibilities of other regional institutions to copy the EU example. They do emphasize, however, the possibilities of successful transnational institutional adaptation (Jordan, 1998).

The central elements of the EU sustainable development strategy can be found in the Lisbon Strategy (2000) and the Gothenburg agenda (2001). The Lisbon process emphasizes some general policy principles of sustainable development and further elaborates the social and economic elements of this strategy. The environmental aspects are less prominent. As a reaction to this lack of an environmental vision, a number of environmental policy objectives of sustainable development were added during the Gothenburg Summit.[4] In addition, there is increased attention to the integration of environmental policy objectives in other policy domains, a serious shift in energy production and use, and in general for what is called decoupling. This means that the goal is to increase the creation of wealth with increasingly less material input and energy use.

More critical voices claim that despite several years of verbal and also institutional commitment to sustainable development, not much has changed in the actual policy dynamics of the EU (Lightfoot and Burchell, 2004). The overall direction of economic and environmental developments in the EU is not exactly characterizeable as a serious turn towards sustainable development (Bruyninckx, 2003). In addition, the positive implementation of EU policies in East Central Europe will probably be countered because of rapid economic development in these new member countries. This basically means increased production and especially consumption, and hence higher levels of material input and pollution outputs.[5]

Some of the other interesting examples of sustainable development dynamics are to be found at the regional level. By 'regional' I mean parts of countries that are connected internationally because they share certain characteristics such as a relevant biophysical region (for example, the Barents region; the Baltic Sea region) or are places of intense economic interaction and hence connected through production and consumption (for example the EU region between Germany, the Netherlands and Belgium). Although the processes are more local or regional in their functioning, there is a clear transnational element present in their organization (Kern and Löffelsend, 2004). Of importance in these cases are the transboundary regional dialogue that takes place and also the fact that actors (stakeholders) are motivated by a sort of connection to a common regional cause and understanding. It is an appeal to the concept of 'belonging' to a certain place that is often present in the discourse.

the state and the institutionalization of sustainable development

One of the most lively and interesting debates of recent years in the academic literature on globalization has been on governance and the environment, namely the role of the state (see Biermann in this volume). Yet many authors still consider the state to be a major catalyst if not the major agent of change in the direction of sustainable development (Jänicke, 2000), in the literature (Eckersley, 2004; Frickel and Davidson, 2004; Lipschutz, 1996) as well as in dynamics at the international level. This point of view can be illustrated by looking at the most visible state-based practices in sustainable development policy-making. As mentioned above, a large number of countries have by now completed planning for national sustainable development polices and have started implementation processes. In a number of them, this has led to changes in the constitution or legislative initiatives on sustainable development (for example, France's national law on sustainable development of 2002). In a number of countries sustainable development is recognized as a policy field at the level of ministerial competence (France, Canada), or state secretary. In a large number of countries sustainable development agencies have been set up.

In addition to these political choices and their institutional translation, conducting sustainable development polices also has its impact on bureaucratic or interagency functioning. Agencies are expected to work together in a more thematic way, which has been against the longstanding bureaucratic custom of functionally organized government.[6]

Sustainable development policy processes have also led to significant changes in the institutional aspects of policy participation (Frickel and Davidson, 2004). In almost all countries, national advisory bodies or councils are functioning. The general tendency is that this has increased the opportunities of environmental and also developmental NGOs and actors to influence governmental policies (at least in principle). Another consequence has been the redrawing of the advisory landscape in some countries with strong neocorporatist traditions. Indeed, labour unions and employers' organizations have been the preferred partners of the government to negotiate governmental policies for several decades. Sustainable development has added other dimensions to socioeconomic policy-making, and hence shifted the debate in different ways. This has meant that traditional bodies have sometimes added the environmental theme to their agenda, or have been enlarged with environmental groups. Sometimes new bodies incorporating the traditional social partners have formed the new arena to discuss issues (see below on state–society relations).

the debate on the local dimension: decentralizing sustainability

A recurring debate in the academic literature goes back to the 'small is beautiful debate' of the 1970s (Schumacher, 1973). In fact, one of the interesting evolutions has been the application of sustainable development at the local level (mostly the municipality). A number of authors describe decentralized and local initiatives as simply a part of a multilevel governance thinking that for reasons of efficiency includes local level institutions and practices (Lafferty and Eckerberg, 1998; Lafferty and Meadowcraft, 1996). The subsidiarity principle is central in this thinking. Local *Agenda 21* initiatives have spread surprisingly fast to all countries and very different types of municipalities. The emphasis is usually on the cooperation between local policy-making institutions and stakeholders to deal with local problems.

For others, it includes a more process-oriented approach. Actors or stakeholders try to formulate problem definitions together and reach consensus about local goals of sustainable development, and the contribution, which they can all bring to the process of change at the local level (Lafferty and Meadowcraft, 1996). A number of these initiatives also explicitly include a more global dimension, either through evaluating their own local contribution to reaching international policy goals, such as the reduction of greenhouse gas emissions, or because they have (if they are in the North) a clear connection with the South. In addition, some seem to link sustainable development to localism and regionalism in a sort of autarchic meaning. This is much in line with traditions such as social ecology (Light, 1998) and bioregionalism (Kretzmann and McKnight, 1993).

It is important to mention that although local *Agenda 21* activities are driven by local dynamics, significant transnational networks and structures (for example, the International Council for Local Environment Initiatives (ICLEI) with an emphasis on developing countries) have developed. These networks provide communities with local approaches to issues, practical toolkits or instruments, good practices and also, more fundamentally, create a space for interaction and discussion, or literally for exchange and linking up, as some municipalities have created small global networks of cooperation. Countries that are particularly strong in this regard are the Scandinavian countries. But we also find important examples in Africa (Kenya, for example), North America, and other places. Academic attention has included a number of studies on the role of cities, sustainability and their role in a globalizing world (Sassen, 1996).

debates on stakeholders and actors

Few elements of the sustainable development discourse have produced such a large social science literature as the participatory and stakeholder elements. One of the strong suggestions in the academic literature is that sustainable development initiatives of various types illustrate the transition from traditional government to governance arrangements that may or may not include the state as key point of reference (Mol et al., 2005). If we regard sustainable development as a process of social change, rather than a policy process, this makes sense. In the absence of strong traditional government in the area of sustainable development, and based on interpretations of the participatory dimensions of sustainable development, a number of innovative new networks on specific environmental issues have emerged that are closely linked to sustainable development practices (Hemmati, 2002). Indeed, economic, social and environmental actors have created networks that influence production and consumption processes in such areas as tropical forest products, agricultural products, and energy consumption. Through labelling networks, for example, sustainable production and consumption are promoted. The fact that state institutions play only a marginal or even negligible role in some of these schemes demonstrates at first sight that sustainable development does have a viable existence outside of formal state politics. We will illustrate this by emphasizing the various roles played by stakeholders in this sort of arrangements.

Environmental and development NGOs have been among the earliest and most enthusiastic supporters of the sustainability concept (Agyeman et al., 2003; Fisher, 1993; Zaccaï, 2002). They have used it to emphasize their older ideas on the essential nature of environmental protection, on the need for solidarity between North and South, and maybe even more, they have found elements in the sustainability discourse in support of more structural changes in our system of production and consumption. In addition, they have used the emphasis on participation to demand more input in policy-making processes at all levels of government and governance (Dower and Williams, 2002). Also during international negotiations on sustainable development, they have gradually gained a more important position. Where they were only marginally represented in the first real side-conference at the Stockholm Conference in 1972, they were represented with literally tens of thousands at Rio and more recently at Johannesburg. Moreover, NGOs representing the social, economic and ecological aspects of sustainable development have started their own more independent processes of international negotiation. The most important example is the World Social Forum, which was held three times in the Brazilian town of Porto Alegre, and once in the Indian city of Mumbai. Thousands of environmentalists, trade unionists, developmentalists, religious leaders and others from North and South discussed possible solutions and cooperation in light of a more sustainable global system and society and also celebrated their common cause and unity in diversity. But also at other levels, this sort of new alliances and bundling of creative powers have led to cooperation between different NGOs. Some label these alliances the formation of global citizenship (Attfield, 2002) and see it as a counter-force to liberal globalization.

Another phenomenon is the emergence of a new sort of NGOs representing sustainable development as such. They reflect the ideas of integration and participation. Often they are umbrella organizations that bring together other NGOs in the three subspheres of sustainable development. In terms of actual involvement in non-statist environmental governance we can refer to eco-labelling regimes in various areas such as clothing, food products and forestry (Cashore and Bernstein, 2004).

In the business world we have witnessed a number of uses of the term sustainable (Zadek, 2001). Some have clearly emphasized the sustainable growth aspects to highlight that regardless of environmental or social concerns, the economy is dependent on businesses that can guarantee long-term employment and profit. Others have linked the term more to technological innovations, in an attempt to stress technocratic interpretations and solutions to environmental problems (Sagoff, 2000). Both of these approaches are rather reductionist in that they are closely linked to the dominant view on business. Other more comprehensive translations of sustainable development into the business community are based on concepts such as triple bottom line management, integrated business management, stakeholder management, and so on. The company is viewed as a social actor with social responsibilities that transcend narrow profit-driven thinking, or companies' economic and technical functioning (Peeters, 2003). As actors in a social context, companies have social and economic responsibilities towards the community in which they operate (Mol et al., 2005).

Critical voices have correctly pointed out that the number of companies that have really incorporated this sort of new approach is rather limited. In addition, once economic crisis is imminent, these new ideas are easily questioned as threats to the bottom line of the company, namely profit. So the critique is that sustainable entrepreneurship seems to be a sort of 'luxury' in times when things are going well. Others talk about a more fundamental trend, and cite companies such as Shell as an example. Global scrutiny by NGOs and other interest groups have, so they claim, changed the environment in which these companies operate so fundamentally that they have made significant changes and are taking public opinion seriously into account and have adapted accordingly (Mol et al., 2005).

Workers' organizations or labour unions have been rather slow to adopt sustainable development wholeheartedly as a concept that could further their claims (Kjaergaard and Westphalen, 2001). They have been especially hesitant because of the dominant position of environmental elements in the discourse. Although some official documents might suggest otherwise, unions have taken a very ambivalent position towards environmental issues in general. As long as they referred to workers' health and safety, they have supported them within specific companies, or in sectoral agreements. As soon as there was reference to more general environmental issues associated with certain industrial sectors, such as the petrochemical or the energy sector, they have been very hesitant to accept any fundamental questioning based on environmental arguments. They have regarded environmental issues in those cases more as a potential threat to employment. The environmental and the labour movement have therefore not always been friendly with each other. This history of mixed feelings partially explains the lukewarm acceptance of sustainable development as a concept (Bruyninckx, 2003).

However, in the last few years things have shifted as the more transnationally organized umbrella organizations of unions have been active during negotiation processes and have produced basic position papers (for example, the European Trade Union Conference (ETUC)). Unions obviously emphasize the more socioeconomic elements of sustainable development. It is fair to say that they have a very strong tradition in these domains as they have been the advocates for strong social rights, social services, fairness in income distribution, and international social rights (for example, through the International Labour Organization). Increasingly, however they are also integrating environmental issues in their vision on sustainable development.

the role of knowledge and instruments

Another academic and practitioners literature has emerged about the knowledge requirements for the transition towards a more sustainable society. In addition, the more practical translation of this debate has been reflected in a sort of 'instruments for sustainable development' approach, often operating at a nuts and bolts level. However, at the most fundamental or meta-level, authors like Homer-Dixon (2000) have described the need for a completely new ability to face the enormous complexity of challenges in the sphere of sustainable development (see also Capra, 2002). The complexity idea is further developed by social scientists in relationship to the need of different scientific approaches and epistemologies to understand social interaction. British sociologist John Urry (2003) pleads for a whole new conceptual framework to look at global society. This new conceptual toolkit and epistemology then forms the basis for social action.

In order to answer some of the major calls for new knowledge based approaches, global networks on partial issues of sustainable development have formed in the scientific and academic communities. The Human Dimensions of Global Environmental Change Programme, for example, involves a significant portion of the top people in international environmental politics and sustainable development. The Intergovernmental Panel on Climate Change (IPCC) could be described as the global epistemic community (Haas, 1992) of the global warming regime. Although generally considered as a key element in global sustainable development efforts, some critical voices are present in the academic debate about the role of the IPCC. Boehmer-Christiansen, for example, is a sharp critic of the large-scale institutionalized and fairly closed arrangements of scientific funding and research. She suggests that the political economy of such arrangements prevents the necessary critical thinking and independence of this knowledge-based arrangement (Boehmer-Christiansen, 1996).

An interesting part of the knowledge and instruments issue is linked to the development of sustainability indicators. Practitioners and scientists alike have engaged in this exercise. One of the most well known exercises has come from Rees and Wackernagel (1996) who developed the ecological footprint metaphor (see also Wackernagel and Yount, 1998). More applied versions can be found in the concrete indicator development projects. Soon after UNCED, the UNCSD started a process to develop a set of indicators for sustainable development. In addition, literally hundreds of cities, regions and countries started similar programmes.

The search for indicators has played a central role in many processes of planning and implementation. Two central elements are important in this debate. One is the need for a sort of 'dashboard' of indicators adapted to the specific policy context. This means that indicators have to reflect the policy conditions and issues of a specific place and policy competence. So we have seen the development of indicators at the level of the municipality, the regional level, the national and the international level. At the same time, there has been a serious debate about the role of indicators in the process of sustainable development. Some feel that disproportionate interest is going to indicators. Others see in indicators an essential precondition to provide accessible knowledge necessary for public debate and participation (Spangenberg, 2000; Spangenberg et al., 1999).

Another academic and intellectual battle on the knowledge base for sustainable development is being fought between believers and non-believers of scenario exercises (and between the defenders of different scenarios as well!). Raskin et al. in the Global Scenario Group (1998) claim to show 'possible pathways for the future' in their *Bending the Curve* report of 1998. *Global Environmental Outlook* makes the same claims. Notorious non-believers such as Lomborg (2001) have defended the position that most scenarios from the past have proven to be far of the mark in many respects. Yet, without making any claims in this debate, it is clear to us that the exercise of bringing together literally hundreds, if not thousands, of variables and datasets is an important attempt to give a state of the art scientific appraisal of the state of the planet (Swart et al., 2003). From social scientists we know that knowledge (claims) potentially play an important role in shifting policy discourses and policy-making dynamics. Sustainability scenarios have the potential to play this role.

One of the latest themes in the knowledge for sustainable development debate is the potential of local or indigenous knowledge (Corell, 1999). The idea that hard science constituted a Western-biased approach to sustainable development lived strongly amongst some. Local knowledge is supposed to be more authentic and more adapted to local demands (Bruyninckx, 2004). In several international regimes we find this type of reasoning, including the Biodiversity Convention, the Desertification Convention and several others.

In this part of the chapter I have looked at different processes that form part of the gradual institutionalization of sustainable development. I have placed those in the context of ongoing academic debates. By doing so it should become clear that many aspects of sustainable development remain open for academic debate. This academic debate is not free-floating, however. It is linked to more essential debates in the social sciences in general and international relations more specifically, about the underlying dynamics and driving forces of global politics.

a new generation of international environmental regimes: the desertification convention

Since the Rio Conference in 1992 a number of environmental regimes have been formed which incorporate innovative elements that are part of the sustainable development discourse. Some of the elements that keep being repeated in these regimes are their clearly global character, a strong North–South dimension, the role of participatory processes, the long-term horizon that is embedded in them, and innovative implementation schemes. The most obvious examples of such regimes are the ones that were on the table for signature during the UNCED Conference in 1992, namely the Biodiversity Convention and the United Nations Framework Convention on Climate Change. The other document open for signature at Rio was the Authoritative Non-legally Binding Declaration of Forest Principles. All three issues demonstrate new elements that are in line with sustainable development. Yet, for the purpose of this chapter, I will use a different regime to illustrate the impact of the sustainable development paradigm on international regimes, the United Nations Convention to Combat Desertification (UNCCD). During the Rio process, the countries of the South demanded a convention on this issue. To make the symbolism even clearer, it was the first international convention with a clear emphasis on Africa.

The UNCCD regime makes ample use of the newer international policy discourses of participatory policy-making and implementation, decentralization as a fundamental policy goal, and the use of local knowledge as an explicit 'good' (Corell, 1999). The sustainable development concept forms the overarching umbrella for these discourses, which represent – at least in policy practices – a fairly recent dimension in international environmental policy-making. Not only is the Desertification Convention setting norms and standards for the behaviour of states, but it also encourages states to reach certain goals, through constitutional and other reforms (for example, decentralization of state power), and/or through the implementation of a different view on state–society relations (for example, participatory policy-making at decentralized policy levels). It is clear that the inclusion of these elements in the UNCCD has its impact on how signatory states try to comply with the regime's requirements. In addition, this normative framework also adds an important challenge to the task of evaluating the performance of actors, and the regime as a whole.

These innovative elements are emanations of policy discourses, which have been gaining in importance with the fairly broad acceptance of sustainable development and *Agenda 21* as guiding conceptual frameworks for international environmental and development initiatives. In Article 9 of the UNCCD, the link with sustainable development is made explicitly. It states that the National Action Programmes should 'be closely interlinked to formulate national policies for sustainable development' (Article 9, section 1). In the convention this new policy discourse is further translated into specific articles, including obligations and recommendations for signatory states.

In the remainder of this section, we take a closer look at three specific discourses, namely participation, decentralization and local knowledge. Each one is strongly represented in the convention and has, at least in theory, significant impact on the policy dynamics in the signatory states.

the participation discourse

Sustainable development is often described as the integration of social, economic and environmental objectives (WCED, 1987). One could add the participatory dimension as the fourth objective. *Agenda 21* devotes a lot of attention to the participatory aspects, both instrumentally and in a more fundamental or normative way. From an instrumental perspective, participation is useful because it makes the implementation of policies more acceptable and hence easier. But there is also a more normative side to the debate. A more participatory society is deemed 'better' than one with less input from citizens and groups, because of a higher democratic and representative nature. The stakeholder approach and policy debate, bring these two elements together: stakeholder involvement in policy-making leads to empowerment as a positive outcome and better policy-making in both planning and implementation stages.[7]

It is obvious from the convention text, that participation should be understood as a policy process, which includes local populations, NGOs and decentralized institutions. The more normative elements connected to the empowerment discourse are less strongly present and more implicit. The actual translation is that both in the policy planning and policy implementation processes, countries in the South have paid much attention to participatory processes and feedback. This has clearly led in some cases to the involvement of a range of actors, including farmers' groups and women's organizations.

the decentralization discourse

The decentralization discourse suggests that decentralized policy-making and implementation has a better chance of reaching the goals of the convention than the traditional centralized command and control style of policy-making, which is dominant in most developing countries. The reason would be that local governments have a better feeling for the 'real' problems; can better decide on the priorities based on more adapted, and hence more relevant, knowledge of the local situation; and are better equipped to allow for stronger inputs from civil society, through participatory processes.

The UNCCD, however, implicitly suggests an even stronger argument for decentralization. The general idea is that decentralization in developing countries will provide a more fertile ground for participatory and more effective policy-making, through the spreading of a *more democratic* political context. This same discourse can be found in documents and policy programmes conducted by the World Bank and the IMF, where it is often partially translated in the conditions for financial support and loans under the heading of 'good governance'.

These are very far-reaching recommendations/obligations. The decentralization of state power in terms of institutional arrangements and policy-making capacity and responsibilities is a fundamental political decision in any country. To request this under an international regime in the African context is very unusual and requires the combination of a number of complex political, bureaucratic and policy processes.

the local knowledge discourse

The knowledge base for most international agreements comes from traditional scientists. Each regime has its specific way of introducing science and scientists into the political and policy cycle. Scientists can be organized in a sort of 'in house' arrangement or through the specific use of knowledge, coming from certain well-defined groups or institutions of scientists. Other regimes are less specific in where they get the scientific underpinnings but overall, the knowledge base can be found in traditional academic scientific research principles. However, an emphasis on local or traditional knowledge as a basis for planning and policy-making is in line with the *Agenda 21* discourse on knowledge. The idea is that local groups – including farmers, women, foresters and others – have a strong 'common sense' and experience-based knowledge of the dynamics of desertification and other problems. This knowledge is until now hardly inventoried, analysed or put to use for policy-making.

It is obvious that the inclusion of local or traditional knowledge does not happen spontaneously. The literature describing the role of this type of knowledge in policy processes emphasizes specific methodologies to make the process work. In addition, the normative bias towards local knowledge – good and authentic – versus scientific knowledge – distant and culturally unadapted – is an element that should be taken into account when analysing the implementation of this type of discourse.

conclusion and directions for future research

Sustainable development has quickly conquered the policy discourse in a number of very important fields of policy-making such as environmental policy, development policy and spatial planning. It has done so in a surprisingly pervasive fashion and at all levels of policy-making. In addition, the concept is used by all sorts of social actors in very varied contexts in both developed or industrialized countries and developing or industrializing countries.

It remains, however a highly contested concept. For some its vagueness is enough to ignore any policy relevance to it. Others emphasize the essential distinction between holistic versus more ecological interpretations of sustainable development. In the North–South debate, sustainable development is seen as a concept that tries to bridge a gap by some, yet by others as the next intellectual and conceptual attempt by the North to control the international development and political economy agenda. The most fundamental critiques of sustainable development claim that it leaves the unsustainable nature of liberal capitalism untouched and hence is a smokescreen concept that avoids more structural debates.

We have also given numerous examples of the institutional consequences of sustainable development at all sorts of policy-making levels. This has made it clear that regardless of the debates, the concept is very much influencing governmental and governance processes. The example of the Desertification Convention finally illustrates how in a 'new generation' of international regimes, several sustainable development foci are incorporated. Participation, decentralization and the social production of knowledge as an interactive process are clearly emphasized in this convention.

In light of the issues discussed in this chapter, there are several suggestions possible for future research. First, the linkages between the sustainable development literature and the governance literature need to be further developed and clarified. Both literatures are rather broad, encompass numerous topics that are not always coherently related, and in addition increasingly refer to each other. Several issues stand out in the academic debates: is there a real possibility of a global governance system based on the current interpretations of sustainable development? What sort of role is there for the United Nations in such a global network and to what extent is this related to the debate about restructuring the UN system?

A second and related issue is the challenge that sustainable development poses to current state–society relations. The emphasis on governance and on the role of stakeholders seems to suggest that the role of the state is becoming less and less important. Yet at the same time, debates about the role of democratizing state–society relations in authoritarian states are closely linked to the concept of the state. There is clearly a need for a better conceptualization of participatory, stakeholder and civil society approaches to sustainable development and their link to debates about sovereignty and the role of the state in global politics.

Third, there is a serious need for scientifically grounded policy evaluation literature on sustainable development. Since the Rio Conference, endless initiatives have been taken under the heading of sustainable development. Although some indicator-development initiatives and UNCSD reports have attempted to substantiate claims about the 'state of sustainable development policies', there is a serious lack of rigorous research on the issue.

Fourth and finally, I think a fundamental set of questions on the linkages between the critical debates on globalization and those on sustainable development need to be asked. Does a transition toward sustainability imply serious threats to current patterns of globalization? If not, what sort of promises does the sustainable development agenda hold for those seeking a different kind of globalization? Is there an outspoken critical political economy of sustainable development? If so, why is it not more explicit in the literature?

notes

1. Because of this type of conclusion, people who followed the reasoning of the Club of Rome and others were called 'doomsday thinkers'. They have, in my opinion, wrongly been accused of being pessimists about the future of the world. I would rather defend the position that they attempted to provide serious analysis of global trends in order to come to more realistic policies for a more sustainable future.
2. This document had far less impact since it was disregarded by President Reagan and thus never really used in policy debates.
3. With the term 'institutionalization', I mean the more political and public policy use of the term. It refers to the framework of concrete, formal institutions, but

also the processes, norms and values that exist within these formal institutions. For this chapter I do not use the more sociological understanding of institution which would refer to the norms and rules that provide the framework for social behaviour of actors: social practices as sociologists refer to them.
4. These objectives are in line with the social democratic foundations of European societies. It was deemed necessary to reinforce those principles, however, as they have been under attack from more market-based and individualistic approaches that have been dominant in the United States (Bruyninckx, 2002).
5. This tendency is already visible in the 'most advanced' countries such as the Czech Republic, Hungary and Slovenia which have started to emit higher levels of greenhouse gases and are already using higher levels of energy.
6. An example would be to work on surface water quality as a theme. This would include numerous agencies in fields ranging from environment to agriculture to infrastructure.
7. This type of reasoning is spreading out into other international policy-making efforts as well; desertification is clearly not an isolated case. A good example is the EU integrated water management directive, which requires member states to change towards more participatory forms of integral water management.

bibliography

Adger, W. Neil, and Timothy O'Riordan (2000) Population, adaptation and resilience. In Timothy O'Riordan (ed.) *Environmental Science for Environmental Management*, 2nd edn, London: Longman, pp. 149–70.
Agyeman, Julian, Robert D. Bullard and Bob Evans (eds) (2003) *Just Sustainabilities in an Unequal World*, Cambridge, Mass.: MIT Press.
Attfield, Robin (2002) 'Global Citizenship and the Global Environment', in Nigel Dower and John Williams (eds) *Global Citizenship: A Critical Introduction*, New York: Routledge.
Barber, J. (2003) 'Production, Consumption and the World Summit on Sustainable Development', in *Environment, Development and Sustainability* 5, pp. 63–93.
Barney, Gerald (1982) *The Global 2000 Report to the President: Entering the 21st Century*, New York: Penguin.
Bartelmus, Peter (1994) *Environment, Growth and Development: The Concepts and Strategies of Sustainability*, London: Routledge.
Beohmer-Christiansen, Sonja (1996) 'The International Research Enterprise and Global Environmental Change: Climate-Change Policy as a Research Process', in John Vogler and Mark F. Imber (eds) *The Environment and International Relations*, London: Routledge, pp. 171–95.
Brandt Commission (1977) *Report of the Independent Commission for International Developmental Issues (the 'North–South Commission')*, Washington DC: World Bank.
Brown, Lester et al. (eds) (1999) *State of the World*, New York: W. W. Norton & Co.
Bruyninckx, Hans (2004) 'The Convention to Combat Desertification and the Role of Innovative Policy Making Discourses: The Case of Burkina Faso', *Global Environmental Politics*, 4, 3, pp. 107–27.
Bruyninckx, Hans (2003) 'De Europese Dimensie van Duurzame Ontwikkeling', in Patrick Develtere (ed.) *Het Draagvlak voor Duurzame Ontwikkeling: Wat het is en zou kunnen zijn*, Leuven: De Boeck Publishing.

Bruyninckx, Hans (2002) *Towards a Social Pact in Sustainability Matters*, Brussels: DWTC.

Bruyninckx, Hans (2001) *Towards a Social Pact in Sustainability Matters: Concluding Research Remarks on Participation in the Belgian Sustainable Development Policy Context*, Brussels: DWTC Publications.

Caldwell, Lynton (1990) *International Environmental Policy: Emergence and Dimensions*, Durham: Duke University Press.

Campos Mello, Valérie de (1999) 'Global Change and the Political Economy of Sustainable Development in Brazil'. Paper presented at International Studies Association, Annual Convention, Washington, DC, 16–20 February, 1999.

Capra, Fritjof (2002) *The Hidden Connections: A Science for Sustainable Living*, New York: HarperCollins.

Cashore, Benjamin and Steven Bernstein (2004) 'Non-State Global Governance: Is Forest Certification a Legitimate Alternative to a Global Forest Convention?', in John Kirton and Michael Trebilcock (eds) *Hard Choices, Soft Law: Combining Trade, Environment, and Social Cohesion in Global Governance*, Aldershot: Ashgate.

Club of Rome (1971) *Limits to Growth. First Report of the Club of Rome*, New York: Club of Rome Penguin.

Cohen, Maurice J., and Joseph Murphy (eds) (2001) *Exploring Sustainable Consumption: Environmental Policy and the Social Sciences*, Oxford: Pergamon Press.

Corell, Elizabeth (1999) *The Negotiable Desert: Expert Knowledge in the Negotiations of the Convention to Combat Desertification*, PhD thesis, Linköping Studies in Arts and Science, No. 191.

Cox, Robert (1987) *Production, Power, and World Order: Social Forces in the Making of History*, New York: Columbia University Press.

Dower, Nigel and John Williams (eds) (2002) *Global Citizenship: A Critical Introduction*, New York: Routledge.

Eckersley, Robyn (2004) *The Green State: Rethinking Democracy and Sovereignty*, Boston: MIT Press.

Ehrlich, Paul (1968) *The Population Bomb*, New York: Ballentine.

Faber, Daniel, and Deborah McCarthy (2003) 'Neo-liberalism, Globalization and the Struggle for Ecological Democracy: Linking Sustainability and Environmental Justice', in Julian Agyeman, Robert D. Bullard and Bob Evans (eds) *Just Sustainabilities in an Unequal World*, Cambridge, Mass.: MIT Press.

Fisher, Judy (1993) *The Road From Rio: Sustainable Development and the Nongovernmental Movement in the Third World*, Westport: Praeger.

Frickel, Scott, and Debra J. Davidson (2004) 'Understanding Environmental Governance: A Critical Review' *Organization & Environment* 17, 4, pp. 471–92.

Glasby, Geoffrey (2002) 'Sustainable Development: The Need for a New Paradigm', *Environment, Development and Sustainability* 4, pp. 333–45.

Gupta, Joyeeta (2002a) 'Global Sustainable Development Governance: Institutional Challenges from a Theoretical Perspective', *International Environmental Agreements: Politics, Law and Economics* 2, pp. 361–88.

Gupta, Joyeeta (2002b) *Our Simmering Planet – What to do about Global Warming?*, London: Zed Books.

Haas, Ernst (1964) *Beyond the Nation-State: Functionalism and International Organization*, Stanford: Stanford University Press.

Haas, Peter M. (ed.) (1992) *Knowledge, Power, and International Policy Coordination*, Columbia, SC: University of South Carolina Press.

Haas, Peter, Robert Keohane and Marc Levy (eds) (1993) *Institutions for the Earth: Sources of Effective International Environmental Protection*, Cambridge, Mass.: MIT Press.

Heinrich Böll Foundation (2001) *The Road to Earth Summit 2002*, Washington, DC: Heinrich Böll Foundation.

Hemmati, Minu (2002) *Multi-stakeholder Processes for Governance and Sustainbility: Beyond Deadlock and Conflict*, London: Earthscan.

Homer-Dixon, Thomas (2000) *The Ingenuity Gap*, Random House.

Hurrell, Andrew, and Benedict Kingsbury (eds) (1992) *The International Politics of the Environment*, Oxford: Clarendon.

Ikeme, Jekwu (2003) 'Equity, Environmental Justice and Sustainability: Incomplete Approaches in Climate Change Politics', *Global Environmental Change* 13, 3, pp. 195–206.

International Union for the Conservation of Nature and Natural Resources (IUCN) (1980) *The World Conservation Strategy*, Gland, Switzerland: IUCN.

Jänicke, Martin (2000) *Umweltplanung im internationalen Vergleich. Strategien der Nachhaltigkeit*, Berlin, Heidelberg, New York: Springer Verlag.

Jordan, Andrew (1998) 'EU Environmental Policy at 25: The Politics of Multinational Governance – European Union', *Environment* 40, pp. 14–18.

Keekok, Lee, Alan Holland and Desmond McNeil (eds) (2000) *Global Sustainable Development in the 21st Century*, Edinburgh: Edinburgh University Press.

Kern, Kristine, and Tina Löffelsend (2004) 'Sustainable Development in the Baltic Sea Region. Governance Beyond the Nation State', *Local Environment* 9, 5, pp. 451–67.

Kjaergaard, Carsten, and Sven-Age Westphalen (2001) *From Collective Bargaining to Social Partnerships: New Roles of the Social Partners in Europe*, Copenhagen: Copenhagen Centre Publications.

Kretzmann, John P., and John L. McKnight (1993) *Building Communities from the Inside Out: A Path Toward Finding and Mobilizing a Community's Assets*, Chicago: Center for Urban Affairs and Policy Research.

Kütting, Gabriela (2000) *Environment, Society and International Relations: Towards More Effective International Environmental Agreements*, London: Routledge.

Lafferty, William M., and Katarina Eckerberg (eds) (1998) *From the Earth Summit to Local Agenda 21: Working Towards Sustainable Development*, London: Earthscan.

Lafferty, William M., and James Meadowcraft (eds) (1996) *Democracy and the Environment: Problems and Prospects*, Cheltenham: Edward Elgar.

Lee, Norman, and Colin Kirkpatrick (eds) (2000) *Sustainable Development and Integrated Appraisal in a Developing World*, Cheltenham: Edward Elgar.

Lélé, Sharachchandra (1991) 'Sustainable Development: A Critical Review', *World Development* 19, 6, pp. 607–21.

Light, Andrew (ed.) (1998) *Social Ecology after Bookchin*, New York: Guilford Publications.

Lightfoot, Simon, and Jon Burchell (2004) 'Green Hope or Greenwash? The Actions of the European Union at the World Summit on Sustainable Development', *Global Environmental Change* 14, 6, pp. 337–44.

Lipschutz, Ronnie, with Judith Mayer (1996) *Global Civil Society and Global Environmental Governance: The Politics Of Nature from Place to Planet*, Albany: SUNY Press.
Lohmann, Larry (1990) 'Whose Common Future?', *The Ecologist* 20, 3, pp. 82–4.
Lomborg, Bjorn (2001) *The Skeptical Environmentalist*, Cambridge: Cambridge University Press.
Lozoya, Jorge Alberto (1980) *Alternative Views of the New International Economic Order*, New York: Elsevier Science Publishing.
Lumley, Sarah, and Patrick Armstrong (2004) 'Some of the Nineteenth Century Origins of the Sustainability Concept', *Environment, Development and Sustainability* 6, pp. 367–78.
McLaren, Duncan (2003) 'Environmental Space, Equity and Ecological Debt', in Julian Agyeman, Robert D. Bullard and Bob Evans (eds) *Just Sustainabilities in an Unequal World*, Cambridge, Mass.: MIT Press.
Mestrum, Francine (2003) 'Poverty Reduction and Sustainable Development', *Environment, Development and Sustainability* 5, 1–2, pp. 41–61.
Mol, Arthur, Fred Buttel and Gert Spaargaren (2005) *Governing Environmental Flows*, Cambridge, Mass.: MIT Press.
Nierynck, Eddy, Anthony Van Overschelde, Tom Bauler, Edwin Zaccai, Luc Hens and Marc Pallemaerts (eds) (2003) *Making Globalisation Sustainable: The Johannesburg Summit on Sustainable Development and Beyond*, Brussels: VUB University Press.
O'Riordan, Tim (1985) 'Future Directions in Environmental Policy', *Journal of Environment and Planning*, 17, pp. 1431–46.
Paelke, Robert (2001) 'Environmental Politics, Sustainability and Social Science', *Environmental Politics* 10, 4, pp. 1–22.
Pallemaerts, Marc (2003) 'Is Multilateralism the Future? Sustainable Development or Globalisation as a Comprehensive Vision of the Future of Humanity', *Environment, Development and Sustainability* 5, pp. 275–95.
Pearce, David (1999) 'Economic Analysis of Global Environmental Issues: Global Warming, Stratospheric Ozone and Biodiversity', in Jeroen C. J. M. van den Bergh (ed.) *Handbook of Environmental and Resource Economics*, Cheltenham and Northampton, Mass.: Edward Elgar.
Peeters, Herwig (2003) 'Sustainable development and the role of the financial world', *Environment, Development and Sustainability* 5, pp. 197–230.
Petrella, Ricardo (2003) 'Sustainable Development in a Globalizing World', in Eddy Nierynck, Anthony Van Overschelde, Tom Bauler, Edwin Zaccai, Luc Hens and Marc Pallemaerts (eds) *Making Globalisation Sustainable: The Johannesburg Summit on Sustainable Development and Beyond*, Brussels: VUB University Press.
Raskin, Paul, Gilberto Gallopin, Pablo Gutman, Al Hammond and Rob Swart (1998) *Bending the Curve: Toward Global Sustainability, Report of the Global Scenario Group*, Stockholm: Stockholm Environment Institute, PoleStar Series Report No. 8.
Redclift, Michael (1987) *Sustainable Development: Exploring the Contradictions*, New York: Methuen Press.
Rees, William, and Mathis Wackernagel (1996) *Our Ecological Footprint: Reducing Human Impact on Earth*, British Columbia: New Society Publishers.
Sagoff, Mark (2000) 'Can Technology Make the World Safe for Development? The Environment in the Age of Information', in Lee Keekok, Alan Holland

and Desmond McNeill (eds) *Global Sustainable Development in the 21st Century*, Edinburgh: Edinburgh University Press.

Sassen, Saskia (1996) 'Globalization and its Impact on Cities', *Public Culture* 8, 2, pp. 49–63.

Saurin, Julian (1996) 'International Relations, Social Ecology and the Globalisation of Environmental Change', in John Vogler and Mark Imber (eds) *The Environment and International Relations*, London: Routledge.

Schumacher, E. F. (1973) *Small is Beautiful*, New York: Blond and Briggs.

Spaargaren, Gert, and B. J. M. van Vliet (2000) 'Lifestyles, Consumption and the Environment: The Ecological Modernisation of Domestic Consumption', *Environmental Politics* 9, pp. 50–77.

Spangenberg, Joachim (2000) 'Towards Sustainability', in Brendan Gleeson and Nicholas Low (eds) *Governing for the Environment: Global Problems, Ethics, and Democracy*, New York: Palgrave.

Spangenberg, Joachim (2004) 'Reconciling Sustainability and Growth: Criteria, Indicators, Policies', *Sustainable Development* 12, pp. 74–86.

Spangenberg, Joachim, A. Femia, F. Hinterberger and H. Schütz (1999) *Material Flow Based Indicators in Environmental Reporting*, Luxembourg: Office for Official Publications of the European Communities.

Stevis, Dimitris, and Valerie Assetto (eds) (2001) *The International Political Economy of the Environment: Critical Perspectives*, Boulder: Lynne Rienner.

Swart, Rob J., Paul Raskin and John Robinson (2003) 'The Problem of the Future: Sustainability and Scenario Analysis', *Global Environmental Change* 14, pp. 137–46.

Tinbergen, Jan (1970) Report of the Tinbergen Commission, Amsterdam: UN Press.

Trittin, Jürgen, Uschi Eid, Sascha Müller-Kraenner and Nika Greger (2001) *From Rio To Johannesburg: Contributions to the Globalization of Sustainability*, Berlin: Heinrich Böll Foundation.

UNCED (1992a) *Agenda 21*, New York: United Nations.

UNCED (1992b) *The Rio Declaration on Environment and Development*, New York: United Nations.

Urry, John (2003) *Global Complexity*, London: Polity Press.

Van Ypersele, Jean-Pascale (2003) 'The 2002 Johannesburg Summit and Global Warming', in Eddy Nierynck, Anthony Van Overschelde, Tom Bauler, Edwin Zaccai, Luc Hens and Marc Pallemaerts (eds) *Making Globalisation Sustainable: The Johannesburg Summit on Sustainable Development and Beyond*, Brussels: VUB University Press.

Velasquez, Jerry (2000) 'Prospects for Rio+10: The Need for an Inter-linkages Approach to Global Environmental Governance', *Global Environmental Change*, 10, pp. 307–12.

von Moltke, Konrad (2002) 'Governments and International Civil Society in Sustainable Development: A Framework', *International Environmental Agreements: Politics, Law and Economics*, 2, pp. 341–59.

Wackernagel, Mathis, and J. David Yount (1998) 'The Ecological Footprint: An Indicator of Progress Toward Regional Sustainability', *Environmental Monitoring and Assessment* 51, pp. 511–29.

Wackernagel, Mathis, and David Yount (2000) 'Footprints for Sustainability: The Next Steps', *Environment Development and Sustainability* 2, pp. 21–42.

Wilk, Richard (2002) 'Consumption, Human Needs and Global Environmental Change', *Global Environmental Change* 12, pp. 5–13.
Wilkinson, David (1997) 'Towards Sustainability in the EU? Steps within the European Commission Towards Integrating the Environment into Other EU Policy Sectors', *Environmental Politics* 6,1, pp. 153–73.
World Commission on Environment and Development (1987) *Our Common Future*, Oxford: Oxford University Press.
Zaccaï, Edwin (2002) *Le Développement Durable: Dynamique et constitution d'un projet*, Brussels: P. I. E. Peter Lang.
Zadek, Simon (2001) *The Civil Corporation: The New Economy of Corporate Citizenship*, London: Earthscan.

annotated bibliography

Agyeman, Julian, Robert D. Bullard and Bob Evans (eds) (2003) *Just Sustainabilities in an Unequal World*, Cambridge, Mass.: MIT Press. This book has a rather pragmatic approach to sustainable development. It discusses possibilities of implementing sustainable development in the South in light of the more fundamental debate about North–South dimensions of global development. As suggested by the 'just' in the title, it also has a normative approach. The main point is that the North is overwhelmingly responsible for the unsustainability of global development and has the moral duty to tackle some of the basic issues in partnerships with the South. This book is excellent because it represents an important subdiscourse of sustainable development and links it to concrete actions in the South. It provides an introduction to elements such as participation, ownership and stakeholder approaches which are particularly relevant in debates about the North–South dimension of sustainable development.

Gupta, Joyeeta (2002) 'Global Sustainable Development Governance: Institutional Challenges from a Theoretical Perspective', *International Environmental Agreements: Politics, Law and Economics* 2, pp. 361–88. This article discusses the different views on global governance for sustainable development. It starts from the hypothesis that in order to reach global sustainability, a system of governance will have to be set up. In an initial section the article gives a nice overview of the different efforts at the international level (especially the UN) to come to functioning institutional arrangements on sustainable development. In a second part, Gupta discusses these efforts in light of the main international relations theories on global institutionalization. The governance debate is placed in the context of neorealism, neo-institutionalism, idealistic supranationalism and historical materialism. In this way it links three important analytical elements: sustainable development as a policy discourse, governance as a sort of normative framework, and IR theories about global dynamics. It is one of the very few publications that does this in such a coherent, concise and systematic way.

Keekok, Lee, Alan Holland and Desmond McNeil (eds) (2000) *Global Sustainable Development in the 21st Century*, Edinburgh: Edinburgh University Press. This book is a great introduction to both the concept of sustainable development and a number of the debates that are attached to it. Several of the most important European authors have contributed to this volume. In separate chapters a description of the main issues of environmental, social and economic aspects of sustainable development are discussed. In addition, significant attention is paid to the North–South dimension of sustainable development. In a final part there is a good discussion on the practical implications of sustainable development as a policy concept.

Paehlke, Robert (2001) 'Environmental Politics, sustainability and Social Science', *Environmental Politics* 10, 4, pp. 1–22. Robert Paehlke, one of the important social science authors on environmental issues and sustainable development, strongly defends the importance of social science in the intellectual debates on sustainable development in this article. He claims that solid social science could have a serious impact on policy practices as well. The inherent complexity of fundamental social change in the direction of sustainable development needs, or requires, social science analysis, according to Paehlke. Another major point is that social scientists should engage more in public debates and advocacy because complex issues require a knowledgeable and policy relevant translation into daily political discourses.

Wackernagel, Mathis, and David Yount (2000) 'Footprints for Sustainability: The Next Steps', *Environment, Development and Sustainability* 2, pp. 21–42. The ecological footprint metaphor is used very often in debates on environmental issues and sustainability. The idea that each individual has some kind of fair share of the earth's resources links rational elements about the availability of these resources to essential debates about normative and distributive issues on a global scale. The concept introduced by Rees and Wackernagel about ten years ago has been used as a heuristic, a pedagogical tool, a normative framework and an analytical tool. In this article, one of the original authors and his co-author look back on the experience with the concept and offer a number of empirical and methodological improvements, which are placed in current policy relevant debates. The article is a great introduction to a much-used concept, and how the debate surrounding its use has evolved.

11
the effectiveness of environmental policies[1]
jørgen wettestad

Much of my work in this field has been conducted as a member of arguably the first major international project on the effectiveness of international environmental regimes, led by Ed Miles and Arild Underdal. The final book from this project was published in 2002 (Miles et al., 2002). Hence it will come as no surprise that I will use the basic framework which was used in this project to structure this chapter also.

More specifically, in the first section of this chapter I will sum up central contributions to the study of the effectiveness of international environmental regimes. I will first focus on the *measuring* of effectiveness. With Underdal's distinction between a problem-solving perspective (that is, 'distance to collective optimum') and a more political and institutional perspective (that is, 'relative improvement') as a conceptual backdrop, I discuss three major 'waves' in the development of the understanding of what constitutes 'effectiveness' in this context. I will then turn to the *explaining* of effectiveness and particularly the identification of promising institutional techniques to enhance regime effectiveness. Also in this venture I take the major perspectives from the Miles et al. project as a point of departure; that is, the distinction between 'institutional and problem-solving capacity' versus 'characteristics of the problem(s)'. I put forward and discuss a 'top five' list of central problematic characteristics of or obstacles to the improvement of effectiveness and some central related institutional cures and techniques.

The second section then seeks to put the theoretical perspectives and insights from the first section into practice by carrying out a brief empirical case study of the Convention on Long-Range Transboundary Air Pollution (CLRTAP). The purpose here is to provide a more systematic empirical illustration of the concepts and perspectives presented in the first section, not to fully explore all aspects of the complex functioning of this fascinating regime.

The third section winds up the chapter with some concluding reflections. Have we managed a significant relative improvement of knowledge? Are we still far from intellectual 'problem-solving'? What are the main interesting topics for further research?

Before I embark upon my reading of what has been written within the field of international environmental regime effectiveness in the recent decade or so, I need to emphasize that I will not be covering the rich and interesting field of studies focusing upon regime formation.[2] Regimes can also be evaluated according to criteria such as fairness, equity, legitimacy and robustness.[3] This line of thinking will not be further pursued in this context. It can just generally be noted that such concerns are of course driving factors behind such developments as differentiated policy commitments and funds for financial and technology transfer from North to South in international environmental politics. These particular developments will be further discussed at the end of the first section. A main point of the literature summaries in section one is to stimulate further reading as I am unable to do these rich studies full justice in this context.

central contributions to the study of the effectiveness of international environmental regimes[4]

measuring effectiveness: the three waves

the first wave: introducing the central perspectives of problem-solving and behavioural change

A very important idea first put forward by Arild Underdal at the beginning of the 1990s (Underdal, 1990, 1992) was that there are at least two crucial and quite different ways to conceptualize and measure the effectiveness of international regimes: 'distance to collective optimum' and 'relative improvement'. With regard to the first *'distance to collective optimum'* perspective, this is 'the appropriate perspective if we want to determine to what extent a collective problem is in fact "solved" under present arrangements' (1992, p. 231). So when authors such as Kütting (1999) maintain that early effectiveness studies ignored the environment and the ecological problem-solving effectiveness, this author disagrees.[5]

But what is then a 'problem solution', that is, what constitutes the collective optimum and the maximum that can be achieved? Whenever we are dealing with collective decisions that can only be made through agreement, this would be the Pareto frontier 'when no further increase in benefits to any party can be obtained without thereby leaving one or more prospective partner(s) worse off' (Underdal, 1992, p. 233). However, realizing that trying to apply this criterion to specific cases was very difficult,[6] Underdal suggested focusing instead upon more easily measurable proxies. As one sensible rule of thumb, he suggested looking for independent expert advice indicating to decision-makers what the (technically) 'perfect' solution would be. When expert advice could not be found (or experts heavily disagreed!), his proposed fallback strategy was to look for some official declaration of a joint goal or purpose (for example limit and reduce transboundary air pollution) to serve as a point of reference.

Turning then to the '*relative improvement*' perspective, this is 'clearly the notion we have in mind when considering whether and to what extent "regimes matter"' (ibid., p. 231). This is basically a counter-factual perspective. What is the hypothetical 'state of affairs' that would have obtained if, instead of the present regime, we were left in a 'no regime' condition? However, similar to the challenge of measuring the distance to the collective optimum, carrying out a regime counter-factual is a tall order as there are a number of factors at the international, national and subnational levels to take into consideration. Underdal advises us to look for whatever predictions we can find in negotiation documents, preferably documents that can be seen as 'non-partisan' inputs. Other than that, the task of determining what would otherwise have happened simply calls for the best judgement that the analyst can produce, on the basis of available sources.

Seen together, these perspectives are clearly complementary. As a key insight, Underdal states that 'even a regime leading to substantial improvement may fall short of being "perfect"' (ibid., p. 231). Moreover, he elaborates the relationship between the perspectives as seen in Figure 11.1.

	Distance to collective optimum	
	GREAT	SMALL
Relative improvement HIGH	Important, but still imperfect	Important and (almost) perfect
LOW	Insignificant and suboptimal	Unimportant, yet (almost) optimal

Figure 11.1 Dimensions of regime effectiveness

With Underdal's seminal concepts as a backdrop, let us then sum up the further evolution of this field. Although I do not think the first wave of effectiveness studies 'ignored the environment', I think it is quite appropriate to say that this first wave of projects and studies was in a sense both overambitious (with regard to measuring problem-solving so far) and gave far too little attention to domestic matters and implementation (see Andresen and Wettestad, 2004). As witnessed by the report published by Steinar Andresen and myself in 1991 (Wettestad and Andresen, 1991), much focus was given to the strength of protocols and international outputs. As we noted in our more recent reflections on the development of the field (Andresen and Wettestad, 2004), the reason for this was partly that important implementation processes had just started in several of the regimes which were studied. And as some of these regimes were less than a decade old, the ink had only just dried on important protocols and declarations. Thus some of the cases were simply premature in terms of tracing behavioural impacts.

So let us then turn to the 'Institutions for the Earth' project, which certainly has been a much cited and influential project (Haas et al., 1993). How did this project handle the crucial challenge of measuring effectiveness? In comparison with the two central dimensions put forward by Underdal, the 'Institutions for the Earth' project group put most emphasis on the 'relative improvement' perspective. The editors noted that

> truly effective international environmental institutions would improve the quality of the global environment. Much of this activity, however, is relatively new, and on none of the issues discussed in this book do we yet have good data about changes in environmental quality as a result of international institutional action. *So we must focus on observable political effects of institutions rather than directly on environmental impacts.* (ibid., p. 7, emphasis added)

Moreover, a central tool was the use of 'hypothetical counterfactual analysis' (ibid., pp. 18, 19). However, the most noted contribution of this project was the describing of three specific mechanisms through which regimes could improve effectiveness, the three C's – that is, increasing governmental *Concern*, for instance, by helping to improve scientific evidence and serving as magnifiers of public pressure; enhancing the *Contractual* environment, for instance, by providing bargaining forums for states and by providing monitoring and verification services; and increasing national *Capacity*, for instance, by providing technical assistance and aid. These were useful, evocative and certainly influential concepts. However, in terms of specific regime design, the concepts were quite vague and hence the conclusions from the project offered only limited precise advice for practitioners and analysts.

So it was another project started about at the same time and led by Oran Young that more forcefully put focus on and elaborated what may be termed the 'behavioural' dimension of regime effectiveness.[7] According to Young and Levy (1999, p. 1), 'a regime that channels behavior in such a way as to eliminate or substantially ameliorate the problem that led to its creation is an effective regime. A regime that has little behavioral impact, by contrast, is an ineffective regime.' Hence they established behavioural impact as a *necessary condition* for a regime to be counted as effective. This perspective was rather implicit in the aforementioned model put forward by Arild Underdal. Hence, spelling out the behavioural part much more explicitly was a significant contribution to the thinking within the field at this point in time (that is, the early 1990s). More specifically, Young and his collaborators offered helpful clarifications of various effects of regimes (for example, internal versus external and direct versus indirect effects). Moreover, as the most significant contribution, they suggested some important *behavioural mechanisms/pathways* through which international regimes could influence actors and processes in the domestic contexts. Compared to the more inductively discovered mechanisms put forward by Haas et al., the Young team's six mechanisms were more firmly theoretically grounded (ibid., p. 21). Let us briefly sum up these mechanisms.

- First, regimes can function as *modifiers of the utility functions* of actors, by increasing the costs and/or benefits related to certain ways of action. For instance, 'there can be no doubt ... that the costs of trying to avoid the use of SBT [segregated ballast tanks] and COW [crude oil washing] technologies have risen sharply with the establishment of rules spelling out equipment standards as part of the oil pollution regime' (ibid., p. 22).
- Second, regimes can function as *enhancers of cooperation* by mitigating the collective-action problems that stand as barriers to the realization of joint gains. For instance, fears of free-riding can be reduced by verification and monitoring and hence enhancement of transparency.
- Third, regimes can function as *bestowers of authority* upon implementing agencies and other central domestic actors. Hence, 'it is the normative status or the authoritativeness of regime rules

and activities that triggers the behavioral response rather than some calculation of the anticipated benefits and costs associated with different options available to decision-makers' (ibid., p. 24).
- Fourth, regimes can function as *learning facilitators*. 'The learning in question can take the form of new perspectives on the nature of a particular problem to be solved, new ideas about measures likely to prove effective in solving the problem at hand, new insights into the process of implementing these measures, or new solution concepts for larger classes of problems to which the specific case belongs' (ibid.).[8]
- Fifth, there is the function of '*role definers*', as actors take on new roles under the terms of institutional arrangements. For instance, 'the enhanced role of coastal states helped Norway and Russia phase out third-party fishing in the Barents Sea in a relatively noncoercive manner, and the growing strength of coastal and port states in contrast to flag states appears to be a factor of some significance in the case of oil pollution' (ibid., p. 26).
- Sixth, regimes can function as '*agents of internal realignments*'. This mechanism relaxes the unitary actor assumption and focuses on how regimes can affect behaviour by creating new constituencies and/or or shifting the political balance among domestic factions or subgroups. For instance, 'the advent of equipment standards under MARPOL [International Convention for the Prevention of Pollution from Ships] in the oil pollution regime has clearly affected the relative strength of various constituencies that seek to influence the actions of tanker owners and operators' (ibid., p. 27).

In their conclusions, Young et al. found that all six mechanisms had some role to play in making regimes effective. But their significance was not uniform across the set of focused regime cases, 'and their operation in specific cases is often more complex than simple models would lead one to believe' (ibid., p. 260).

the second wave: specifying the behavioural part by studies of domestic implementation

The next wave of research, with projects starting in the mid-1990s, helped to sort out some of this complexity by carrying out a number of case studies on the domestic implementation of international environmental commitments. The following brief summary will present some important characteristics and findings of these projects, but, as stated earlier, it is not possible really to do these comprehensive and rich studies full justice in this context.

In the project led by Edith Brown-Weiss and Harold K. Jacobson (1998), a key idea was 'engaging countries'.[9] As they stated: 'engaging countries means engaging *all relevant actors* to promote compliance ... A strategy of compliance must look *beyond governments* to provide incentives and pressures for all relevant actors to comply with the environmental agreements' (Brown-Weiss and Jacobson, 1999, p. 44, emphasis added). In the light of the six behavioural mechanisms put forward by Young and Levy (1999), this project particularly highlighted how regimes can function as 'agents of internal realignments'. Hence, considerable attention was given to the role of 'the international environment' (for example, major international conferences and INGOs) and domestic NGOs in the strengthening of compliance/implementation.[10] Moreover, Brown-Weiss and Jacobson emphasized how international financial and technical assistance could modify the utility function and enhance implementation in weak parties. It should also be noted that the project was one of the first to contribute systematic knowledge on environmental policy implementation processes in key developing countries such as Brazil, China and India (and also a small developing country, Cameroon). The Soviet Union/Russia was also included in the set of country cases.

With a focus on the Convention on Long-Range Transboundary Air Pollution (CLRTAP), the project led by Ken Hanf and Arild Underdal launched three models of how to understand both policy formation and implementation performance: the Unitary Rational Actor (URA) model; the 'Domestic Politics' model; and the 'Social Learning and Policy Diffusion' model (Underdal and Hanf, 2000).[11] If we again compare with the Young project's mechanisms, these three models elaborated several of the mechanisms. The URA model brought together elements both from the Young project's 'utility modification' and 'cooperation enhancement' mechanisms, as it focused on actors' cost-benefit calculations and how monitoring and transparency could increase the costs of defection and non-compliance. The domestic politics model considerably enriched and systematically fleshed out elements hinted at in the Young project's 'agents of internal realignments' mechanism. As the model also drew attention to how the signing and ratifying of international agreements empowered certain governmental agencies vis-à-vis other governmental and societal actors, there was also a link to Young et al.'s idea of regimes as 'bestowers of authority'. The social learning and policy diffusion model then touched upon the 'learning facilitation' mechanism. Among other things, this model focused upon the potential role of transnational networks of experts ('epistemic communities') in building consensual knowledge and shared social norms among the regime parties.

When interpreting the empirical evidence, Underdal and Hanf found patterns overall consistent with both the URA model and the domestic politics models. However, the latter enabled a deeper penetration of the processes. Hence 'the more specific the aspect of behaviour that we want to predict or explain, the greater the marginal utility of moving beyond the narrow confines of the unitary, rational actor model' (ibid., p. 377). With regard to the third model (that is, social learning), it also made a useful contribution, as 'it is abundantly clear that knowledge and ideas played very important roles in the development of the LRTAP regime' (ibid.). As to empirical evidence on the URA and domestic politics models in the environmental policy context, the dissertation project on North Sea cooperation carried out by Jon B. Skjærseth should clearly also be mentioned as an outstanding example of a deep-diving investigation of the causal chain from international commitments right down to (among other things) farmers' practices in the UK, Netherlands and Norway (Skjærseth, 2000).

The International Institute for Applied Systems Analysis (IIASA)-based project led by David G. Victor and Eugene Skolnikoff improved knowledge on these issues on three accounts (Victor et al., 1998): first, specifying elements in the domestic politics model described above, several in-depth case studies shed light on how access procedures and participation patterns affected domestic implementation processes. Second, building upon the emphasis placed on transparency and implementation review by, among others, the 'Institutions for the Earth' group and the Young group, several analytical and empirical contributions on implementation review enhanced knowledge of this issue. Third, several chapters on Russian and Eastern European environmental policy implementation considerably enhanced knowledge on these important actors.[12]

Finally, as the last of the big projects initiated in this phase, the project led by Bill Clark focused on the concept of social learning (Clark et al., 2001a, 2001b). Hence, it eventually produced more in-depth knowledge on the (social) learning issue included in the Young and Hanf and Underdal projects. This included some knowledge on implementation processes, but the project covered the whole policy cycle, from early problem framing to implementation and evaluation.

the third wave: broadening the methodological palette, rediscovering problem-solving, and bringing in institutional interaction

The most recent wave of research is very much methodologically driven. There is a drive to refine the earlier largely qualitative case evidence by developing and carrying out more quantitatively oriented studies. In a way, Miles et al. (2002) can be seen as part of this drive, as the final phase of the project augmented the existing qualitative evidence within the project with a certain quantitative element – for the first time in this field.[13] The purpose was to enable a systematic examination of patterns across cases. Two techniques were applied: first, a dichotomized 'truth table' was produced and analysed in accordance with the principles of Boolean logic. The other data file produced contained interval and ordinal scale data analysed by simple statistical techniques.[14]

Others seek to go down the quantitative avenue more firmly. Detlef Sprinz has been a driving force in this effort (for example, Helm and Sprinz, 2000; Sprinz, 2003). Helm and Sprinz's (2000) measure of regime effectiveness establishes both an empirical lower bound of performance and an upper bound, and then relates the actual level of performance to both of them – thereby producing a simple coefficient of regime effectiveness. Ron Mitchell (2002) suggests replacing the systemic level of analysis with the analysis of yearly country-level performance. Analysing the country-level data, he suggests using regression analysis. Mitchell winds up by indicating that 'intermediate models specified to explain the variation in the dependent variable across a set of regimes that are selected for similarity in their predicted impacts may reach the right balance between … too-generic and too-specific extremes' (2002, p. 80). In 2003, Hovi et al. launched the 'Oslo–Potsdam solution' to measuring regime effectiveness (Hovi et al., 2003a). What the 'Oslo' (that is, Miles et al., 2002) and 'Potsdam' (for example, Helm and Sprinz, 2000) approaches have in common 'is that the basic components of the analysis are conceptually identical, namely measures of the no-regime counterfactual, actual performance, and the collective optimum' (Hovi et al., 2003a, p. 77).[15] This has initiated an interesting debate (see Hovi et al., 2003b; Young, 2001, 2003).

There have also been some quite recent studies seeking to 'bring the environment back in', i.e. giving more weight to environmental conditions and problem-solving as a measuring rod for the effectiveness of international collaborative efforts (for example, Kütting, 1999). Among other things, Kütting ends up with a plea for placing agreements and regimes more clearly within the social and ecological contexts in which they operate (ibid., p. 134). This can perhaps be seen as a counter-reaction to the number of implementation projects and studies summed up above, which have very much flowed from the standpoint that we cannot measure regimes' effects on environmental conditions because the causal chain is too long and complicated. Hence, we must concentrate on behavioral change.

With regard to the plea for placing regimes more clearly within the social context in which they operate, the upsurge of studies on institutional interaction can be seen as a promising development.[16] Hence, increasing attention has been given to the phenomenon that measures taken in one collaborative context may counteract – or strengthen – the effect of measures taken in other collaborative contexts (for example, Oberthür and Gehring, 2003, forthcoming; Rosendal, 2001; Stokke, 2001; Young, 1996). This is of course related to the generally increasing institutional density which has taken place in the field of international environmental and resource cooperation – with one problem after another being addressed by specific regimes. A prominent example is the relationship between the World Trade Organization (WTO) that promotes free international trade and several multilateral environmental agreements that establish trade restrictions, such as the 1973 Convention on International Trade in Endangered Species of Wild Fauna and Flora (CITES) and the 1987 Montreal Protocol on Substances that Deplete the Ozone Layer. Although the initial attention within this field of studies was focused on such cases of problematic interaction, much (and perhaps most) interaction is positive and synergistic (Oberthür and Gehring, 2003).

shedding light on differing levels of effectiveness and identifying promising institutional techniques

With regard to the explaining of effectiveness and the particularly intriguing question of institutional techniques for improving effectiveness, as indicated in the introduction, I always find the two fundamental perspectives of 'problem characteristics' and 'problem-solving capacity' helpful. Problem characteristics are fundamental aspects of the environmental problems addressed by the regimes. For instance, is the underlying collaborative problem one of complicated transboundary effects and competition over collective goods, implying the need for painful and often costly behavioural changes, or is it a more simple coordination problem where modest behavioural adjustments will do? Moreover, to what extent is knowledge about the problem uncertain and disputed? Problem-solving capacity is then a combination of the institutional efforts established and the entrepreneurial efforts made to address and hopefully solve the environmental or resource problems.[17] A core idea is that some regimes are more effective than others either because the problems they deal with are more benign – or because they are addressed by more effective problem-solving instruments and efforts.

Hence, institutional design and 'techniques' form a central part of problem-solving capacity – and such techniques and smart ways of designing international institutions and commitments can obviously be seen as ways to overcome cooperative obstacles flowing from malign problem characteristics. For instance, establishing a fund for the transfer of technological know-how from North to South can be seen as an institutional technique and response to the wide variance in capabilities to establish effective abatement policies between Northern and Southern participants within a global regime. On this background, what can then be seen as the 'top five' list of central obstacles to the improvement of effectiveness and related important institutional cures and techniques?

First, a central obstacle is often that of marked differences in capabilities among the actors to establish or implement abatement policies. Hence, as indicated above, an important institutional 'cure' and technique is the establishment of a *funding mechanism*. This approach has clearly played an important role in global cooperation to protect the ozone layer. The establishment of a specific Fund in 1990 to pay for developing countries' incremental costs and some technology transfer can be seen as a watershed development within the regime in terms of securing support among developing countries for the Vienna Convention and Montreal Protocol (Benedick, 1991; DeSombre and Kaufman, 1996; Parson 2003; Wettestad, 1999). More generally, the funding and technology transfer issue was discussed in the background of a number of case studies in a project reported in Keohane and Levy (1996). Another institutional technique to deal with the problem of differing capabilities is the possibility of *differentiated commitments*. CLRTAP and the 1999 Gothenburg Protocol are very good examples, as will be further elaborated in the next section.

A second central obstacle to effectiveness is marked differences among the parties with regard to perceptions of the (seriousness of the) environmental problems. An important institutional cure and technique in this connection is the establishment of a well-functioning *international knowledge-improvement effort*. In this connection, CLRTAP is of course a very good example. But the promises – and pitfalls! – of organizing a good science–politics dialogue in the international regime context have been discussed in a number of studies (for example, Andresen et al., 2000; Parson, 2003; Skodvin, 2000). A central dilemma pointed out in Andresen et al. (2000) is to find the right balance between scientific integrity and political involvement. The establishment of a specific buffer body (or several bodies) was put forward as a promising technique by Andresen and his collaborators.

A third central obstacle to regime effectiveness is undoubtedly differences among the parties with regard to positions on how to deal with the problems. An important institutional technique to beat the 'law of the least ambitious program' (Underdal, 1980) and come up with strong and ambitious policies and protocols is the introduction of *majority-voting*. CLRTAP is not a good example in this regard, and nor are many other international regimes, in fact. Within international environmental politics, the most relevant example is probably the European Union (for example, Wallace and Wallace, 2000). But there are also other and less-demanding 'fast track' options (Sand, 1990), including the possibility of *opt-out provisions* and the establishment of smaller clubs where not all regime parties participate. The North Sea cooperation regime is a good case in this regard (for example, Skjærseth, 2000).

A fourth central obstacle to effectiveness is uncertainty among the parties with regard to domestic regulatory bite and domestic abatement possibilities more generally. A standard institutional technique here is the establishment of specific *clauses on regular renegotiations of commitments*. Both the North Sea regime and the ozone regime are good examples of this, as these regimes have developed through several stages and amendments (Skjærseth, 2000; Parson, 2003). Within the ozone regime, for instance, the 1987 Montreal Protocol came first. The requirements were then strengthened and further substances were added in the 1990 London and 1992 Copenhagen amendments.

A fifth central obstacle to effectiveness is inadequate knowledge of parties' 'real' implementation and follow-up. The general cure is of course the establishment of a reporting and verification system. However, a more specific institutional technique is *the establishment of a specific implementation/compliance committee*. CLRTAP is a good example here, as the operation of a specific Implementation Committee (IC) from 1997 on has improved reporting procedures and regime debates on implementation considerably, as will be further elaborated below. But the ozone regime established a specific IC in 1990 (Greene, 1996; Parson, 2003; Parson and Greene, 1995; Victor, 1998), and this has functioned as an important model for the climate regime (Wettestad, 2005). However, let us at this point take a break in the summary of concepts and ideas so far and see how these insights can be utilized to make sense of the empirical progress of one specific regime: the CLRTAP.

putting theory into practice: measuring and shedding light on the effectiveness of the convention on long-range transboundary air pollution

In 1968, Swedish scientist Svante Oden published a paper in which he argued that precipitation over Scandinavia was becoming increasingly acidic, thus inflicting damage on fish and lakes (Oden, 1968). Moreover, he maintained that the acidic precipitation was to a large extent caused by sulphur compounds from British and Central European industrial emissions. This development aroused broader Scandinavian concern and diplomatic activity related to acid pollution. The specific background for formal negotiations on an air pollution *convention* was the East–West détente process in the mid-1970s, in which the environment was identified as one potential area for cooperation. Due to the East–West dimension, the United Nations Economic Commission for Europe (UNECE) was chosen as the institutional setting for the negotiations.

The Economic Commission for Europe's CLRTAP was signed by 33 Contracting Parties (32 countries and the EC Commission) in Geneva in November 1979. Four main aspects of the 1979 Convention may be discerned: first, the recognition that airborne pollutants were a major problem; second, the declaration that the Parties would 'endeavour to limit and, as far as possible, gradually reduce and prevent air pollution, including long-range transboundary air pollution' (Article 2); third, the commitment of Contracting Parties 'by means of exchange of information, consultation, research and monitoring, develop without undue delay policies and strategies which should serve as a means of combating the discharge of air pollutants, taking into account efforts already made at the national and international levels' (Article 3); and fourth, the intention to use 'the best available technology which is economically feasible' to meet the objectives of the Convention.

The Convention did not specify any pollutants, but stated that monitoring activity and information exchange should start with sulphur dioxide (SO_2). The Convention has been in force since 1983 with a membership in January 2005 of 49 Parties. Moreover, the Convention was to be overseen by an 'Executive Body' (EB), which included representatives of all the Parties to the Convention as well as the EC. Furthermore, the UNECE secretariat was given a coordinating function. The institutional structure has also included several Working Groups, Task Forces and 'International Cooperative Programmes'. Rooted in the Convention's strong initial focus on knowledge improvement and monitoring, a specific financing protocol for the Cooperative Programme for Monitoring and Evaluation of Long-Range Transmissions of Air Pollutants in Europe (EMEP) monitoring programme was established in 1984.

clrtap: 'medium' – or 0.39 effectiveness?

How effective has CLRTAP been? First, in terms of policy development through the establishment of specific protocols, it is clear that the record is impressive indeed. In this connection, it is interesting to note that the initial Convention did not mention subsequent protocols at all! The protocol development can be summed up in the following manner:

- *The 1985 Protocol on the Reduction of Sulphur Emissions*. In Helsinki, July 1985, 21 countries and the EC signed this legally binding protocol. The Protocol stipulated a reduction of emissions/transboundary fluxes of SO_2 by at least 30 per cent as soon as possible, and by 1993 at the latest, with 1980 levels as a baseline. However, some major emitter states failed to join the agreement, among them the UK, the US, and Poland. The protocol entered into force in September 1987 and has been ratified by 22 Parties.
- *The 1988 Sofia Protocol on Nitrogen Oxides (NO_x)*. Here, the signatories pledged to freeze NO_x emissions at the 1987 level from 1994 onwards and to negotiate subsequent reductions. Twenty-five countries signed the protocol, including the UK and the US. Moreover, 12 European signatories went a step further and signed an additional (and separate) joint declaration committing them to a 30 per cent reduction of emissions by 1998. The protocol entered into force in February 1991 and has been ratified by 28 Parties.
- *The 1991 Geneva Protocol on Volatile Organic Compounds (VOCs)*. VOCs are a group of chemicals which are precursors of ground-level ozone. The protocol called for a reduction of 30 per cent in VOC emissions between 1988 and 1999, based on 1988 levels – either at national levels or within specific 'tropospheric ozone management areas'. Some countries were allowed to opt for a freeze of 1988 emissions by 1999.[18] Twenty-one Parties signed the protocol in 1991 and it entered into force in September 1997. It has been ratified by 21 Parties.
- *The 1994 Second Sulphur Protocol* was then signed in Oslo in June 1994 by 28 Parties. This Protocol was based on the critical loads approach. The aim of this approach was that emissions reductions should be negotiated on the basis of the (varying) effects of air pollutants, rather than by choosing an equal percentage reduction target for all countries involved.[19] Hence the Protocol set out individual and varying national reduction targets for the year 2000 for half of the countries, and additional 2005 and 2010 targets for the other half – with 1980 as the base year. The protocol entered into force in August 1998 and has been ratified by 18 Parties.
- Two new protocols on transboundary air pollution by *heavy metals* and *persistent organic pollutants* (POPs), which were signed by 34

Parties in Århus in June 1998. They have been ratified by 24 and 22 Parties respectively and entered into force in late 2003.
- *The 1999 Protocol to Abate Acidification, Eutrophication and Ground-level Ozone* was signed in Gothenburg in December 1999 by 31 Parties. This Protocol is by far the most advanced within the regime so far and covers four substances (NOx, VOCs, NH_3 [ammonium] and SO_2) and three environmental effects (acidification, ground-level ozone and eutrophication). The Protocol establishes varying reduction targets for all Parties involved (Wettestad, 2002a). Fourteen countries had ratified the treaty by January 2005 and it is not yet in force.

Both in my 1999 book (Wettestad, 1999) and the Miles et al. project (Wettestad, 2002b) I landed on an overall 'medium' effectiveness score for the regime. The basic reasoning goes like this: on the one hand, as has been summed up above, a substantial international regulatory progress has taken place. There has been a steady development of protocols, covering more substances with regulations gradually becoming both binding and specific and more fine-tuned to ecological and economic variations between the countries. Moreover, national compliance with these protocols must be characterized overall as high. Particularly the work on reducing sulphur emissions has been marked by substantial overcompliance. In fact, sulphur emissions in Europe have been reduced by 71 per cent in the period 1980–2000!

Then there is the (social) learning dimension (cf. above; also Clark et al., 2001a, 2001b; Underdal and Hanf, 2000; Young and Levy, 1999). It is clear that CLRTAP has been very important as a forum for organizing the production and dissemination of economic and natural scientific knowledge. Moreover, it has also been important as a forum where bureaucrats, researchers and NGOs from various countries have met and learnt from each other. It is very tricky to assess the exact value of this. It has clearly been an important factor for driving the regime forward and producing scientific reports and protocols. But, as elaborated more below, in the implementation processes, other and more substantial economic and social forces have generally come more to the forefront. Still, as has been pointed out by one experienced regime analyst, 'it is hard to imagine being where we are today regarding transboundary air pollution in the absence of the LRTAP process'.[20] So the regime has clearly contributed to the reductions in emissions witnessed, and the promising steps taken towards problem-solving.

On the other hand, a closer scrutiny of available national and sub-national knowledge indicates that many forces other than the CLRTAP regime have been involved in bringing about policy changes and emissions reductions in this field. Much behavioural change would probably have happened 'anyway' – due to energy policy changes, more fundamental economic and industrial changes, European Community/Union processes, and so on – even if the picture clearly varies between countries.[21] Take, for instance, the sulphur and NO_x implementation processes in the UK, Germany, the Netherlands and Norway. Aided not least by knowledge produced within the context of the Hanf and Underdal project discussed earlier, it can be concluded that the majority of the initial reductions and compliance levels achieved in three of the four countries (the UK, Netherlands, and Norway) are apparently explained by processes that are not primarily related to environmental protection, at least with regard to sulphur reductions (Underdal and Hanf, 2000; Wettestad, 1996; 1998). In the UK, industrial recession and reduced energy demand in the 1980s were important factors. Moreover, the privatization and the switch from coal to gas were also important factors. In the Netherlands, a gradual conversion to domestic natural gas related to domestic political and financial reasons was clearly important. In Norway, much was achieved by, among other things, reducing consumption of heavy fuel oil on land (Wettestad, 2004). The exception is Germany, where environmental regulations were – at least initially – the main driving forces.

Focusing more closely on the CLRTAP contribution to these processes, the direct, easily detectable influence has been moderate, at least for these particular countries. In a situation without the CLRTAP, significant initial reductions would probably have taken place anyway, due to other economic and political processes and domestic political pressure motivated by environmental damage. Take, for instance, Germany. It is important to remember that Germany was almost as reluctant as the UK at the Convention negotiations in the late 1970s. Without the rapidly increasing concern about forest damage/*Waldsterben* in the early 1980s, both German and European acid rain politics would have looked very different today. Moreover, with regard to German and British acid rain politics, the European Community decision-making arena has possibly been more important than the CLRTAP.[22]

However, there is still a possibility that the CLRTAP has been more important for the other Western and not least East European countries. This is for example indicated by Levy (1993, pp. 118–21) who suggests that countries like Austria, Finland, the Netherlands and Switzerland were influenced by the CLRTAP through increased awareness of domestic acid rain damage. Moreover, countries like Denmark, the UK, and the

Soviet Union were influenced by the CLRTAP through various types of linkage effects.

Turning then briefly to the 'distance to collective optimum' and hence the question of environmental optimality and problem-solving, there was a widespread feeling in the scientific community that the targets in the first rounds of protocols were ecologically ineffective, but were, however, steps in the right direction.[23] The adoption of the most recent and most ambitious protocol in CLRTAP history – that is, the 1999 Gothenburg Protocol – points towards a considerable improvement. If it is implemented faithfully, substantial environmental improvements are within reach. With regard to acidification, critical loads modelling indicate that critical levels were being exceeded in around 32.5 million hectares of ecosystem area in 1990. In comparison, implementation of the Gothenburg Protocol and the National Emissions Ceilings (NEC) Directive will reduce this area to around 4.4 million hectares in 2010. Likewise, with regard to ozone, instances where World Health Organization (WHO) guidelines for protecting human health are exceeded will be reduced by around 70 per cent between 1990 and 2010. Moreover, the other side of the coin of these figures is, of course, that some vulnerable areas will *still* suffer from acidification in 2010, and WHO guidelines will still be exceeded many places.

All in all, although clearly highly complex and complicated, it is still tempting to give the regime a 'medium' score – not least as relative improvement brought about *specifically by the regime* seems only moderate so far, and there is still a significant distance to optimal air quality and environmental conditions in Europe. Compared to this, Helm and Sprinz's assessment of 0.39 effectiveness for the SO_2 regime and 0.31 for the NOx regime is certainly eye-catching and attractive. As discussed in section one, the quantitative approach to the study of regime effectiveness clearly has its merits. But given the considerable uncertainty with regard to the true impact of CLRTAP in relation to a number of other social forces even in well-studied countries such as Norway, I still think that such figures function best as contributions to the internal debate within the field. For external purposes, I fear that the use of such figures may convey to policy-makers a far too optimistic message about the state of knowledge within this field.[24]

Shedding light on medium success: malign problems – but softened by some effective institutional techniques

In this section, I give a summary overview of how the central explanatory perspectives of problem characteristics and problem-solving capacity can shed light on the 'medium' effectiveness of CLRTAP, then I turn to the specific institutional 'cures' and techniques outlined earlier.

My general assessment is that CLRTAP's medium success so far has much to do with an initially strongly malign case to deal with: first, emissions of the pollutants involved stem from societally important activities related to energy production and consumption, industrial processes and transport, with related powerful target groups for regulators to deal with (see also Wettestad, 1999, 2002b). Moreover, there were generally technologically complicated and quite expensive abatement options, and a lack of technological breakthroughs. Third, and not least important, there was a strong asymmetry in the transboundary flow of pollutants, with some nations being net importers and some net exporters. This situation was furthermore worsened by an asymmetrical vulnerability to air pollutants, with unfortunate combinations like the cases of Norway and Sweden, both being considerable net importers of pollutants and having particularly vulnerable soil characteristics. In addition, one must not forget the East–West context which increased the need for delicate diplomatic balancing acts and consensual and 'non-intrusive' processes. However, for understanding the degree of success the CLRTAP regime has after all achieved, not least symbolized by an impressive regulatory development in the 1990s, the catalytic event is definitely Germany's turnabout in 1982 related to the domestic '*Waldsterben*' uproar over forest damage. Germany's shift from laggard to leader represented a crucial and symbolic lasting shift in the power balance between reluctants and pushers within the regime.

Turning then briefly to the issue of problem-solving capacity, this must overall be characterized as moderate, although increasing over time. The moderate element is clearly reflected in institutional aspects like a limited and stable secretarial capacity (in a period where several protocols and tasks have been added to the regime) and a consensual decision-making style. However, it should be noted that the consensus requirement has been exercised with some flexibility. Reluctant countries have simply not signed the protocols and hence not held back the rest of the countries. Perhaps the most important institutional contribution to effectiveness so far has been the evolution of the scientific working groups and hence the 'scientific-political complex'. As indicated earlier, this has contributed to considerable social learning among the parties. More about these institutional aspects follow below.

The strength of entrepreneurial leadership has increased over time, primarily related to the catalytic change in German acid rain policies as described above. German leadership has added considerable political weight to these processes, and not least leadership continuity at the point in the regime development process where the interests of several Nordic countries got much more complicated and the initial Nordic leadership coalition broke down (that is, from the mid-1980s on). This breakdown symbolizes the fact that even if procedural leadership has been strengthened over time and general problem-solving capacity improved, the basic interests of many countries have continued to be complicated in this issue area, both with regard to domestic regulatory capacities and international competitive aspects.

Let us then elaborate a little bit more the relevance of the more specific institutional techniques described at the end of section one. As can be recalled, a central obstacle is often marked differences in capabilities among the actors to establish or implement abatement policies, resulting in the establishment of a specific *funding mechanism*. Within CLRTAP, this difference has been most prominent in the difference between Western and Eastern actors within the regime. Although this mechanism has not played a central role within CLRTAP, it has in fact been used there also. In order to facilitate full involvement in the negotiations of countries with economies in transition, a Trust Fund for Assistance to Countries in Transition (TFACT) was established by the CLRTAP Executive Body in 1994 (Selin, 2000, p. 137). Donor countries, mainly from Western Europe, deposited money into TFACT, the Executive Body decided on its use, and the secretariat was authorized to offer funding to one government-designated expert from each qualified country. According to well-informed sources, this arrangement contributed to wide participation in the negotiations leading up to the 1999 Gothenburg Protocol.[25]

Another institutional technique to deal with the problem of differing capabilities is the possibility of *differentiated commitments*. This is clearly very relevant in the CLRTAP context, as CLRTAP commitments over time have become steadily more differentiated. As indicated in the earlier overview of CLRTAP policy development, the first step away from the 'common cuts' approach was taken in the 1991 VOC Protocol. The differentiation was taken a significant step further in the 1994 Second Sulphur Protocol, where the concept of critical loads for the first time in CLRTAP's history was used as a foundation for policy-making. Then this whole exercise was developed considerably in the work on the 1999 Gothenburg Protocol. Here, the outcome was a myriad of differing commitments for the countries involved, related to four substances (Wettestad, 2002b).

Let us now turn to the obstacle to effectiveness stemming from marked differences among the parties with regard to perceptions of the (seriousness of the) environmental problems. An important institutional cure and technique in this connection is the establishment of a well-functioning *international knowledge-improvement effort*. In this connection, CLRTAP is of course a very good example (Wettestad, 2000). What are then the main aspects of the CLRTAP model? First, the basic flexibility of the system must be noted. As pointed out by Levy (1993), the fact that the CLRTAP has been 'consistently science- and ecosystem-driven' means that working groups have progressively been organized around potential environmental damage, and permitted transfrontier pollutants to enter onto the diplomatic agenda: 'This accounts for the ease with which VOCs entered the agenda, as well as for the current investigations into mercury and persistent organic compounds' (Levy, 1993, p. 111).

Second, the formally advanced (that is, financing based on a separate, specific protocol) and well-functioning EMEP system has represented a strong scientific foundation and 'core' in the development of the regime. A third interesting element in the CLRTAP model is the establishment of a permanent negotiating forum in the Working Group on Strategies (WGS). This body may be seen as a mediating buffer between science and politics – a 'not too formal' meeting-place for scientists and administrators, allowing the building of consensual knowledge on both scientific and political strategic matters (cf. the buffer idea put forward in Andresen et al., 2000). Regime participants emphasize the flexibility in frequency of meetings and generally much less time-consuming formalities as an advantage of the WGS style of functioning compared to the Executive Bureau meetings.

The next central obstacle to regime effectiveness with relevance for CLRTAP is uncertainty among the parties with regard to domestic regulatory bite and domestic abatement possibilities. As indicated, an interesting institutional technique here is the establishment of specific *clauses on regular renegotiations of commitments*. CLRTAP is a good example of this, where SO_2, NO_x and VOCs have been negotiated and renegotiated in light of improved knowledge on environmental conditions and regulatory instruments. In this process, the initial basically flat-rate reduction commitments have been replaced with the much more ambitious and differentiated commitments in the 1999 Gothenburg Protocol.

Finally, there is the obstacle to effectiveness consisting of missing and diffuse knowledge about Parties' 'real' implementation and follow-up, addressed by the establishment of a reporting and verification system and specifically *the establishment of a specific implementation/compliance committee*. As indicated, CLRTAP is an interesting case in this regard

(Wettestad, 2005). The first step was taken in connection with the 1991 VOC Protocol, which required the parties to establish a mechanism for monitoring compliance with the Protocol. This call was followed up and made more specific in the 1994 Second Sulphur Protocol. Article 7 formally established a specific Implementation Committee. The committee's mandate was to review implementation and compliance, including decisions on 'action to bring about full compliance with the protocol'.

The more specific and practical establishment of the Implementation Committee took place in 1997. At this point, the mandate was broadened to cover the review of compliance with all the CLRTAP protocols. Within environmental politics in general, the implementation committee established within the ozone-layer regime clearly served as the institutional model. So this is an example of interinstitutional learning. Apart from 'institutional diffusion' from the ozone regime, important background factors for tougher compliance procedures in CLRTAP are the fundamental changes in the East–West relationship and greater openness in the East. These changes have provided a much more beneficial setting for critical follow-up discussions, both at the international and national levels, than was the case in the 1970s and 1980s. In addition, the increased regulatory sophistication over time has created an increased need for improved institutional procedures.

The CLRTAP Implementation Committee is composed of eight legal experts from the Parties. At its first meeting, it was decided that it would take all decisions by consensus and any report to the Executive Body on a specific Party would first be shown to 'and if necessary discussed with' that Party. The important role of the Secretariat was also indicated, and given the committee's limited own resources, the function of the Committee was to reach its conclusions on the basis of analyses carried out by the secretariat or experts. Initially, the Implementation Committee concentrated on reviewing reporting procedures and practices. Part of this work consisted of publishing overview tables of reporting 'scores'. So a main part of the work, and hence what may be characterized as its 'compliance strategy', has been to increase transparency.

The committee has adopted a strategy of gradually increasing highlighting in cases of non-compliance with reporting obligations: the first time, the case is noted without highlighting the Party's name; if it happens again, the Party's name is revealed; and the third time around, the committee includes the Party in a recommendation to the Executive Body urging it for action to achieve reporting compliance. The committee has also systematically reviewed compliance with the various protocols. With regard to the functioning of the Implementation

Committee, experienced negotiators emphasize the value of having both technical and legal expertise represented. All in all, the Implementation Committee has improved reporting procedures and regime debates on implementation considerably.

concluding comments: significant relative improvement of knowledge, but far from intellectual problem-solving?

Summing up, many core insights with regard to the study of effectiveness were launched in the first part of the 1990s. One core insight was the distinction between behavioural change and the 'relative improvement' brought about by the operation of a regime – and the extent to which the behavioural change and the regime really solve the fundamental environmental problems and hence close the gap to the collective optimum. More implementation and behavioural change within the regimes have provided richer empirical evidence. This has spurred a number of large-scale projects on compliance and implementation. It can be noted that this topic is quite unique in its research history, in the way it has been dominated by some big projects and a fairly tight-knit group of scholars in close debate with each other. More recently, a phase of reappraisal of concepts and methodological tools has started (see Hochstetler and Laituri in this volume). Among other things, a more comprehensive and 'contextual' perspective has been introduced, where various forms of interaction between regimes and international institutions are the focus of attention. Moreover, the need for a broadening of the methodological toolbox is increasingly emphasized, with more attention given to more quantitatively based methods. So in terms of 'relative improvement' of regime effectiveness knowledge I think we have not done that badly.

But looking back I think no one realized the tall order of the analytical challenges involved. Tracing the effect of a protocol established within an international regime through the national and subnational processes and through to the effects in terms of environmental improvement is extremely complicated. This is probably also a contributing factor to the fact that other interesting and important issues in this context related to concepts such as fairness and equity have not been much explored (see Parks and Roberts in this volume). Hence, in terms of knowledge on problem-solving and distance to the 'collective learning optimum', I think we are still far from a truly broad and comprehensive state of effectiveness knowledge. For instance, although we are becoming increasingly certain that regimes do matter, we really do not know that much about *how* they matter. True, some promising institutional techniques and 'cures' have been identified, as summed up in the previous sections. And as shown by the CLRTAP case study in this chapter, these techniques have contributed positively to the improving effectiveness of CLRTAP. But we need to know far more about under what conditions these techniques and cures really work. Hence, it could very well be that the debate over methods and research strategies as witnessed in the August 2003 edition of *Global Environmental Politics* can be seen as a kind of necessary reflective step back in order to make a big jump forward in the identification of effective regime design.

Finally, let me briefly outline what I see as some other interesting topics for further research. With regard to *actors*, over time, increasing attention has been given to the role of environmental NGOs in the policy-making, implementation and effectiveness phases. Although the role of industry has been given some attention too, it can be argued that these actors should be given (even) more attention ahead, not least in light of the increasing attention given to market-based flexible policy instruments (see below) and corporate social responsibility. After all, industry is a very important target group, controlling the effectiveness of policies to a far higher degree than NGOs.

With regard to *policy instruments*, a very interesting development is the increasing attention and weight given to so-called flexible policy instruments, including voluntary agreements and not least emissions trading. This development raises the question of how and to what extent this development requires adjusted and changed regime design. For instance, existing monitoring and verification systems are geared towards checking governmental actions. It is clear that emissions trading will mean a much messier picture, with a flurry of transboundary transactions and multilevel games. Computer-based registries at the national and international levels are being established to meet this challenge. But how will this function in practice? It is also clear that the meaning of *national* environmental policies and commitments will change in such a flexible and fundamentally transnational context. These are interesting themes for further research.

A related theme has to do with the *interaction of instruments* and finding the right and most effective policy mix. In this connection, an interesting subtheme has to do with the 'limits of trading'. It is becoming increasingly clear that although well-functioning emissions trading may be a necessary condition for developing an efficient and effective response to problems such as climate change, trading is not a sufficient condition.

Other instruments must accompany trading in order to obtain the right mix of incentives. This 'optimal instrument mix' at international and national levels should be explored further.

Finally, with regard to *implementation research*, as was noted in earlier sections, a wave of research was started in the mid-1990s and ended just after the turn of the millennium. As several years have now passed, is it perhaps time for a new wave of in-depth country case studies? With respect to CLRTAP, the regime I know best, it is clear that very little implementation research has been carried out both with regard to VOC commitments and also the more recent heavy metals and POPs commitments.

notes

1. In addition to helpful comments from the editors and the project group I have received very useful comments on a previous draft from Steinar Andresen and Jon B. Skjærseth.
2. For good overviews of this literature, see for instance Young and Osherenko (1993); Underdal (1995); Levy et al. (1995); Young (1998). For instance, there are good reasons to assume that it is important to actively engage central policy target groups such as industry in the policy-making and regime formation stage in order to achieve faithful implementation and high effectiveness. For those who are particularly interested in such links, I recommend the works of Victor et al. (1998); Underdal and Hanf (2000); Young (2002). See also Biermann in this volume.
3. See Young (1994; ch. 6) for an overview of potential ways to understand the effectiveness concept. With regard to the discussion of legitimacy, see Stokke and Vidas (1996).
4. There are several good and interesting overview articles of the development of this field. See, for example, Bernauer (1995); Levy et al. (1995); Zurn (1998).
5. As further elaborated below, the environmental problem-solving perspective was also explicitly launched and commented upon both by the 'Institutions for the Earth' project (Haas et al., 1993) and the Young effectiveness project (Young and Levy, 1999).
6. As noted by Underdal (1992, p. 234): 'The Pareto frontier can be determined only for a given negotiation setting, including a given set of actors and a certain set of issues and issue linkages. A change in any of these elements may affect the range of politically feasible solutions.'
7. Due to various reasons, although most of the work within the project was carried out in the first part of the 1990s, the concluding book from the project was not published until 1999 (Young and Levy, 1999). The cases studied included oil pollution, Barents Sea fisheries, and transboundary air pollution in Europe and North America.
8. This relates to the literature on 'epistemic communities', primarily by Peter M. Haas. See for instance Haas (1990, 1992).

the effectiveness of environmental policies   323

9. The following country cases were conducted within this project: Brazil, Cameroon, China, Hungary, India, Japan, the Soviet Union/Russian Federation and the United States. In addition, the EU was studied. These cases were studied within the context of five regimes: the World Heritage Convention; the Convention on International Trade in Endangered Species of Wild Fauna and Flora (CITES); the International Tropical Timber Agreement; the London Dumping Convention, and the Vienna Convention and Montreal Protocol on Substances that Deplete the Ozone Layer.
10. With compliance/implementation as the main dependent variable, their intermediate variable was 'factors involving the country' and the three independent variables were 'characteristics of the activity involved', 'characteristics of the accord', and 'the international environment'.
11. Nine country cases were conducted within this project: Finland, France, Germany, Italy, Norway, Spain, Sweden, Switzerland and the UK.
12. As indicated earlier, the project also contributed valuable evidence on the relationship and links between the policy-making and implementation phases.
13. Arild Underdal was the driving force in this work.
14. More about this exercise in ch. 2 ('Methods of Analysis'), by Arild Underdal, in Miles et al. (2002, pp. 47–63).
15. But there are also differences: 'A main difference is that Miles et al. use ordinal metrics, whereas Helm and Sprinz use interval scales – which give rise to the use of a sensitivity coefficient for the effectiveness score ... Perhaps most important, there is a difference in strategy of analysis. Whereas Helm and Sprinz develop a single measure, Miles et al. derive two independent measures that are not directly related to each other' (Hovi et al., 2003a, p. 77).
16. However, as pointed out by Steinar Andresen, the interaction studies are by nature focused on institutions and politics and they do not contribute to a higher emphasis on ecological context and problem-solving.
17. See Miles et al. (2002, ch. 1), for a further elaboration of these two seminal perspectives.
18. Among the signatories, 15 countries and the EC committed themselves to the regular 30 per cent reduction; four chose the freeze option; and three chose the Tropospheric Ozone Management Area (TOMA) option. See Gehring (1994, p. 180).
19. For an analysis of the negotiations and content of the 1994 sulphur protocol, see, for example, Gehring (1994); Churchill et al. (1995).
20. Communication with Oran Young, October 1997.
21. According to Levy (1993, p. 126), 'the sulfur protocol *probably* had significant effects on the emission reductions in seven countries, including the largest and fourth-largest emitters in Europe (USSR and United Kingdom). A protocol that affects only these seven *probably* counts as a success' (emphasis added).
22. For instance, the effect of the 1988 Large Combustion Plant Directive has been significant in the UK.
23. More about this in my CLRTAP chapter in Miles et al. (2002).
24. I fear that policy-makers will easily ignore the sophisticated caveats provided by analysts along with the figures.
25. Communication with Lars Nordberg, former Head of the CLRTAP Secretariat, 16 October and 30 November 2000.

bibliography

Andresen, Steinar, Tora Skodvin, Arild Underdal and Jørgen Wettestad (2000) *Science and International Environmental Regimes – Combining Integrity with Involvement*, Manchester: Manchester University Press.

Andresen, Steiner, and Jørgen Wettestad (2004) 'Case Studies of the Effectiveness of International Environmental Regimes: Balancing Textbook Ideals and Feasibility Concerns', in Arild Underdal and Oran R. Young (eds) *Regime Consequences – Methodological Challenges and Research Strategies*, Dordrecht: Kluwer, pp. 49–71.

Benedick, Richard E. (1991) *Ozone Diplomacy*, Cambridge, Mass.: Harvard University Press.

Bernauer, Thomas (1995) 'The Effect of International Institutions: How We Might Learn More', *International Organization* 49, 2, pp. 351–77.

Brown-Weiss, Edith, and Harold Jacobson (eds) (1998) *Engaging Countries: Strengthening Compliance with International Accords*, Cambridge, Mass.: MIT Press.

Brown-Weiss, Edith, and Harold Jacobson (1999) 'Getting Countries to Comply with International Agreements', *Environment* 41, 6, pp. 16–20, 37–45.

Churchill, Robin, Gabriela Kütting and Lynda M. Warren (1995) 'The 1994 UN ECE Sulphur Protocol', *Journal of Environmental Law* 7, 2, pp. 169–97.

Clark, William C., Nancy Dickson, Jill Jäger and Josee van Eijndhoven (eds) (2001a) *Learning to Manage Global Environmental Risks, Vol. 1: A Comparative History of Social Responses to Climate Change, Ozone Depletion, and Acid Rain (Politics, Science, and the Environment)*, Cambridge, Mass.: MIT Press.

Clark, William C., Nancy Dickson, Jill Jäger and Josee van Eijndhoven (eds) (2001b) *Learning to Manage Global Environmental Risks, Vol. 2: A Functional Analysis of Social Responses to Climate Change, Ozone Depletion, and Acid Rain (Politics, Science, and the Environment)*, Cambridge, Mass.: MIT Press.

DeSombre, Elizabeth and Joanne Kaufman (1996) 'The Montreal Protocol Multilateral Fund: Partial Success', in Robert Keohane and Marc Levy (eds) *Institutions for Environmental Aid. Pitfalls and Promise*, Cambridge, Mass.: MIT Press, pp. 89–126.

Gehring, Thomas (1994) *Dynamic International Regimes: Institutions for International Environmental Governance*, Berlin: Peter Lang Verlag.

Greene, Owen (1996) 'The Montreal Protocol: implementation and development in 1995', in John B. Poole and Richard Guthrie (eds), *Verification 1996: Arms Control, Environment and Peacekeeping*, Boulder: Westview Press, pp. 407–26.

Haas, Peter M. (1990) *Saving the Mediterranean: The Politics of International Environmental Cooperation*, New York: Columbia University Press.

Haas, Peter M. (1992) 'Introduction: Epistemic Communities and International Policy Coordination', *International Organization* 46, 1, pp. 1–37.

Haas, Peter M., Robert O. Keohane and Marc A. Levy (eds) (1993) *Institutions for the Earth – Sources of Effective International Environmental Protection*, Cambridge, Mass.: MIT Press.

Helm, Carsten, and Detlef Sprinz (2000) 'Measuring the Effectiveness of International Environmental Regimes', *Journal of Conflict Resolution* 45, 2, pp. 630–52.

Hovi, Jon, Detlef Sprinz, and Arild Underdal (2003a) 'The Oslo-Potsdam Solution to Measuring Regime Effectiveness: Critique, Response and the Road Ahead', *Global Environmental Politics* 3, 3, pp. 74–97.

Hovi, Jon, Detlef Sprinz and Arild Underdal (2003b) 'Regime Effectiveness and the Oslo–Potsdam Solution', *Global Environmental Politics* 3, 3, pp. 105–8.

Keohane, Robert O., and Marc A. Levy (eds) (1996) *Institutions for Environmental Aid. Pitfalls and Promise*, Cambridge, Mass.: MIT Press.

Kütting, Gabriela (1999) *Environment, Society and International Relations – Towards More Effective International Agreements*, London: Routledge.

Levy, Marc (1993) 'European Acid Rain: The Power of Tote Board Diplomacy', in Peter M. Haas, Robert O. Keohane and Marc A. Levy (eds) *Institutions for the Earth*, Cambridge, Mass.: MIT Press, pp. 75–133.

Levy, Marc A., Oran R. Young and Michael Zurn (1995) 'The Study of International Regimes', *European Journal of International Relations* 1, 3, pp. 267–330.

Miles, Edward L., Arild Underdal, Steinar Andresen, Jørgen Wettestad, Jon Birger Skjærseth and Elaine M. Carlin (2002) *Environmental Regime Effectiveness – Confronting Theory with Evidence*, Cambridge, Mass.: MIT Press.

Mitchell, Ronald (2002) 'A Quantitative Approach to Evaluating International Environmental Regimes', *Global Environmental Politics* 2, 4, pp. 58–83.

Oberthür, Sebastian, and Thomas Gehring (2003) 'Investigating Institutional Interaction: Toward a Systematic Analysis'. Paper presented at the 2003 International Studies Association Annual Convention, Portland, Oregon, February–March.

Oberthür, Sebastian, and Thomas Gehring (eds) (forthcoming 2005) *Institutional Interaction: How to Enhance Synergies and Prevent Conflicts between International and EU Institutions*, Cambridge, Mass.: MIT Press.

Oden, Svante (1968) 'The Acidification of Air and Precipitation and its Consequences in the Natural Environment', *Ecology Committee Bulletin* 1.

Parson, Edward A. (2003) *Protecting the Ozone Layer – Science and Strategy*, Oxford: Oxford University Press.

Parson, Edward A., and Owen Greene (1995) 'The Complex Chemistry of the International Ozone Agreements', *Environment* 37, 2, pp. 16–20, 35–43.

Rosendal, G. Kristin (2001) 'Impacts of Overlapping International Regimes: The Case of Biodiversity, *Global Governance* 7, pp. 95–117.

Sand, Peter (1990) *Lessons Learned in Global Environmental Governance*, Washington, DC: World Resources Institute.

Selin, Henrik (2000) *Towards International Chemical Safety – Taking Action on Persistent Organic Pollutants (POPs)*, Linköping: Linköping Studies in Arts and Science 211.

Skjærseth, Jon Birger (2000) *North Sea Cooperation – Linking International and Domestic Pollution Control*, Manchester: Manchester University Press.

Skodvin, Tora (2000) *Structure and Agent in the Scientific Diplomacy of Climate Change – An Empirical Case Study of Science–Policy Interaction in the Intergovernmental Panel on Climate Change*, Dordrecht: Kluwer.

Sprinz, Detlef (2003) 'The Quantitative Analysis of International Environmental Policy'. Paper presented at the 2003 International Studies Association Annual Convention, Portland, Oregon, February–March.

Stokke, Olav Schram (ed.) (2001) *Governing High Seas Fisheries: The Interplay of Global and Regional Regimes*, Oxford: Oxford University Press.

Stokke, Olav S., and Davor Vidas (eds) (1996) *Governing the Antarctic: The Effectiveness and Legitimacy of the Antarctic Treaty System*, Cambridge: Cambridge University Press.

Underdal, Arild (1980) *The Politics of International Fisheries Management: The Case of the Northeast Atlantic*, Oslo: Universitetsforlaget.

Underdal, Arild (1990) 'Negotiating Effective Solutions: The Art and Science of Political Engineering'. Unpublished paper, University of Oslo.

Underdal, Arild (1992) 'The Concept of Regime "effectiveness"', *Cooperation and Conflict* 27, 3, pp. 227–40.

Underdal, Arild (1995) 'The Study of International Regimes', *Journal of Peace Research* 32, 1, pp. 113–19.

Underdal, Arild, and Kenneth Hanf (eds) (2000) *International Environmental Agreements and Domestic Politics: The Case of Acid Rain*, Aldershot: Ashgate.

Victor, David G. (1998) 'The Operation and Effectiveness of the Montreal Protocol's Non-Compliance Procedure', in David G. Victor, Kal Raustiala and Eugene B. Skolnikoff (eds) *The Implementation and Effectiveness of International Environmental Commitments*, Cambridge, Mass.: MIT Press, pp. 137–77.

Victor, David G., Kal Raustiala and Eugene B. Skolnikoff (eds) (1998) *The Implementation and Effectiveness of International Environmental Commitments*, Cambridge, Mass.: MIT Press.

Wallace, Helen and William Wallace (eds) (2000) *Policy-Making in the European Union*, Oxford: Oxford University Press.

Wettestad, Jørgen (1996) *Acid Lessons? Assessing and Explaining LRTAP Implementation and Effectiveness*, WP-96–18 March, IIASA Working Paper. A revised version was published as 'Acid Lessons? Assessing and Explaining LRTAP Implementation and Effectiveness', *Global Environmental Change* 7, 3, 1997, pp. 235–49.

Wettestad, Jørgen (1998) 'Participation in NOx Policy-Making and Implementation in the Netherlands, UK, and Norway: Different Approaches, but Similar Results?', in David Victor, Kal Raustiala and Eugene Skolnikoff (eds) *The Implementation and Effectiveness of International Environmental Commitments*, Cambridge, Mass.: MIT Press, pp. 381–431.

Wettestad, Jørgen (1999) 'More "Discursive Diplomacy" than "Dashing Design"? The Convention on Long-Range Transboundary Air Pollution (LRTAP)', in *Designing Effective Environmental Regimes – The Key Conditions*, Cheltenham: Edward Elgar, pp. 85–125.

Wettestad, Jørgen (2000), 'From Common Cuts to Critical Loads: The ECE Convention on Long-range Transboundary Air Pollution (CLRTAP)', in Steinar Andresen, Tora Skodvin, Arild Underdal and Jørgen Wettestad (eds) *Science and Politics in International Environmental Regimes – Between Integrity and Involvement*, Manchester: Manchester University Press, pp. 95–122.

Wettestad, Jørgen (2002a) *Clearing the Air – European Advances in Tackling Acid Rain and Atmospheric Pollution*, Aldershot: Ashgate.

Wettestad, Jørgen (2002b) 'The Convention on Long-Range Transboundary Air Pollution (CLRTAP)', in Edward L. Miles et al. (eds) *Environmental Regime Effectiveness – Confronting Theory with Evidence*, Cambridge, Mass.: MIT Press, pp. 197–223.

Wettestad, Jørgen (2004) 'Air Pollution: International Success, Domestic Problems', in Jon Birger Skjærseth (ed.) *International Regimes and Norway's Environmental Policy: Crossfire and Coherence*, Aldershot: Ashgate, pp. 85–111.

Wettestad, Jørgen (2005) 'Enhancing Climate Compliance: What are the Lessons to Learn from Environmental Regimes and the EU?', in Olav Schram Stokke, Jon Hovi and Geir Ulfstein (eds) *Implementing the Climate Regime – International Compliance*, London: Earthscan, pp. 209–233.

Wettestad, Jørgen and Steinar Andresen (1991) 'The Effectiveness of International Resource Cooperation: Some Preliminary Findings', R:007-1991, Fridtjof Nansen Institute.

Young, Oran R. (1994) *International Governance – Protecting the Environment in a Stateless Society*, Ithaca: Cornell University Press.

Young, Oran R. (1996) 'Institutional Linkages in International Society: Polar Perspectives', *Global Governance*, 2, 1, pp. 1–24.

Young, Oran R. (1998) *Creating Regimes – Arctic Accords and International Governance*, Ithaca: Cornell University Press.

Young, Oran R. (2001) 'Inferences and Indices: Evaluating the Effectiveness of International Environmental Regimes', *Global Environmental Politics* 1, 1, pp. 99–121.

Young, Oran R. (2002) *The Institutional Dimensions of Environmental Change: Fit, Interplay, and Scale*, Cambridge, Mass.: MIT Press.

Young, Oran R. (2003) 'Determining Regime Effectiveness: A Commentary to the Oslo-Potsdam Solution', *Global Environmental Politics* 3, 3, pp. 97–105.

Young, Oran R., and Marc A. Levy (eds) (1999) *The Effectiveness of International Environmental Regimes: Causal Connections and Behavioral Mechanisms*, Cambridge, Mass.: MIT Press.

Young, Oran R., and Gail Osherenko (eds) (1993) *Polar Politics: Creating International Environmental Regimes*, New York: Cornell University Press.

Zurn, Michael (1998) 'The Rise of International Environmental Politics – A Review of Current Research', *World Politics* 50, 4, pp. 617–49.

annotated bibliography

Brown-Weiss, Edith, and Harold Jacobson (eds) (1998) *Engaging Countries: Strengthening Compliance with International Accords*, Cambridge, Mass.: MIT Press. Considerable attention was given to the role of 'the international environment' (for example, major international conferences and INGOs) and domestic NGOs in the strengthening of compliance/implementation. Moreover, Brown-Weiss and Jacobson emphasized how international financial and technical assistance could modify the utility function and enhance implementation in weak parties. The project was one of the first to contribute systematic knowledge on environmental policy implementation processes in key developing countries and the Soviet Union/Russia.

Haas, Peter M., Robert O. Keohane and Marc A. Levy (eds) (1993) *Institutions for the Earth – Sources of Effective International Environmental Protection*, Cambridge, Mass.: MIT Press. This was the first published book in this field summing up a number of empirical case studies. The most noted contribution of this project was the describing of three specific mechanisms through which regimes could improve effectiveness, the three C's – that is, increasing governmental Concern, for instance, by helping to improve scientific evidence and serving as magnifiers of public pressure; enhancing the Contractual environment, for instance, by providing bargaining forums for states and by providing monitoring and verification services; and increasing national Capacity, for instance, by providing technical assistance and aid.

Miles, Edward L., Arild Underdal, Steinar Andresen, Jørgen Wettestad, Jon Birger Skjærseth and Elaine M. Carlin (2002) *Environmental Regime Effectiveness – Confronting Theory with Evidence*, Cambridge, Mass.: MIT Press. This project led by Ed Miles and Arild Underdal put forward and explored the seminal concepts of 'relative improvement' and 'distance to collective optimum' in the measuring of effectiveness, and 'problem characteristics' versus 'problem-solving capacity' as explanatory perspectives. Based on a high number of empirical case studies, the project was the first to systematically combine quantitative and qualitative methods in the analysis of regime effectiveness.

Underdal, Arild and Ken Hanf (eds) (2000) *International Environmental Agreements and Domestic Politics: The Case of Acid Rain*, Aldershot: Ashgate. With a focus on the Convention on Long-Range Transboundary Air Pollution (CLRTAP), the project led by Ken Hanf and Arild Underdal launched and explored three major perspectives on how to understand both policy formation and implementation performance: the Unitary Rational Actor (URA) model; the 'Domestic Politics' model; and the 'Social Learning and Policy Diffusion' model. This was explored in a number of country case studies.

Victor, David G., Kal Raustiala and Eugene B. Skolnikoff (eds) (1998) *The Implementation and Effectiveness of International Environmental Commitments*, Cambridge, Mass.: MIT Press. The IIASA-based project led by David G. Victor and Eugene Skolnikoff improved knowledge on especially three accounts: first, several in-depth case studies shed light on how access procedures and participation patterns affected domestic implementation processes. Second, with regard to transparency and implementation review, several analytical and empirical contributions on implementation review enhanced knowledge of this issue. Third, there were several chapters on Russian and Eastern European environmental policy implementation.

Young, Oran R., and Marc A. Levy (eds) (1999) *The Effectiveness of International Environmental Regimes: Causal Connections and Behavioral Mechanisms*, Cambridge, Mass.: MIT Press. This project was the first to forcefully put focus on and elaborate what may be termed the 'behavioural' dimension of regime effectiveness. Moreover, as a significant contribution, they suggested six important behavioural mechanisms/pathways through which international regimes could influence actors and processes in the domestic contexts. Regimes can hence function as: modifiers of the utility functions, enhancers of cooperation, bestowers of authority, learning facilitator, role definers, and agents of internal realignments.

12
environmental and ecological justice[1]

bradley c. parks and j. timmons roberts

It has become painfully clear over the last three decades that the causes and consequences of global environmental degradation cannot be addressed without tackling inequality and injustice. The 'pollution of the rich and poor' was a charged sub-theme of the 1972 United Nations Stockholm Conference (raised first by Indira Gandhi) and has steadily gained force at subsequent gatherings: Rio in 1992 and Johannesburg in 2002. With economic globalization and the increasing awareness of global warming's devastating potential have come new discourses to address inequality on these vaster scales. Environmental issues such as climate change are being 'reframed' as issues of global justice, and as this happens, new potential alliances between poor nations and environmental social movements are emerging.

The term 'environmental racism' was coined in 1982 by Benjamin Chavis, then the head of the National Association for the Advancement of Colored People (NAACP), at a landmark protest in the black town of Afton, in Warren County, North Carolina (Bullard, 1990, 1994; Cole and Foster, 2001). Citizens and civil rights activists from around the United States attempted to block the dumping of contaminated soil in the county, which had the highest concentration of blacks and among the highest poverty rates in the state. Expecting groundwater contamination from this largest polychlorinated biphenyl (PCB) dump in US history, one speaker commented that 'the depositing of toxic wastes within the black community is no less than attempted genocide' (Dr Charles E. Cobb, director of the United Church of Christ's Commission for Racial Justice in 1972, cited in Bullard 1990, p. 31).

The concept of environmental racism was soon broadened to 'environmental justice', to include unequal exposures by class, race and ethnicity: poor Latino and Native American communities were quickly seen to face the same types of 'disproportionate impacts' of pollution as blacks in the US South. The concept and the social movement by the same name gained considerable momentum among scholars, policy-makers and citizen activists, evolving with the entrance and exit of different actors and issues. The phrase, however, is used in many different senses, creating a rich but sometimes bewildering and inconsistent array of literature. In fact, the very definition of 'justice' is hotly contested. One commentator has written that outsiders to the scholarly discussion on justice are usually left with an impression of 'philosophical pandemonium ... a cacophony of discordant philosophical voices ... incommensurability' (Cullen, 1992, p. 60, cited in Harris, 1999). Similarly, another observer suggests that the pursuit of definitional consensus is a 'hopeless and pompous task' (1992, p. 177, cited in Harris, 1999). But a social movement does not need a seamless definition of its core conceptual frame: it needs one that motivates people to act, and one which puts pressure on policy-makers who are for a number of reasons averse to being tagged as racist.

It is not entirely clear when environmental justice as a concept and a social movement took to the international stage. The American environmental justice movement certainly influenced international discourse during the 1980s and 1990s. However, earlier North–South debates over 'sustainable development', which often boiled down to issues of distributive justice, were also enormously important (see Bruyninckx in this volume). For example, at the 1972 Stockholm Conference, rich and poor states agreed to 'marry' environment and development, but late developers – who feared future restrictions on their economic growth – had to threaten non-cooperation and appeal to socially shared norms of social justice in order to achieve this outcome.[2] Faced with the possibility of a North–South standoff and Southern opportunism, architects of the Stockholm Declaration designed a 'Resolution on Institutional and Financial Arrangements' and included an 'Environment Fund' to assist developing nations in their efforts toward sustainability.

Over the course of the 1970s and 1980s, environmental justice gradually secured its position as the dominant rallying cry for poor- and middle-income countries. With the rise of the New International Economic Order (NIEO) as an influential demand in international politics,[3] less developed country (LDC) policy-makers grew increasingly strident in their criticism of Northern environmentalism – an environmentalism that they perceived as 'pull[ing] up the development ladder' (Najam, 1995, p. 249). The NIEO, in fact, provided the intellectual foundation for poor nations' participation in international environmental negotiations. Whether the issue was population growth, seabed mining, oil pollution, ozone depletion and global climate change, the South's ability to articulate clear and coherent reasons why the North stood to gain by discriminating in their favour made it exceedingly difficult for their concerns to be marginalized (Krasner, 1985; Najam, 2004; Sebenius, 1991).

By the beginning of the 1990s, the 'compensatory justice' principle had been enshrined in the Montreal Protocol Fund and it was clear that any failure to honour environmental aid commitments would jeopardize international cooperation across multiple issue domains (Albin, 2001; Najam, 2002; Sell, 1996).[4] Realizing their position of leverage vis-à-vis industrialized nations, Southern leaders turned up the rhetorical volume at the 1992 Rio Earth Summit. Malaysian Prime Minister Mahathir Mohamed's remarks capture the adversarial tone taken:

When the rich chopped down our forests, built their poison-belching factories and scoured the world for cheap resources, the poor said nothing. Indeed, they paid for the development of the rich. Now the rich claim a right to regulate the development of the poor countries. And yet any suggestion that the rich compensate the poor adequately is regarded as outrageous. As colonies we were exploited. Now as independent nations we are to be equally exploited. (Mohamed, 1995, p. 288)

Such posturing and stated unwillingness to cooperate sent a clear message: there would be no progress without a 'Rio Bargain'. The North responded with a proposed financial package of US$141.9 billion a year in 'new and additional' concessional funding for sustainable development and global environmental problems,[5] but OECD nations delivered only a pitiful fraction of what was originally promised.[6]

Justice debates have since then raged on in negotiations over biodiversity, ozone, desertification, deforestation and international waters. However, the issue of global climate change has without a doubt created the sharpest divide between North and South. Poor nations are least responsible for climate change, but stand to lose most from its effects. If sea levels rise as expected, the small island states and impoverished low-lying nations like Bangladesh are expected to suffer human casualties 'of biblical proportions', an ethnicide some say approaches genocide.[7] Unsurprisingly, the Kyoto Protocol has foundered upon requests by some rich nations that poor nations set binding limits on their carbon emissions, and the term 'climate justice' has emerged as a logical fit.

In the remainder of this chapter we first review some of the difficulties of applying the concept of justice to environmental issues, especially on the side of assigning 'rights' to non-human actors. We then briefly describe a few environmental justice struggles and explore different international relations theories that may shed light on these outcomes. We find several key elements missing from these theories which make them unable to account for environmental injustice broadly: attention to the colonial history of poor nations, their current (disadvantaged) insertions into the world economy, their weakness in the face of growing transnational corporations, and their feeble domestic institutions. We therefore draw upon world-systems theory, an analytical framework which we see as uniquely situated to explain global environmental injustice. As a case study, we examine the complex set of injustices underlying the current debate over climate change. We peel away ten layers of inequality and injustice in how the benefits and costs of climate change are being distributed. By way of conclusion, we offer a summary, several policy recommendations, and some proposals for future research. The environmental justice field is beginning to blossom, but we believe it would benefit from some synthesis and direction.

applying justice to the environment

ecological justice, social ecology, and other senses of the term

Much of the existing theoretical work on justice has focused on human–human relationships. The justice literature has, in other words, concerned itself with the righting of some distribution of burdens or benefits in society that is perceived as being unfair. But one must distinguish between fairness and wrongness. Brian Barry explains that 'we would not in normal usage describe murder or assault as unjust, even though they are paradigmatically wrong. Rather, we reserve terms from the "justice" family for a cause in which some distributive consideration comes into play' (1999, p. 94).

It should therefore come as no surprise that many justice scholars have serious reservations about extending notions of justice to nature, future generations, and abstract entities like nation-states. As Barry puts it, 'justice and injustice can be predicated only on relations who are regarded as equals in the sense that they weigh equally in moral scales' (1999, p. 67). For most scholars and casual observers, only humans can justifiably be considered equal in moral terms. But there are those who disagree with such 'antiquated' notions of justice. Low and Gleeson (1998) are thought to have coined the term of 'ecological justice', which refers to a fair distribution of environmental goods and bads among different species (see also see Baxter, 2000; Cooper and Palmer, 1995). But many other ecocentrists including Benton had argued for 'do[ing] justice to non-human inhabitants' (1993, p. 212) before their 1998 publication. Much earlier, ecocentrism and deep ecology proposed that nature had rights and value entirely separate from human interests (Devall and Sessions, 1985; Naess, 1973; Tokar, 1988).

Wilfred Beckerman (1999) has argued that these newer, more radical understandings of justice are impractical and unconstructive. The 'intergenerational question', according to him, is a case in point.[8] Since the Brundtland Commission's 1987 report, sustainable development has more or less meant economic growth that addresses the needs of the present generation without jeopardizing the ability of future generations to also do so (WCED, 1987). But how one goes about pursuing this objective is a matter of deep contestation. Beckerman argues that 'the best way to provide decent societies for future generations is to improve the institutions that are partly or largely responsible for the humiliating circumstances in which many people live today' (Beckerman, 1999, p. 91), but this remains an open question. Some social scientists and ethicists might apply a similar logic to plants and animals (for example, Singer, 1999), and many ecologists and natural scientists have suggested that what future generations demand goes far beyond what those of us concerned with the present human condition can even fathom. Still others argue that if we open up this can of worms, we quickly become responsible not only for doing justice to animal species and plant life (whose moral status is unclear) today, but also the descendants of interminable generations of plants and animals (Ferry, 1992).

To the contrary, social ecologists have argued that environmental and ecological justice may actually be two sides of the same coin (Low and Gleeson, 1998). They reject the strict dichotomy that places society and nature forever at odds. Social theorist Murray Bookchin writes that 'the divisions between society and nature have their deepest roots in divisions within the social realm, namely deep-seated conflicts between human and human that are often obscured by our broad use of the word "humanity"' (1990, p. 32). Thus, exploitation of the human and the non-human are somehow inextricably linked, providing a more encompassing rationale for action. Ecofeminists see parallels and even causal connections between our mistreatment of women and the subordination of nature, and social ecologists make the broader point that subordinating classes, races and ethic minorities perpetuates the mistreatment of nature (Brulle, 2000; Mellor, 1992).

environmental justice struggles as 'ecological distributional conflicts'

From its origin in the civil rights movement of the US South, the term 'environmental justice' (EJ) has opened up something of a Pandora's box, being used to label countless cases of similar injustice around the world. While there have been ups (especially around 1992) and downs in global environmentalism, the movements are diverse, arising 'from social conflicts on environmental entitlements, the burdens of pollution, the sharing of uncertain environmental risks, and the loss of access to natural resources and environmental services' (Martinez-Alier, 2003, p. 201). Joan Martinez-Alier (1994) refers to these historical and contemporary incidents as 'ecological distributional conflicts'.

Guha and Martinez-Alier (1997) insist that conflicts like these have been particularly acute throughout the developing world and paved the way for new Third World environmental justice perspectives. The emerging terminology in academia illustrates this trend: 'livelihood ecology' (Gari, 2000, cited in Martinez-Alier, 2000), 'liberation ecology' (Peet and Watts, 1996), 'subaltern environmentalism' (Pulido, 1996), 'biopiracy' (Shiva, 1997), 'environmentalism of survival' (Guha and Martinez-Alier, 1997) and the 'environmentalism of the poor' (Gadgil and Guha, 1995; Martinez-Alier, 2003) are among a much longer list of new EJ frames.[9] These new approaches to environmental problems suggest an attempt to shift environmental attention toward human issues and inequality and away from what many in the South perceive as excessive attention to 'green' issues. So-called luxury goods such as habitat preservation, for many Southern environmental groups, connote a certain elitism (Bullard, 1990; Martinez-Alier, 1995).

That said, the green–brown divide has been overcome in some cases. In the 1990s, for example, the US environmental movement made allies with the civil rights-based environmental justice movement, reaching out to EJ communities and organizations (for example, Cole and Foster, 2001; Roberts and Toffolon-Weiss, 2001). However, these new solidaristic links have only begun to develop globally.

Historically, this type of coalition-building typically begins with localized struggles that pitch a globalizing corporation against a local community which is attempting to gain the support of an international social movement. For examples, one can think of the struggle led by rubbertapper and labour leader Chico Mendes in Brazil's Amazon forest against ranchers, Nigeria's Ogoni people's struggle against Shell Oil, ongoing lawsuits and protests by victims of the terrible accident at the Union Carbide factory in Bhopal, India, protests against mining giant

Freeport McMoRan's Grasburg mine in Irian Jaya, Indonesia, or the displacement of indigenous populations associated with huge dams and other megaprojects. This list is continually extended.

applying theories of international relations to environmental injustice

Scholars interested in issues of *international* environmental justice issues face a unique set of challenges (Harris, 2001b). For decades, the international relations (IR) literature has ignored questions of justice. Realists and neorealists have argued that the application of moral principles or notions of justice to the interstate system is irrelevant because foreign policy will always be about acquiring and maintaining power in a world where there is no supranational authority to enforce the rules (Waltz, 1979; see also Paterson in this volume). And since theories of justice assume a relational precondition of reciprocity, these two literatures are believed to be fundamentally incompatible (Bull, 1977, p. 95). To put it very plainly, the struggle over power leads to injustice, not justice.

Liberal institutionalists advance a more optimistic view of international relations and have suggested important ways that fairness may 'matter'. They argue that by negotiating mutually acceptable 'rules of the game', states can increase information, reduce uncertainty, lower transaction costs, stabilize expectations, constrain opportunism, increase the credibility of their commitments and promote collective action (Abbott and Snidal, 2000; Keohane, 1984; Martin, 2000). Fairness rules and norms are therefore said to reduce the costs of negotiating, monitoring and enforcing agreements under certain conditions.[10]

In coordination dilemmas, where multiple equilibria exist along the Pareto frontier, establishing *shared* principles, norms, rules, and decision-making procedures can enable states to zero in on a limited range of possible equilibria and enhance their prospects for cooperation. Shared understandings of *fairness* therefore provide what game theorists call 'focal points'. By isolating one point along the contract curve that every party would prefer over a non-cooperative outcome, states can stabilize expectations for future behaviour and reduce the costs of arriving at a mutually acceptable agreement (Garrett and Weingast, 1993; Keohane, 2000; Mitchell, 2002; Müller, 1999; Schelling, 1960; Snidal, 2002, p. 85; Young, 1999).

In collaboration games, where states have mixed motives for cooperation and face powerful free-rider incentives, fairness principles may affect the costs of monitoring and enforcing agreements. The difficulty of climate-change treaty-making is a prime example. Since every state faces a strong temptation to free-ride on others' climate stabilization efforts because asymmetric information reduces the 'observability' of deviant behaviour and the benefits of a stable climate are non-excludable and non-rival, it is in every state's self-interest to disguise their preferences and misrepresent their level of contribution to the collective good. Therefore, demandeurs must make compliance economically-rational for more reluctant nations through financial compensation, issue linkage and other incentive-restructuring schemes.[11] Doing so dramatically weakens the incentives for cheating and defection at the monitoring and enforcement stage of cooperation.[12]

Finally, 'fairness' norms can lower the costs of distributive bargaining in situations where some parties are risk-averse. Developing countries, for example, are extremely sensitive to distributional concerns because of their 'structural vulnerability' to changes in the international system (Abbott and Snidal, 2000; Krasner, 1985). As one author explains, the South,

> as a self-professed collective of the weak ... is inherently risk-averse and seeks to minimize its losses rather than to maximize its gains; ... its unity is based on a sense of shared vulnerability and a shared distrust of the prevailing world order ... [and] because of its self-perception of weakness [it] has very low expectations. (Najam, 2004, p. 128)

In short, for fear of widening power asymmetries and Northern opportunism, developing countries may require special signals of confidence, solidarity, empathy and kindness (Abbott and Snidal, 2000; Keohane, 2000; Koremenos et al., 2001; Roberts and Parks, n.d.; Shadlen, 2004).

new ir approaches to global environmental injustice

While liberal institutionalists have persuasively argued that embedding principles of fairness in environmental regimes may be as a propulsive force for 'deep' and long-term cooperation, such optimism may be misplaced: liberal theory appears to be disastrously unsupported by the historical record of persistent, sometimes worsening, underdevelopment and environmental degradation in the South (Redclift and Sage, 1999; Sachs, 1999; Wade, 2004).

Critics have therefore complained that institutional 'solutions' are many times only 'institutional bandage[s] applied to a structural hemorrhage' (Vogler and Imber, 1996, p. 16). Rather than adopting a case-by-case 'problem-solving' approach to global environmental problems, careful social scientists, they argue, should explain why highly asymmetric distributions arise in the first place. Rationalist theories are notoriously bad at explaining such outcomes. By their own admission, liberal institutionalists cannot account for what they call 'discrete trends', such as 'population pressures, unequal resource demands, and reliance on fossil fuel and chemical products' (Paterson, 2000, p. 26). A passage from the conclusion of Haas et al.'s *Institutions for the Earth* volume is particularly telling: 'Each set of [environmental] issues has been considered separately', they submit, 'independently of possible underlying causes such as population growth, patterns of consumer demand, and practices of modern industrial production' (Haas et al., 1993, p. 423).[13]

In attempting to marry the proximate political causes of environmental injustice to its deeper social and historical determinants, we have found in our own research that the historical-materialist, globalist and structuralist traditions offer many important insights (Roberts et al., 2004; Roberts and Parks, n.d.). Here we briefly explore world-systems theory – an approach that addresses many factors left untouched in the realist and liberal institutionalist traditions, including the colonial legacy of poor nations, the 'structural vulnerability' that many countries experience in the current world economy, and the power and influence of transnational corporations.

According to world-systems theory, there is a global stratification system that places nations on one of three levels: core, semi-periphery and periphery. In a relationship that has shifted but not reversed since colonial times, the wealthy core nations import low-priced raw or intermediate materials from the poor, 'periphery' nations (a process often referred to as 'peripheralization', for example, Martinez-Alier, 2003; Wallerstein, 1974). Wealthy nations export higher-value industrial manufactures or services, and contain the headquarters of massive financial institutions and commodities markets. Unlike poorer nations who have accumulated oppressive burdens of debt to these institutions, wealthy states have the ability to impose monetary policies to stabilize their economies in times of crisis. And while manufacturing has shifted more to the periphery, these are often the low value-added phases of a product's lifecycle (Dicken, 1998; Gereffi and Korzeniewitz, 1994). Further, with the imposition of environmental laws and greater enforcement in wealthier nations, increasingly the poorer nations are becoming the location of the most polluting parts of the 'commodity chain' by which raw materials become finished products. What remains then in the core is increasingly only the research, development, design, marketing and management phases of the cycle. Dispersed low-wage firms bid to contract the materials supply and assembly phases.

The overall pattern of environmental burdens, then, is that they fall most on the poorest nations, and the poorest and most disempowered people in all nations. These people reap ridiculously few benefits from the very processes that create the burdens. Waste flows downhill, in the literal and social structural sense. This is perhaps the core insight of the environmental justice movement and scholarship. In an attempt to point a way that international environmental politics might move ahead, we will explore three additional theoretical advantages of structuralist theories: their ability to bridge the 'domestic–international divide', their treatment of transnational corporations (TNCs) as actors in their own right, and their historical explanation for the weak domestic institutions that plague many poor nations.

explaining internal heterogeneity

The first critical improvement that world-systems theory makes is its rejection of the 'state as a black box' assumption often found in both realism and liberal institutionalism. Traditional IR theorists are arguably overly parsimonious in their assumption of the state's internal homogeneity (Paterson, 1996b). World-systems theory, by contrast, assumes a deeply entrenched domestic and international class structure and struggle. World-systems theory specifically emphasizes export elites: those who control the main money-makers for a nation, which Vernon (1993) usefully referred to as the 'polluting elites' (see also Roberts and Grimes, 2002). These groups have significant power over fiscal and trade policies, and these can create significant weaknesses and penetration of the state. Terry Karl (1997), for example, describes how nations heavily dependent upon petroleum exports end up overdependent on revenues from oil exports, leaving an inability to muster other taxes when oil prices fall. She also observes that export elites can penetrate state institutions to the point where public officials are no longer able to make autonomous decisions in the public interest. Indeed, in such settings, corruption almost inevitably follows. We believe that these elements of a nation's productive structure are crucial to any understanding of international environmental justice. Such elites are insulated and distanced by export-related wealth from the dire concerns of their nation's poor.

Environmental inequalities within states also require examination. In 1989–90, the Indira Gandhi Institute for Development Research found that the carbon emissions of the top 10 per cent of the urban population in India was 13 times greater than that of the bottom 50 per cent of the rural population (Karunakaran, 2002). Wolfgang Sachs also observes a deep divide: 'You have a Germany sitting right in India. Germany has 82 million inhabitants, not all of them are really rich ... [but] there are easily 70 million middle class in India' (*EcoEquity*, 2001). Clearly, this middle class – part of which is in the elite 'comprador' class in the global South which is tightly linked to foreign power – may have just as much ability as the industrialized North to consume resources at the expense of marginalized majorities.

bringing tncs back in

Both realism and liberal institutionalism also fail to offer adequate explanation for the plurality of actors influencing international environmental justice outcomes. Though early liberal thought placed great emphasis on the multiplicity of actors affecting world politics (for example, Keohane and Nye, 1977), subsequent theory-building has been rooted in an uncompromisingly 'state-centric' view (Paterson, 1996a). While principal-agent and constructivist authors have successfully brought international organizations, non-governmental organizations, 'epistemic communities' and transnational advocacy networks into analytical focus (Barnett and Finnemore, 1999; Haas, 1990; Nielson and Tierney, 2003; Willets, 1999), and notably with an eye toward global environmental justice (Keck and Sikkink, 1998; Wapner, 1996), most extant IR theories are sadly unable to accommodate what are probably the most influential actors in international environmental relations: corporations (Paterson, 1996a, p. 130).

One report calculated that just 122 corporations are responsible for 80 per cent of the total global CO_2 emissions (Bruno et al., 1999, p. 6). Exxon Mobil emissions alone are roughly equivalent to 80 per cent of all emissions in Africa or South America (Bruno et al., 1999, p. 7). And the *New York Times* refers to Exxon and Mobil as 'rich in cash, aggressive in style ... [and] effective in pursuing their agenda ... at the highest level of government and through arm-twisting in Congress' (cited in Bruno et al., 1999, p. 6). Indeed, in the run-up to climate change negotitations, US and EU energy firms waged a massive and vociferous campaign of disinformation to kill all proposals for reform. Often states simply feel helpless next to some large corporations, who can threaten to move jobs and tax revenues if states become too restrictive in their regulations. One often-cited statistic is Anderson and Cavanaugh's (2000) calculation that of the world's largest 100 economies, 51 of them are companies. De Grauwe and Camerman (2002), who rank countries by value added, not by total sales, put corporations as 29 and nations as 71 of the top 100 economic entities.[14] Asked about what he was going to do about the overwhelming power of corporations, Bill Clinton answered, 'What am I going to do about it? Nothing, I am only the President of the United States. I can't do anything about these companies' (Khor, 2002). Clearly then, to explain how states behave and why they may not respond to environmental justice issues, the power of transnational firms has to be included. Even beyond their influence on governments, corporations seem to deserve analytical attention in their own right (Risse-Kappen, 1995).

World-systems theory prides itself in its emphasis upon economic forces as the primary determinants of international outcomes. Economic elite classes, and the transnational corporations some of them control, are understood as the most important actors in the international system, but states are still the formal political units. The dominant social class is thus embodied in states, corporations, international organizations, and local bourgeoisies around the world (Sklair, 2001). As mentioned above, world-system studies of 'commodity chains' trace the source of products back to their component raw materials, and follow their transformation and assembly to the point of sale. This global 'sourcing' is conducted by or within transnational corporations, through a series of complex social networks. In exploiting the natural resources of peripheral nations for the benefit of wealthy classes in wealthy core nations, transnational mining and oil corporations are at the centre of a system that generates few benefits to the bulk of the world's nations and people, while despoiling their environments. To keep resources and labour cheap, poor nations can scarcely afford the luxury of noisy and intrusive labour unions and environmental activists (Roberts and Grimes, 2002; Roberts and Thanos, 2003). Mining corporations are associated with abuses of environmental activists and indigenous peoples around the world (for example, Gedicks, 2001). World-system and dependency/structuralism theories provide theoretical frameworks for understanding the origins and continuing unequal relations that drive this unjust exposure to environmental burdens.

constructing a theory of the state

A final weakness inherent in the realist and liberal institutionalist tradition is the absence of any plausible explanation for 'weak' and 'strong' states. Whereas both of these theories derive their state-centric ontology from some sort of pre-theoretical intuition, world-systems theory attempts to provide a historical explanation for state formation. Wallerstein (1974) argues that with the emergence of economic surplus during the Middle

Ages and eventually colonialism, an international division of labour was created where some regions produced primary products and other regions processed those products and took part in more advanced economic activities. Strong states in the core colonial nations applied force to create favourable terms of trade between them and weak states in the periphery (Wallerstein, 1974, p. 355). Today, 'colonial overhang' manifests itself in the noticeable exporting sector elite that depends on the state and its rents and acts as a drain on local resources. And because of the long legacy of extraction and exploitation for external markets, many of these so-called 'weak states' are also notorious for their feeble domestic institutions, which Rodrik et al. (2004) find to be the most decisive factor determining a nation's long-term economic development (Acemoglu et al., 2001, 2002; Rodrik, 2000, 2003). Likewise, we have found that the narrowness of a nation's export structure has strong effects on both the degree of voice and accountability and mobilization of civil society within a nation (Roberts et al., 2004). These findings are especially relevant since repressive, unaccountable governments typically find it easier to ignore the demands of environmental justice activists.

theorizing global environmental injustice

To review, we must reconcile our theories of international environmental relations and justice with the emerging empirical realities. A multiplicity of actors do indeed exist, including states, international organizations, transnational corporations, epistemic communities and civil society groups. All exert influence on environmental justice outcomes at the international level, but some certainly more than others. As Paul Harris notes, 'these actors relate to one another, again much like in domestic society, in a myriad of complex, cross-cutting, voluntary and involuntary, cooperative or competitive ways' (1999).

World-systems theory is only one example of a theory offering leverage on questions of global environmental justice and it certainly has its drawbacks.[15] A pressing need remains for globally minded EJ scholars to find more flexible theories that allow us to think systematically about the broad patterns and roots of global environmental injustice and solutions for the future. A welcome addition has been constructivist theory, which rather than emphasizing unfair social structures, has focused on the agency that non-state actors and networks exercise in global environmental politics. As Keck and Sikkink (1998, p. x) note,

> where the powerful impose forgetfulness, [transnational activist] networks can provide alternative channels of communication ... .

Transnational networks multiply the voices that are heard in the international and domestic policies. These voices argue, persuade, strategize, document, pressure and complain. ... By overcoming the deliberate suppression of information that sustains any abuses of power, networks can help reframe international and domestic debates, changing their terms, their sites, and the configuration of participants.

Theory-building of this type clearly offers an important complement to scholarship focusing on the social structures that create and re-create environmental injustice, and we suspect efforts to explain such agency can only increase our collective understanding of international environmental politics and better inform future policy-making.

peeling the onion: ten layers of climate injustice

The injustice of climate change is a crucially important example for the future of our planet, and a worthy test for the value of these political economy approaches. Without understanding claims of injustice arising from the Third World, Northern policy-makers will find cooperative solutions ever more elusive (Roberts and Parks, n.d.; Young, 1994). Among the many environmental issues we could examine in depth, we believe that climate change is unique in that Southern nations sometimes hold considerable political leverage. As Neumayer describes, 'the biggest bargaining power of developing countries – especially of big ones like China, India, Brazil and Indonesia – is their ability to obstruct'. Taking this to its logical conclusion, Neumayer argues that 'as their current emissions and populations grow faster than the ones in developed countries, any comprehensive treaty in the early next century will be futile without the cooperation of these countries' (2000, p. 191).

As a way to explore the value of the environmental justice perspective in understanding international environmental politics, we begin here to 'peel the onion' of injustices revealed by global climate change.[16] We have identified ten layers of this injustice; each opens a complex debate. However, given our limited space here, our goal is simply to suggest the complexity and importance of a series of claims of an emerging pattern of global climate injustice. To do so we present these layers of injustice as ten provocative possibilities for further examination.

Layer 1 – Who is most responsible for climate change? With only 4 per cent of the world's population, the US is responsible for 21 per cent of all global emissions. Compare that to 136 developing countries that together are only responsible for 24 per cent of global emissions (Marland et al., 2000). Clearly, poor nations remain far behind the US in terms of emissions per person. The average American citizen dumps as much greenhouse gas into the atmosphere as eight Chinese and as much as 20 citizens of India. Overall, the richest 20 per cent of the world's population is responsible for over 60 per cent of its current emissions of greenhouse gases. That figure surpasses 80 per cent if our past contributions to the problem are considered. They probably should be considered, since CO_2, the main contributor to the greenhouse effect, remains in the atmosphere for over 100 years.

Layer 2 – Unequal Vulnerability I: Which nations will suffer worst and first? Certainly global warming threatens everyone on the planet, but some places and some people in those places will suffer much sooner and much more profoundly than others (Adger and Brooks, 2003; Meyer-Abich, 1993). Kasperson and Kasperson (2001) explain that 'developing countries, and particularly the least developed countries, are the most vulnerable regions to climate change. They will experience the greatest loss of life, the most negative effects on economy and development, and the largest diversion of resources from other pressing needs.' As stated above, island nations and those with large populations in low-lying areas such as Bangladesh, are facing devastating ecological disasters if the sea level rises as much as is predicted. Africa will face devastating droughts, which may destabilize governments and bring even greater strife and suffering to the region, according to the same report by 2,000 international scientists (IPCC, 2001a, 2001b).

Layer 3 – Unequal Vulnerability II: Within nations, which cultures, ecosystems and segments of society will be hit hardest? Climate scientists expect the natural impacts of global warming to be highly differentiated across cultures, ecosystems and social classes within nations (Kasperson and Kasperson, 2001). Those threatened most by global climatic threats are almost invariably the same people that have the fewest adaptive resources at their disposal to deal with the problem. Whether one points their analytic lens at the poorest and most marginalized communities, classes, cultures, races, genders or ethnicities, the story seems always the same. Those with crucial assets such as education, health, technology, non-climate dependent income, social insurance and infrastructure, tend to be those with the necessary buffers to protect against dangerous climatic change.

Layer 4 – Is anyone responsible for the extraordinarily high levels of social, economic and environmental vulnerability in the South besides the developing countries themselves? Although much of the extant literature accepts these vulnerabilities as given, we would argue that there is an 'historical and contemporary production and reproduction of vulnerability' (Ribot, 1996, p. 3) that is often overlooked and which deserves further academic enquiry. This type of dialogue is far from being considered politically viable within the IPCC. However, we believe that in order to fully understand the 'structural vulnerability' of developing countries, we must pay close attention to how they have been 'inserted' in the world economy through their colonial and postcolonial history (Acemoglu et al., 2001, 2002; Engerman and Sokoloff, 2002; Roberts and Parks, n.d.; Roberts et al., 2004). Put crudely, 'there is an underlying and binding cement to be found in their common experience of imperialism and colonialism together with the common disadvantage they suffer under the present world economic order' (Manley, 1991, p. 4).

Layer 5 – Do any nations stand to gain from the greenhouse effect? Not only will the burdens of global warming fall primarily on the poor and vulnerable, but certain industries and some nations in the North may actually benefit. As Bhaskar (1995) points out, climate change is the quintessential example of market failure, where some nations enjoy tremendous economic benefit at the expense of the larger international community. Robert Mendelsohn (2001) of Yale University, a prominent environmental economist, predicts that North America and parts of Northern Europe will actually enjoy many economic gains associated with longer growing seasons, less frost, and thus increased agricultural output (see also Mendelsohn and Nordhaus, 1996). Of course, the North will also face many adverse effects, especially if we surpass dangerous emission thresholds, but the positive impacts are most likely to go to the North.

Layer 6 – Are all emissions created equal? Common sense suggests that all emissions are not created equal. Surely there is a qualitative distinction that must be made between those emissions coming from the 'gas-guzzling, air-polluting automobiles in Europe and North America' and those emanating from the 'methane emissions of flatulent cattle and the fermenting rice fields of subsistence farmers in West Bengal' (Agarwal and Narain, 1992). We use the terms 'lifestyle emissions' and 'livelihood emissions' to capture this crucial difference (Mwandosya, 2000; Shue, 1993). Although Southern NGOs argue this point repeatedly, Henry Shue may have best expressed it when he noted that justice 'does not permit that poor nations be told to sell their blankets in order that rich nations may keep their jewelry' (1992, p. 397). In other words, the poor are concerned with basic survival: having enough to eat, a safe place to sleep, a way to take care of children. Is it fair for us to 'pull up the development ladder' behind ourselves and insist that they grow 'cleaner' than we did in the early stages of our own economic development (Najam, 1995)? Climate scientists can barely fathom a world in which the families of China and India will drive their own cars. Must they then forgo development? Contrary to the 'pollution prevention pays' principle, sometimes pollution prevention costs more, and nations forced to not develop because of a climate treaty will forever have outsiders to blame for keeping them from lost opportunities. For these nations to be faced with such massive costs because of other nations' profligacy is injust.

Layer 7 – Is it environmentally possible for poor nations to pursue economic development as rich nations did? There is growing concern among Southern nations that the 'catch-up' model of development put forth by the West after World War II has lost credibility due to global climate change (Sachs, 1999; Najam, 1995). If, for example, all 6 billion people on earth emitted at the level of Germany, 67 billion tons of CO_2 would be emitted every year. Yet the planet can only tolerate about ~~16~~=17 billion tons (Sachs, 1999). Again, this highlights the justice–sustainability paradox. Mahatma Gandhi seems then to have posed the appropriate question five decades ago:. 'God forbid that India should ever take to industrialism after the manner of the West ... It took Britain half the resources of the planet to achieve this prosperity. How many planets will a country like India require?' (*The Economist*, 2002, p. 5). Thus, living under the assumption that we can live in a limitless world where we can pursue limitless growth with limitless confidence for our limitless needs has become not just a matter of ignorance, but of justice.

Layer 8 – Has procedural justice been provided to the whole international community? Shue (1992) argues that 'if background injustices have produced the weak bargaining position of the poor nations, it is doubly unfair to exploit that bargaining weakness in order to insist that the poor nations sacrifice the interest in question'. He continues, '[I]f the rich nations, have caused, albeit unintentionally, the impending harms [of climate change] that co-operation would help to prevent, it is doubly unfair to leave poor nations that have pitched in on the prevention effort to cope on their own with what the effort fails to prevent.' No doubt we must question the process by which countries participate in the Intergovernmental Panel on Climate Change, the United Nations Framework Convention on Climate Change, and the Kyoto Treaty decision-making process. The first question that arises is whether all parties are respected and included in decision formulation, and the answer clearly is no. Beyond that, however, we must ask the critical question of how large, marginalized populations within nations with non-transparent and unaccountable governments will receive meaningful representation from their diplomatic envoys. The IPCC brought this additional layer of voicelessness to light in their 2001 Working Group III report with a specific example: by highlighting that 'damage suffered primarily by poor farming communities in developing countries generates a less vigorous political response than damage that hits the infrastructure of the "modern" sectors of the economy'(IPCC, 2001b).

Layer 9 – Does the North owe the South an 'ecological debt'? The 'ecological debt' argument, advanced most forcefully by Spanish economist Joan Martinez-Alier and the Ecuadorean environmental group Acción Ecológica, is that wealthy nations have been running up a huge debt over the centuries through the exploitation of the raw materials and ecosystems in poor nations (Martinez-Alier, 2003). It is said, for example, that decades of oil spills, gas flaring, human displacement (to secure pipelines), human rights abuses, damage to human health, and disregard for lands of sacred significance requires some indemnification (Gedicks, 2001; Martinez-Alier, 2003). The debt includes not only the historical exploitation of non-Western natural resources, but the current use of global 'environmental space' for dumping waste. Scholars have recently subjected this argument to strict empirical disconfirmation tests and concluded that the 'ecologically unequal exchange' is indeed an observable empirical regularity (Andersson and Lindroth, 2001; Bringezu, 2002; Bringezu et al., 2003; Cabeza-Gutés and Martinez-Alier, 2001; Damian and Graz, 2001; Giljum, 2003, 2004; Giljum and Eisenmenger, 2004; Giljum and Hubacek, 2001, 2004a, 2004b; Heil and Selden, 2001; Hornborg, 1998a, 1998b, Machado et al., 2001; Martinez-Alier, 2003; Muradian et al., 2002; Muradian and Martinez-Alier, 2001a, 2001b; Russi and Muradian, 2003). Poorer nations export large quantities of underpriced products whose value does not include the environmental costs of their extraction, processing or shipping.[17] As Giljum explains, 'low prices for primary commodities allow industrialized countries of the capitalist core to appropriate high amounts of biophysical resources from the peripheral economies in the South, while maintaining external trade relations balanced in monetary terms. ... [W]hat within the system of prices appears as reciprocal and fair exchange masks a biophysical inequality of exchange in which one of the partners has little choice but to exploit and possibly exhaust his natural resources and utilize his environment as a waste dump, while the other partner may maintain high environmental quality within its own borders' (Giljum 2003, p. 17).

Layer 10 – Are states the relevant units of analysis in the study of climate justice? As we described earlier, the notion of the nation-state contributing to, being vulnerable to, and responding to global climate change may obscure crucially important intra-country distinctions. Many developing nations now have a sizeable middle class that affects and is affected by the warming of the earth's atmosphere much differently than the rest of society. Ott and Sachs (2002) have labelled these people as the 'omnivores' of society, since 'they are in [a] position to capture resources at the expense of the social majority'. They go on to argue that 'if the "polluter pays" principle were applied not to states, but to members of the global middle class, then most of the Southern middle classes would have to accept Kyoto reduction commitments already'. Further, this same global middle class will likely have the resources to insulate themselves from the effects of climate change. Therefore, making judgements concerning the justice or injustice of specific climate change outcomes requires opening up the 'black box' of the state.

conclusions

We conclude with a summary, some policy recommendations, and a few possible directions this new field might go, suggesting also where researchers might find their comparative advantage in the study of global environmental (in)justice. Beginning with the narrower topic of climate change and justice, we build from there back to the broader issue of environmental justice.

Climate change is increasingly being understood as an important global precedent for justice in international environmental politics. The prospects for the use of the concept are very good: there are new networks and a widening horizon of possibilities. Climate injustice appears to be a concept with political traction in a manipulated landscape. With the broader concept of an ecological debt owed by the wealthy nations to the poor, climate injustice provides the prospect of addressing environmental issues by empowering and redistributing to the worst off. There are crucial implications for international environmental politics in this case: changes in international treaty negotiations and the potential to reverse or change the direction of the flows of investment, debt, trade, aid, and so on, away from the exploitation of cheap resource and 'pollution havens'. This suggests the potential for some revision of power balances. However there are very many 'ifs' in whether this will occur.

The pitfalls, even with the greater adoption of the climate justice frame, are many. Climate change is a seemingly intractable problem, and given the distribution of power in the international system, it is very possible that most claims of injustice will continue to be marginalized. There is difficulty within the environmental justice movement in the US in establishing priorities, and since the movement arrived on the scene after the mainstream environmentalists, no solid legislative, administrative or legal framework has been instituted (Foreman, 1998; Roberts and Toffolon-Weiss, 2001). As with all social movements, there is the danger of burnout and/or cooptation among its leaders. And like the US environmental justice movement, the international movement needs accountability and recourse mechanisms. There is a further problem of different paradigms within the international movement, between those holding distributive paradigms, those clamoring for cultural recognition and political participation and those maintaining that sustainability should come first. And most fundamentally, there are multiple cultural meanings of both terms: environmental and justice. In the 1990s, Lélé (1991, p. 613) described sustainable development as having a far too large a 'tent' and becoming a '"metafix" that ... unite[d] everybody from the profit-minded industrialist and risk minimising subsistence farmer to the equity seeking social worker, the pollution-concerned or wildlife-loving First Worlder, the growth-maximising policy maker, the goal-oriented bureaucrat, and therefore, the vote-counting politician'.[18] We therefore remain sceptical of environmental justice becoming too inclusive. Clearly, in trying to make the concept acceptable to everyone, those fighting for global environmental justice may succeed to the point of diluting their concept to the point of impotence and analytical worthlessness. It is important, therefore, that the concept of environmental justice be carefully specified and applied.

In terms of strategies for environmental justice movements, there are some implications that can be drawn from our analysis. First, we believe that we need to strengthen the international institutions which are addressing environmental issues. We need to make international law 'harder' – giving 'teeth' to the environmental chamber of the International Court of Justice, for example. We need to fund and respect the rulings of the chamber and the United Nations Human Rights Commission in Geneva, and other UN agencies such as the United Nations Environment Programme. Most urgently, the dozens of existing environmental treaties need functional enforcement mechanisms (Dunoff, 1995; Kalas, 2001; Palmer, 1992). Together, these efforts support the broader need for us to pursue our multiple community loyalties, rather than just our national identities (Sen, 1999; Wapner, 1996).

Somewhat ironically, we believe that these stronger international institutions may be crucial to national environmental justice movements making any further progress, including in the nation where it was born, the United States. There the environmental justice movement has suffered serious setbacks in the last five years and appears to be foundering, precisely at the time the global movement is taking off (Agyeman et al., 2003; Cole and Foster, 2001; Roberts and Toffolon-Weiss, 2001).

To conclude, there are a number of possible directions in which this area of research might head. First of all, there are three areas of action for environmental justice groups: agenda-setting, negotiation and implementation. We should divide the labour among researchers. Analysts of international relations, for example, could study negotiations, sociologists and anthropologists focusing more on agenda-setting, and economists examining the track record of implementation. This division of labour, however, should not be seen as restrictive or exhaustive. The questions addressed in the ten layers provide a useful template for future research on international environmental justice.

International relations scholarship – the specific focus of this book – is unfortunately found wanting in terms of its focus on inequality and injustice. But all hope is not lost. A handful of areas where future research may be focused are outlined here, and there are surely many more. No matter where the literature evolves, we believe a balance must be struck between studying the social structures and human agency that together explain international environmental justice outcomes.[19]

notes

1. The views expressed in this chapter are the authors' own and do not necessarily represent the views of the Millennium Challenge Corporation.
2. The resolution remained vague on operational details and virtually no action was taken for two decades (Haas et al., 1992).
3. As Sebenius explains, the New International Economic Order (NIEO) 'involved a series of proposals advocated by LDCs during the 1970s which included significant wealth redistribution, greater LDC participation in the world economy, and greater Third World control over global institutions and resources' (1991, p. 128).
4. As DeSombre and Kauffman (1996, p. 126) put it, 'Like it or not, the [Montreal Protocol] Fund has set a precedent for dealing with global environmental issues with North–South equity problems. It has created expectations that developing countries will be compensated for the foregone development opportunities or the added burdens required by environmental cooperation.'
5. See the Rio Declaration of Environment and Development, Section 4, Chapter 33. Rich nations responded by pledging modest amounts of bilateral environmental assistance and a new multilateral funding mechanism dubbed

the Global Environmental Facility (GEF). Their insistence upon prioritizing global environmental problems over issues of local concern sowed seeds of even greater discord (Parks et al., n.d.).
6. To be fair, donors face considerable challenges in transferring environmental aid effectively in many recipient countries (Keohane and Levy, 1996; Parks et al., n.d.).
7. Hardly surprising, then, was Bangladeshi Atiq Rahman's threat that 'if climate change makes our country uninhabitable we will march with our wet feet into your living rooms' (Athanasiou and Baer, 2002, p. 23).
8. For an introduction to the various perspectives relating to intergenerational justice, see Beckerman and Pasek (2001), Brown Weiss (1989), Norton (1999).
9. As Wolfgang Sachs has put it, 'After nearly everybody – heads of state and heads of corporation, believers in technology and believers in growth – has turned environmentalist, the conflicts in the future will not center on who is or who is not an environmentalist, but on who stands for what kind of environmentalism' (1993, p. xvi).
10. Others take serious issue with the alleged causal significance of principled beliefs. Some cynical analysts, for example, have suggested that '[e]quity is merely a word that hypocritical people use to cloak self-interest' (H. Young, 1994, p. xi). David Victor writes that 'for most states most of the time, the decision-making process is mainly a selfish one. Consequently, there exists very little evidence that fairness exerts a strong influence on international policy decisions' (Victor, 2001, p. 3).
11. Oran Young (1994, p. 134) summarizes why collaboration dilemmas often require addressing issues of fairness '[I]t is virtually impossible to achieve high levels of implementation and compliance over time through coercion. Those who believe that they have been treated fairly and that their core demands have been addressed will voluntarily endeavour to make regimes work. Those who lack any sense of ownership regarding the arrangements because they have been pressured into pro forma participation, on the other hand, can be counted on to drag their feet in fulfilling the requirements of governance systems.'
12. Additionally, norms and principles of fairness can help cement a collaborative equilibrium and reduce monitoring and enforcement costs through their impact on the domestic ratification process. Benito Müller lays much emphasis on this point. '[A sceptic] might ... concede that equity has a role to play in the selection of initial allocation proposals. But surely, he is bound to interject, the outcome of the negotiations will be determined by good old-fashioned strategic bargaining, reflecting only the bargaining powers of the parties and the bargaining skills of the negotiators. [What the sceptic has overlooked is that] an agreement has to be implemented. This, in turn, requires political ratification which normally is beyond the power of mere negotiating agents' (Müller 1999, pp. 12–13). He bluntly states that 'It would be foolish to assume ... that bodies such as the US Congress or the Indian Lok Sabha could be ... bullied into ratifying an agreement [because] parties may refuse to ratify an agreement if they feel it deviates unacceptably from what they perceive to be the just solution' (Müller, 1999, pp. 12–13).

13. From a policy perspective, then, the dominance of the rational institutionalist research program in international environmental politics may be problematic. However, from a theoretical standpoint, we should probably be more sanguine in our criticism. Like the drunk who searches for his keys beneath the lamp post 'because that's where the light is', rational choice institutionalism's inability to explain the deeper social and historical determinants of environmental degradation is more a reflection of its epistemological and ontological limitations than some egregious oversight on the part of its proponents. As Snidal (2004, p. 227) reminds us, '[models] are descriptively incomplete and even inaccurate, yet they are tremendously valuable'. Indeed, a 'good model is a radically simplified description that isolates the most important considerations for the purpose at hand' (Snidal, 2002, p. 231).
14. We believe even this is a startlingly large number and await more discussion of the most valid way to do these comparisons.
15. In particular, world-systems theory has come under criticism for being overly economistic (Roberts and Grimes, 2002; Shannon, 1996; Sklair, 2002). Another shortcoming is its underestimation of political factors contributing to and mitigating against environmental degradation (Roberts et al., 2004).
16. We examine these issues in greater depth in our forthcoming book *A Climate of Injustice: Global Inequality, North–South Politics, and Climate Policy*.
17. According to Røpke (1999, p. 45), 'prices are distorted not only because of the present [environmental] externalities, but also because such externalities have existed for nearly two centuries and have been built into the social and physical structures of society as accumulated externalities'. Also see Cabeza-Gutés and Martinez-Alier (2001).
18. For more on the 'sustainable development' debate, see Dobson (1996), Jacobs (1999), Redclift (1987).
19. Finally, we must remind ourselves that justice should be a core standard for international environmental politics, but it cannot be the standard (Dobson, 1998, 1999), since ultimately sustainability must be (see also Sutcliffe, 2000).

bibliography

Abbott, Kenneth, and Duncan Snidal (2000) 'Hard and Soft Law in International Governance', *International Organization* 54, 3, pp. 421–56.

Acemoglu, Daron, Simon Johnson and James A. Robinson (2002) 'Reversal of Fortune: Geography and Institutions in the Making of the Modern World Income Distribution', *Quarterly Journal of Economics* 117, pp. 1231–94.

Acemoglu, Daron, James A. Robinson and Simon Johnson (2001) 'The Colonial Origins of Comparative Development: An Empirical Investigation', *American Economic Review* 91, pp. 1369–401.

Adger, W. Neil, and N. Brooks (2003) 'Does Environmental Change Cause Vulnerability to Natural Disasters?', in Mark Pelling (ed.) *Natural Disasters and Development in a Globalising World*, London: Routledge.

Agarwal, Anil, and Sunita Narain (1992) 'The Fridge, the Greenhouse and the Carbon Sink', *The New Internationalist* 230.

Agyeman, Julian, Robert D. Bullard and Bob Evans (eds) (2003) *Just Sustainabilities: Development in an Unequal World*, Cambridge, MA: The MIT Press.

Albin, Cecilia (2001) *Justice and Fairness in International Negotiation*, Cambridge: Cambridge University Press.

Anderson, Sarah, and John Cavanagh (2000) *Top 200: The Rise of Global Corporate Power*, Washington, DC: Institute for Policy Studies.

Andersson, Jan Otto, and Mattias Lindroth (2001) 'Ecologically Unsustainable Trade', *Ecological Economics* 37, 1, pp. 13–122.

Athanasiou, Tom, and Paul Baer (2002) *Dead Heat: Global Justice and Climate Change*, New York: Seven Stories Press.

Barber, James P., and Anna K. Dickson (1995) 'Justice and Order in International Relations: The Global Environment', in David E. Cooper and Joy Palmer (eds) *Just Environments: Intergenerational, International and Interspecies Issues*, London: Routledge.

Barnett, Michael N., and Martha Finnemore (1999) 'The Politics, Power, and Pathologies of International Organizations', *International Organization* 54, 4 (Autumn), pp. 699–732.

Barry, Brian (1999) 'Sustainability and Intergenerational Justice' in Andrew Dobson (ed) *Fairness and Futurity: Essays on Environmental Sustainability and Social Justice*, Oxford: Oxford University Press, pp. 93–117.

Baxter, Brian (2000) 'Ecological Justice and Justice as Impartiality', *Environmental Politics* 9, pp. 43–64.

Beckerman, Wilfred (1999) 'Sustainability and Intergenerational Equity', in Andrew Dobson (ed.) *Fairness and Futurity: Essays on Environmental Sustainability and Social Justice*, Oxford: Oxford University Press, pp. 71–92.

Beckerman, Wilfred, and Joanna Pasek (2001) *Justice, Posterity and the Evironment*, Oxford: Oxford University Press.

Benton, Ted (1993) *Natural Relations: Ecology, Animal Rights and Social Justice*, London: Verso Press.

Bhaskar, V. (1995) 'Distributive Justice and the Control of Global Warming', in V. Bhaskar and Andrew Glyn (eds) *The North, the South and the Environment*, London: Earthscan, pp. 102–17.

Bookchin, Murray (1990) *Remaking Society: Pathways to a Green Future*, Boston: South End Press.

Bringezu, Stefan (2002) 'Material Flow Analysis: Unveiling the Physical Basis of Economies', in Peter Bartelmus (ed.) *Unveiling Wealth: On Money, Quality of Life and Sustainability*, Dordrecht: Kluwer, pp. 109–34.

Bringezu, Stefan, Helmut Schütz and Stephan Moll (2003) 'Rationale for and Interpretation of Economy-wide Materials Flow Analysis and Derived Indicators', *Journal of Industrial Ecology* 7, 2, pp. 43–64.

Brown Weiss, Edith (1989) *In Fairness to Future Generations: International Law, Common Patrimony, and Intergenerational Equity*. Dobbs Ferry, NY: Transnational Publishers.

Brulle, Robert J. (2000) *Agency, Democracy, and Nature*, Cambridge, Mass.: MIT Press.

Bruno, Kenny, Joshua Karliner and China Brotsky (1999) *Greenhouse Gangsters vs. Climate Justice* San Francisco: Transnational Resource & Action Center.

Bull, Hedley (1977) *The Anarchical Society*, New York: Columbia University Press.

Bullard, Robert D. (1990) *Dumping in Dixie: Race, Class and Environmental Quality*, Boulder: Westview Press.
Bullard, Robert D. (1994) 'Overcoming Racism in Environmental Decision Making', *Environment* 36, 4, pp. 10–44.
Cabeza-Gutés, M., and J. Martinez-Alier (2001) 'L'echange Ecologiquement Inegal', in Michel Damian and Jean Christophe Graz (eds) *Commerce International et Développement Soutenable*, Paris: Economica.
Cole, Luke W., and Sheila R. Foster (2001) *From the Ground Up: Environmental Racism and the Rise of the Environmental Justice Movement*, New York: New York University Press.
Cooper, D., and J. Palmer (eds) (1995) *Just Environments: Intergenerational, International and Interspecies Issues*, London: Routledge.
Cullen, Bernard (1992) 'Philosophical Theories of Justice', in Klaus R. Sherer (ed.) *Justice: Interdisciplinary Perspectives*, Cambridge: Cambridge University Press, pp. 15–64.
Damian, Michel, and Jean Christophe Graz (eds) (2001) *Commerce International et Développement Soutenable*, Paris: Economica.
De Grauwe, Paul, and Filip Camerman (2002) *How Big are Multinational Corporations?*, Research Report, University of Leuven and Belgian Senate, Leuven.
DeSombre, Elizabeth R., and Joanne Kaufman (1996) 'The Montreal Protocol Multilateral Fund: Partial Success Story', in Robert O. Keohane and Marc A. Levy (eds) *Institutions for Environmental Aid: Pitfalls and Promise*, Cambridge, Mass.: MIT Press.
Devall, Bill, and George Sessions (1985) *Deep Ecology: Living as if Nature Mattered*, Layton, Utah: Peregrine Smith Books.
Dicken, Peter (1998) *Global Shift*, New York: Guilford Press.
Dobson, Andrew (1996) 'Environmental Sustainabilities: An Analysis and a Typology', *Environmental Politics* 5, pp. 401–28.
Dobson, Andrew (1998) *Justice and the Environment: Conceptions of Environmental Sustainability and Dimensions of Social Justice*, Oxford: Oxford University Press.
Dobson, Andrew (ed.) (1999) *Fairness and Futurity: Essays on Environmental Sustainability and Social Justice*, Oxford University Press: Oxford.
Dunoff, Jeffrey L. (1995) 'From Green to Global: Toward Transformation of International Environmental Law', *Harvard Environmental Law Review* 19, 2, pp. 241–69.
EcoEquity (2001) Interview with Wolfgang Sachs. May 18, 2001. Available online at <www.ecoequity.org/ceo/ceo_3_4.htm>.
The Economist (2002) 'How many planets? A Survey of the Global Environment', 6 July, pp. 3–16.
Engerman, Stanley L., and Kenneth L. Sokoloff (2002) 'Factor Endowments, Inequality, and Paths of Development Among New World Economies', *Economia* 3, 1, pp. 41–88.
Ferry, L. (1992) *Le Nouvel Ordre Ecologique*, Paris: Grasset.
Foreman, Christopher H. (1998) *The Promise and Peril of Environmental Justice*, Washington, DC: Brookings Institution.
Gadgil, Madhav, and Ramachandra Guha (1995) *Ecology and Equity: The Use and Abuse of Nature in Contemporary India*, London: Routledge.
Garrett, Geoffrey, and Barry Weingast (1993) 'Ideas, Interests and Institutions: Constructing the European Community's Internal Market', in Judith Goldstein

and Robert Keohane (eds) *Ideas and Foreign Policy: Beliefs, Institutions and Political Change*, Ithaca: Cornell University Press, pp. 173–206.

Gereffi, Gary, and Miguel Korzeniewitz (1994) *Commodity Chains and Global Capitalism*, Westport: Praeger.

Gedicks, Al (2001) *Resource Rebels: Native Challenges to Mining and Oil Corporations*, Cambridge, Mass.: South End Press.

Giljum, Stefan (2003) 'Biophysical Dimensions of North–South Trade: Material Flows and Land Use'. PhD thesis, University of Vienna.

Giljum, Stefan (2004) 'Trade, Material Flows and Economic Development in the South: The Example of Chile', *Journal of Industrial Ecology* 8, 1–2, pp. 241–61.

Giljum, Stefan, and Nina Eisenmenger (2004) 'North–South Trade and the Distribution of Environmental Goods and Burdens: A Biophysical Perspective', *Journal of Environment and Development*, 13, 1, pp. 73–100.

Giljum, Stefan, and Klaus Hubacek (2001) International Trade, Material Flows and Land Use: Developing a Physical Trade Balance for the European Union'. Interim Report, International Institute for Applied Systems Analysis (IIASA), Laxenburg, Austria.

Giljum Stefan, and Klaus Hubacek (2004a) 'New Approaches of Physical Input–Output Analysis to Estimate Overall Material Inputs of Production and Consumption Activities', *Economic Systems Research* 16, 3, pp. 301–10.

Giljum, Stefan, and Klaus Hubacek (2004b) 'International Trade and Material Flows: A Physical Trade Balance for the European Union'. Under review at *Journal of Industrial Ecology*.

Gramsci, Antonio (1971) 'Fordism and Americanism', in *Selections from the Prison Notebooks*. London: Lawerence & Wishart.

Guha, Ram, and Juan Martinez-Alier (1997) *Varieties of Environmentalism: Essays North and South*, London: Earthscan.

Haas, Peter M. (1990) *Saving the Mediterranean: The Politics of International Environmental Cooperation*, New York: Columbia University Press.

Haas, Peter, Marc Levy and Ted Parson (1992) 'Appraising the Earth Summit: How Should We Judge UNCED's Success?', *Environment* 34, 8, pp. 6–11, 26–33.

Haas, Peter, Robert. O. Keohane and Marc Levy (eds) (1993) *Institutions for the Earth: Sources of Effective Environmental Protection*, Cambridge, Mass.: MIT Press.

Harris, Paul G. (1999) 'What's Fair? – International Justice from an Environmental Perspective'. Paper presented at the 40th Annual International Studies Association Convention, Washington, DC, 16–20 February.

Harris, Paul G. (2001a) 'Defining International Distributive Justice: Environmental Considerations', *International Relations* 15, 2, pp. 51–66.

Harris, Paul G. (2001b) *International Equity and Global Environmental Politics: Power and Principles in US Foreign Policy*, Aldershot: Ashgate.

Harvey, David (1990) *The Condition of Postmodernity*, Oxford: Blackwell.

Heil, Mark T., and Thomas M. Selden (2001) 'International Trade Intensity and Carbon Emissions: A Cross-country Econometric Analysis', *Journal of Environment and Development* 10, pp. 35–49.

Holland, Alan (1999) 'Sustainability: Should We Start Here?', in Andrew Dobson (ed.) *Fairness and Futurity: Essays on Environmental Sustainability and Social Justice*, Oxford: Oxford University Press, pp. 46–70.

Hornborg, Alf (1998a) 'Towards an Ecological Theory of Unequal Exchange: Articulating World systems Theory and Ecological Economics', *Ecological Economics* 25, pp. 127–36.
Hornborg, Alf (1998b) 'Ecosystems and World Systems: Accumulation as an Ecological Process', *Journal of World Systems Research* 4, pp. 169–77.
IPCC (2001a) *Climate Change 2001: Impacts, Adaptation and Vulnerability*, Geneva, Switzerland: Intergovernmental Panel on Climate Change. Available online: <www.ipcc.ch/pub/tar/wg2/>.
IPCC (2001b) *Climate Change 2001: Mitigation*, Contribution of Working Group III to the Third Assessment Report of the Intergovernmental Panel on Climate Change (IPCC), edited by Bert Metz, Ogunlade Davidson, Rob Swart, Jiahua Pan. Cambridge: Cambridge University Press.
Jacobs, Michael (1999) 'Sustainable development as a Contested Concept', in Andrew Dobson (ed.) *Fairness and Futurity: Essays on Environmental Sustainability and Social Justice*, Oxford: Oxford University Press, pp. 21–45.
Jessop, Bob (1990) *State Theory: Putting Capitalist States in their Place*, Cambridge: Polity Press.
Kalas, Peggy Rodgers (2001) 'International Environmental Dispute Resolution and the Need for Access by Non-State Entities', *Colorado Journal of International Environmental Law and Policy* 12, 1, pp. 191–244.
Karl, Terry (1997) *The Paradox of Plenty: Oil Booms and Petro-States*, Berkeley: University of California Press.
Karunakaran, C. E. (2002) 'Clouds Over Global Warming', *Tamilnadu Science Forum*. (24 October). Available online: <www.corpwatch.org/campaigns/PCD.jsp?articleid=4548>.
Kasperson, Roger E., and Jeanne X. Kasperson (2001) *Climate Change, Vulnerability, and Social Justice*, Stockholm: Risk and Vulnerability Programme, Stockholm Environment Institute.
Keck, Margaret, and Kathryn Sikkink (1998) *Activists Beyond Borders: Advocacy Networks in International Politics*, Ithaca: Cornell University Press.
Keohane, Robert (1984) *After Hegemony: Cooperation and Discord in the World Political Economy*, Princeton: Princeton University Press.
Keohane, Robert (2000) 'Governance in a Partially Globalized World', *American Political Science Review* 95, 1, pp. 1–13.
Keohane, Robert, and Marc A. Levy (eds) (1996) *Institutions for Environmental Aid: Pitfalls and Promise*, Cambridge, Mass.: MIT Press.
Keohane, Robert and Joseph Nye (1977) *Power and Interdependence: World Politics in Transition*, 2nd edn, Boston: Little Brown & Company.
Khor, Martin (2002) 'Much More than a Memo', in Isabelle Reinery (ed.) *Comments on the Jo'burg Memo*, Heinrich Böll Foundation. Available online: <www.worldsummit2002.org/download/wsp18.pdf>.
Koremenos, Barbara, Charles Lipson and Duncan Snidal (2001) 'The Rational Design of International Institutions', *International Organization* 55, 4, pp. 761–99.
Krasner, Stephen D. (1985) *Structural Conflict: The Third World Against Global Liberalism*, Berkeley: University of California Press.
Lélé, Sharachandram M. (1991) 'Sustainable Development: A Critical Review', *World Development* 19, 6, pp. 607–21.
Low, Nicholas, and Brendan Gleeson (1998) *Justice, Society and Nature: An Exploration of Political Ecology*, London: Routledge.

Machado, Giovani, Roberto Schaeffer and Ernst Worrell (2001) 'Energy and Carbon Embodied in the International Trade of Brazil: An Input–output Approach', *Ecological Economics* 39, 3, pp. 409–24.

Manley, Michael (1991) *The Poverty of Nations*, London: Pluto Press.

Marland, G., T. A. Boden and R. J. Andres (2000) 'Global, Regional, and National Fossil Fuel CO_2 Emissions', in *Trends: A Compendium of Data on Global Change*, Oak Ridge, Tenn.: Oak Ridge National Laboratory, United States Department of Energy. Available online: <http:cdiac.esd.ornl.gov/trends/emis/em_cont.htm>.

Martin, Lisa (2000) *Democratic Commitments: Legislatures and International Cooperation*, Princeton: Princeton University Press.

Martinez-Alier, Joan (2000) 'Environmental Justice, Sustainability and Valuation'. Paper delivered at Harvard Seminar on Environmental Values, 21 March.

Martinez-Alier, Joan (1994) 'Distributional Conflicts and International Environmental Policy on Carbon Dioxide Emissions and Agricultural Biodiversity', in Jeroen C. J. M. van den Bergh and Jan van der Straaten (eds) *Toward Sustainable Development*, Washington, DC: Island Press.

Martinez-Alier, Joan (1995) 'The Environment as a Luxury Good, or "Too Poor to be Green"?' *Ecological Economics* 13, pp. 1–10.

Martinez-Alier, Joan (2002a) 'Embodied Pollution in Trade: Estimating the "Environmental Load Displacement" of Industrialized Countries', *Ecological Economics* 41, pp. 51–67.

Martinez, Alier, Juan (2002b) *The Environmentalism of the Poor: A Study of Ecological Conflicts and Valuation*, Cheltenham: Edward Elgar.

Martinez-Alier, Juan (2003) 'Environmental Conflicts, Environmental Justice, and Valuation', in Julain Agyeman, Robert D. Bullard, and Bob Evans (eds) (2003) *Just Sustainabilities: Development in an Unequal World*, Cambridge, Mass.: MIT Press.

Mellor, Mary (1992) *Breaking the Boundaries: Towards a Feminist Green Socialism*, London: Virago.

Mendelsohn, Robert (2001) *Global Warming and the American Economy: A Regional Assessment of Climate Change Impacts*, Northampton, Mass.: Edward Elgar.

Mendelsohn, Robert, and William Nordhaus (1996) 'The Impact of Global Warming on Agriculture: Reply', *American Economic Review* 86, 5, pp. 1312–15.

Meyer-Abich, Klaus M. (1993) 'Winners and Losers in Climate Change', in Wolfgang Sachs (ed.) *Global Ecology*, London: Zed Books.

Mitchell, Ronald B. (2002) 'International Environment', in Thomas Risse, Beth Simmons and Walter Carlsnaes (eds) *Handbook of International Relations*, Thousand Oaks, Calif.: Sage, pp. 500–16.

Mohamed, Mahathir (1995) 'Statement to the U.N. Conference on Environment & Development', in Ken Conca, Michael Alberty and Geoffry D. Dabelko (eds) *Green Planet Blues: Environmental Politics from Stockholm to Rio*, Boulder: Westview Press, pp. 287–89.

Muradian, Roldan, and Joan Martinez-Alier (2001a) 'Trade and the Environment: From a "Southern" Perspective', *Ecological Economics* 36, pp. 281–97.

Muradian, Roldan, and Joan Martinez-Alier (2001b) 'South–North Materials Flow: History and Environmental Repercussions', *Innovation: The European Journal of Social Science Research* 14, 2, pp. 171–87.

Muradian, Roldan, Martin O'Connor and Joan Martinez-Alier (2002) 'Embodied Pollution in Trade: Estimating the "Environmental Load Displacement" of Industrialized Countries', *Ecological Economics* 41, pp. 51–67.

Müller, Benito (1999) *Justice in Global Warming Negotiations: How to Obtain a Procedurally Fair Compromise*, Oxford: Oxford Institute for Energy Studies.

Müller, Benito (2001) 'Varieties of Distributive Justice in Climate Change: An Editorial Comment', *Climatic Change* 48, pp. 273–88.

Mwandosya, Mark J. (2000) *Survival Emissions: A Perspective from the South on Global Climate Change Negotiations*, Dar es Salaam: Dar es Salaam University Press and the Centre for Energy, Environment, Science and Technology.

Naess, Arne (1973) 'The Shallow and the Deep, Long-Range Ecology Movement', *Inquiry* 16, pp. 95–100.

Najam, Adil (1995) 'An Environmental Negotiating Strategy for the South', *International Environmental Affairs* 7, 3, pp. 249–87.

Najam, Adil (2002) 'The Unraveling of the Rio Bargain', *Politics and the Life Sciences* 21, 2, pp. 46–50.

Najam, Adil (2004) 'The View from the South: Developing Countries in Global Environmental Politics', in Regina Axelrod, David Downie and Norman Vig (eds) *The Global Environment: Institutions, Law, and Policy*, 2nd edn, Washington, DC: CQ Press, pp. 225–43.

Neumayer, Eric (2000) 'In Defence of Historical Accountability for Greenhouse Gas Emissions', *Ecological Economics* 33, 2, pp. 185–92.

Nielson, Daniel L., and Michael J. Tierney (2003) 'Delegation to International Organizations: Agency Theory and World Bank Environmental Reform', *International Organization* 57 (Spring), pp. 241–76.

Norton, Bryan (1999) 'Ecology and Opportunity: Intergenerational Equity and Sustainable Options', in Andrew Dobson (ed.) *Fairness and Futurity: Essays on Environmental Sustainability and Social Justice*, Oxford: Oxford University Press, pp. 118–50.

Ott, Hermann and Wolfgang Sachs (2002) 'Ethical Aspects of Emissions Trading', in Luiz Pinguelli-Rosa and Mohan Munasinghe (eds) *Ethics, Equity and International Negotiations on Climate Change*, Cheltenham: Edward Elgar.

Palmer, Geoffrey (1992) 'New Ways to Make International Environmental Law', *American Journal of International Law* 86, 2, pp. 259–83.

Parks, Bradley, Michael Tierney, J. Timmons Roberts and Robert Hicks (n.d.) 'Greening Aid? Understanding Environmental Assistance to Developing Countries'. Unpublished manuscript.

Paterson, Matthew (1996a) *Global Warming and Global Politics*, London: Routledge.

Paterson, Matthew (1996b) 'IR Theory: Liberal Institutionalism, Neorealism and the Climate Change Convention', in Mark Imber and John Vogler (eds) *The Environment and International Relations*, London: Routledge, pp. 59–76.

Paterson, Matthew (2000) *Understanding Global Environmental Politics: Domination, Accumulation and Resistance*, London: Macmillan.

Peet, Richard, and Michael Watts (1996) *Liberation Ecologies: Environment, Development, Social Movements*, New York: Routledge.

Pulido, Laura (1996) *Environmentalism and Economic Justice: Two Chicano Struggles in the Southwest*, Tucson: University of Arizona Press.

Redclift, Michael (1987) *Sustainable Development: Exploring the Contradictions*, London and New York: Methuen.
Redclift, Michael, and Colin Sage (1999) 'Resources, Environmental Degradation, and Inequality', in Andrew Hurrell and Ngaire Woods (eds) *Inequality, Globalization, and World Politics*, Oxford: Oxford University Press, pp. 122–49.
Ribot, Jesse C. (1996) 'Climate Variability, Climate Change and Vulnerability: Moving Forward by Looking Back', in Jesse C. Ribot, Antonio Rocha Magalhães and Stahis S. Panagides (eds) *Climate Variability, Climate Change and Social Vulnerability in the Semi-Arid Tropics*, Cambridge: Cambridge University Press, pp. 1–10.
Risse-Kappen, Thomas (ed.) (1995) *Bringing Transnational Relations Back In: Non-State Actors, Domestic Structures and International Institutions*, Cambridge: Cambridge University Press.
Roberts, J. Timmons and Peter E. Grimes (2002) 'World-System Theory and the Environment: Toward a New Synthesis', in Riley E. Dunlap, Frederick H. Buttel, Peter Dickens and August Gijswijt (eds) *Sociological Theory and the Environment: Classical Foundations, Contemporary Insights*, Lanham, Md: Rowman and Littlefield, pp. 167–95.
Roberts, J. Timmons and Bradley C. Parks (n.d.) 'A Climate of Injustice: Global Inequality, North–South Politics, and Climate Policy'. Unpublished manuscript.
Roberts, J. Timmons, Bradley C. Parks and Alexis A. Vásquez (2004) 'Who Ratifies Environmental Treaties and Why? A World System Analysis of Participation in 22 Treaties by 192 Nations'. Working Paper.
Roberts, J. Timmons and Nikki D. Thanos (2003) *Trouble in Paradise: Globalization and Environmental Crises in Latin America*, London: Routledge.
Roberts, J. Timmons, and Melissa M. Toffolon-Weiss (2001) *Chronicles from the Environmental Justice Frontline*, New York: Cambridge University Press.
Rodrik, Dani (2000) 'Institutions for High-Quality Growth: Why They Are and How to Acquire Them?', *Studies in Comparative International Development* 35, 3, pp. 3–31.
Rodrik, Dani (ed.) (2003) *In Search of Prosperity: Analytic Narratives on Economic Growth*, Princeton: Princeton University Press.
Rodrik, Dani, Arvind Subramanian and Francesco Trebbi (2004) 'Institutions Rule: The Primacy of Institutions Over Geography and Integration in Economic Development', *Journal of Economic Growth* 9, 2, pp. 131–65.
Røpke, Inge (1999) 'Prices are not Worth that Much', *Ecological Economics* 29, pp. 45–6.
Russi, D. and R. Muradian (2003) 'Gobernanza global y responsabilidad ambiental', *Ecologia Política* 24, pp. 95–105.
Sachs, Wolfgang (ed.) (1993) *Global Ecology: A New Arena of Political Conflict*, London: Zed Books.
Sachs, Wolfgang (1999) *Planet Dialectics: Explorations in Environment and Development*, London: Zed Books.
Schelling, Thomas (1960) *The Strategy of Conflict*, Cambridge, Mass.: Harvard University Press.
Sebenius, James K. (1991) 'Designing Negotiation Toward a New Regime: The Case of Global Warming', *International Security* 15, pp. 110–48.

Sell, Susan (1996) 'North–South Environmental Bargaining: Ozone, Climate Change, and Biodiversity', *Global Governance* 2, 1, pp. 97–118.

Sen, Amartya K. (1999) 'Global Justice: Beyond International Equity', in Inge Kaul, Isabelle Grunberg and Marc A. Stern (eds) *Global Public Goods: International Cooperation in the 21st Century*, New York: Oxford University Press, pp. 116–25.

Shadlen, Ken (2004) 'Patents and Pills, Power and Procedure: The North–South Politics of Public Health in the WTO', *Studies in Comparative International Development* 39, 3, pp. 76–108.

Shannon, Thomas (1996) *An Introduction to the World System Perspective*, Boulder: Westview Press.

Shiva, Vandana (1997) *The Plunder of Nature and Knowledge*, Boston: South End Press.

Shue, Henry (1992) 'The Unavoidability of Justice', in Andrew Hurrell and Benedict Kingsbury (eds) *The International Politics of the Environment*, Oxford: Oxford University Press, pp. 373–97.

Shue, Henry (1993) 'Subsistence Emissions and Luxury Emissions', *Law and Policy* 15, pp. 39–59.

Singer, Peter (1999) 'Ethics Across the Species Boundary', in Nicholas Low (ed.) *Global Ethics and Environment*, London: Routledge, pp. 146–57.

Sklair, Leslie (2001) *The Transnational Capitalist Class*, Oxford: Blackwell.

Sklair, Leslie (2002) *Globalization: Capitalism and its Alternatives*, Oxford: Oxford University Press.

Spangenberg, Joachim (2002) 'Environmental Space and the Prism of Sustainability: Frameworks for Indicators Measuring Sustainable Development', *Ecological Indicators* 2, pp. 295–309.

Snidal, Duncan (2002) 'Rational Choice and International Relations', in Walter Carlsnaes, Thomas Risse and Beth Simmons (eds) *Handbook of International Relations*, New York: Sage, pp. 73–94.

Snidal, Duncan (2004) 'Formal Models of International Politics', in Detlef F. Sprinz and Yael Wolinsky-Nahmias (eds) *Models, Numbers, and Cases. Methods for Studying International Relations*, Ann Arbor: University of Michigan Press, pp. 227–64.

Sutcliffe, Bob (2000) 'Development After Ecology', in J. Timmons Roberts and Amy Hite (eds) *From Modernization to Globalization: Perspectives on Development and Social Change*, Malden, Mass.: Blackwell, pp. 328–39.

Tokar, Brian (1998) 'Social Ecology, Deep Ecology and the Future of Green Political Thought', *Ecologist* 18, 4/5, pp. 132–41.

Vernon, Raymond (1993) 'Behind the Scenes: How Policymaking in the European Community, Japan and the United States Affects Global Negotiations', *Environment* 35, 5, pp. 12–20, 35–42.

Victor, David (2001) *The Collapse of the Kyoto Protocol and the Struggle to Slow Global Warming*, Princeton; Princeton University Press.

Vogler, John, and Mark F. Imber (eds) (1996) *The Environment and International Relations*, New York: Routledge.

Wade, Robert Hunter (2004) 'Is Globalization Reducing Poverty and Inequality?', *World Development* 32, 4, pp. 567–89.

Wallerstein, Immanuel (1974) *The Modern World System, Vol. 1*, New York: Academic.

Waltz, Kenneth (1979) *Theory of International Politics*, New York: McGraw-Hill.
Wapner, Paul (1996) *Environmental Activism and World Civic Politics*, Albany: SUNY Press.
Willets, Peter (1999) 'Transnational Actors and International Organizations in Global Politics', in John Baylis and Steve Smith (eds) *The Globalizations of World Politics*, Oxford: Oxford University Press.
World Commission on Environment and Development (WCED) (1987) *Our Common Future*, Oxford: Oxford University Press.
Young, Oran R. (1994) *International Governance: Protecting the Environment in a Stateless Society*, Ithaca: Cornell University Press.
Young, H. Peyton (1994) *Equity in Theory and Practice*, Princeton: Princeton University Press.

annotated bibliography

Albin, Cecilia (2001) *Justice and Fairness in International Negotiation*, Cambridge: Cambridge University Press. This book explores the question 'How and why do principled beliefs matter in international politics?' Albin offers some initial answers to these questions through four illustrative case studies: European efforts to reduce acid rain, international trade negotiations, the Israeli-PLO conflict, and the Nuclear Non-Proliferation Treaty.

Athanasiou, Tom, and Paul Baer (2002) *Dead Heat: Global Justice and Climate Change*, New York: Seven Stories Press. The authors argue that the Kyoto Protocol is deeply flawed and that developed countries are largely responsible for bringing the global atmosphere perilously close to a tipping point. In their view, persistent global inequality is at the heart of the crisis and policy-makers must redouble their efforts to address fairness issues in order to stabilize the climate.

Müller, Benito (2001) 'Varieties of Distributive Justice in Climate Change: An Editorial Comment', *Climatic Change* 48, pp. 273–88. Müller tries to disconfirm the notion that states will naturally and spontaneously arrive at socially shared understandings of what is fair or ethical in order to lower their bargaining costs. He argues that the behaviour of all players will not converge around a single equilibrium unless state explicitly identify potential focal points – that is, reconsider and negotiate their own beliefs about 'what is fair'.

Sell, Susan (1996) 'North–South Environmental Bargaining: Ozone, Climate Change, and Biodiversity', *Global Governance* 2, 1, pp. 97–118. Sell argues that bargaining power is the key determinant of outcomes in global environmental politics and that the 'compensatory justice' principle is most often institutionalized when the South's threat of non-participation in an environmental regime is credible.

Victor, David (2001) *The Collapse of the Kyoto Protocol and the Struggle to Slow Global Warming*, Princeton: Princeton University Press. This monograph explores the merits and shortcomings of the Kyoto Protocol and predicts its eventual disintegration. Victor argues that countries' principled beliefs typically have a trivial effect on outcomes in international environmental politics. In his view, national self-interest is the real driver of pro-environmental action.

13
general conclusion

michele m. betsill, kathryn hochstetler and dimitris stevis

In this conclusion, we draw on the book's chapters to briefly reflect on the status of the field of international environmental politics (IEP) as a whole. We return to the cross-cutting themes and queries raised in the introduction, finding that some of them are reinforced and illustrated by the chapters while others need to be reformulated. Many of the field's most important developments can be seen only by examining a variety of substantive and normative issues together. Some of the conclusions are especially well-grounded in individual chapters of the book, however, and they are indicated here with the chapter authors' names in parentheses.

This book clearly shows that the study of IEP has become broader and deeper over time in terms of research agendas, substantive concerns, theoretical approaches, and the geographical and disciplinary origins of researchers (Hochstetler and Laituri, Paterson, Stevis). The evidence of the proliferation of IEP research is demonstrated by the sheer number of publications and publishing venues. Equally important, however, there has been a broadening in terms of research agendas and research areas. Since the early 1990s, for instance, methodological concerns, societal politics and environmental justice have joined the other research areas. The study of IEP has also become deeper in two senses. Not only have additional theoretical views, particularly constructivist and structuralist, joined the fray, but the quality of the theoretical exchanges has also become more sophisticated. The study of IEP has also spread geographically well beyond the US and Europe. A desirable project for the future would be to bring together reviews of the study of IEP from a number of countries not only to ascertain convergences and divergences but, also, in order to move beyond the boundaries of the Anglo-American literature.

The diversification of approaches is reflected in a second overall conclusion, which is that the field of IEP continues to lack a single normative core idea of the kind that efficiency represents for the discipline of economics. The now-venerable concept of sustainable development has been useful in both academic and policy debates, but carries too many meanings to work well as a conclusive evaluative criterion (Bruyninckx). The search for alternatives is reflected in the recent development of normative criteria that capture only some parts of the sustainability concept, such as environmental justice (Parks and Roberts) or effectiveness (Wettestad). Even in the more circumscribed domains of these alternative evaluative criteria, multiple definitions compete with each other. In addition, one can easily imagine environmental policies and projects that might be effective, but not just, or vice versa. Such observations suggest that the ambiguities and contradictions of the sustainable development criterion are not the result of inadequate analytical effort, but may reflect inevitable tensions among the diverse concerns of IEP.

cross-cutting themes

In the introduction we identified three specific cross-cutting themes that we expected to be important across the book's chapters. The first, North–South relations, turned out to be every bit as central as anticipated and is a factor in nearly every chapter. The second theme, international–domestic linkages, needs reformulating in order to adequately capture the changing nature of environmental governance in theory and in practice. Finally, the cross-cutting theme of identifying phases of the policy process does appear, but in a smaller subset of IEP than we had expected.

The North–South dimension has become increasingly important in the practice and study of IEP (Stevis). For example, it is relevant to all of the different candidates for normative evaluation of IEP. The sustainable development concept was forged to bridge the concerns and futures of the so-called developed and developing worlds (Bruyninckx). The North–South dimension is also fundamental in discussions of environmental and ecological justice, especially in the context of globalization (Kütting, Parks and Roberts). We see a North–South dimension in the regime effectiveness literature as well, where differentiated commitments, technology/financial resource transfers, and capacity issues are central in regime development and implementation (Wettestad). The North–South dimension also appears in virtually every case study in this book, from climate change (Betsill, Parks and Roberts) to transboundary environmental security concerns (Swatuk) to trade in toxic wastes (Clapp). These case studies show that North–South relations dominate government negotiations on the environment. Non-governmental actors, whether economic (Clapp) or civil society based (Betsill), also reflect the North–South dimension. Despite the centrality of North–South relations, it must be pointed out that there is no single Southern or Northern voice in IEP or the study of IEP. Instead, the North-South dimension is dynamic and constantly reconstructed, sometimes a divide and sometimes a bridge.

The significance of the North–South dimension does not mean that intra-North and intra-South issues are marginal. As the politics of climate change demonstrate, there are important differences within the North while Southern environmental politics is a growing and increasingly important and diverse phenomenon (Swatuk). Yet as the various chapters collectively indicate, this geographic heuristic should not lead us to reify the country as a unit. Many of the issues discussed cross boundaries, whether as a result of natural or social processes, while important stakeholders may operate at various scales other than that of nation-state, North or South.

With respect to our second proposed cross-cutting theme, the various chapters clearly show that the international–domestic linkage is more complicated than this dichotomy can capture. Analytical concepts such as two-level games and second image reversed are no longer sufficient to capture the complex, cross-border dynamics of international environmental politics. With increasing global integration and environmental policies, concepts such as multilevel governance are better for understanding the interplay of the various levels and kinds of international environmental politics (Betsill, Biermann, Bruyninckx). To speak of *multilevel* governance is to note that there may be considerably more than two levels engaged in shaping IEP and de-emphasizes the special analytical boundary of the nation-state. Similarly, the conceptual move from government to *governance* stresses the growing number and variety of kinds of authority relations that are part of IEP (Betsill, Biermann, Clapp, Kütting). This does not imply that the state is withering away but, rather, that the meaning and roles of the state and interstate agencies require renewed attention.

Finally, we expected that research across issue areas would explicitly address the various phases of the policy process (for example, agenda-setting, negotiation and implementation). To our surprise the various chapters did not find that the phases of the policy process approach were characteristic or central to the study of the research areas that they covered. There are a number of explanations for this finding. First, this is partly due to the individual authors' sense that other concepts or organizational arrangements were more pertinent. Second, it reflects the observation above about the growing importance of multilevel governance of IEP. Environmental policy-making increasingly occurs in a variety of spheres that cross scales of social organization. While the policy phases approach is suitable to intergovernmental policy-making processes (the central focus of liberal institutionalism), it is less useful where policy making takes place in the private sphere and/or through transnational networks. In other words, it is consistent with an analytical shift from government to governance. Third, it reflects the fact that for many analysts IEP encompasses much more than the policy-making process. IEP also includes the origins and framing of environmental issues, the meaning and operationalization of fundamental concepts and the power relations involved in environmental practices. Finally, the lack of emphasis on phases of the policy process reinforces our finding about the breadth of theoretical approaches used in the study of IEP. While the phases approach is central in liberal institutionalist studies of IEP, this is but one small, though prominent, research programme in the field.

where next?

The concluding section of each of the book's chapters outlines the most important emerging issues for the chapter's topic. Here we examine a set of overarching substantive themes and broader research areas or agendas that emerge from the chapters as a whole.

new substantive themes

With respect to substantive environmental issues, we see a great deal of debate emerging over the nature of specific environmental problems, such as climate change, biotechnology, and so on. These query the likely causes and impacts of these problems as well as their solutions. The book's chapters introduce actors and analytical constructs which must be understood to advance many of these debates. We think the debates within the field are more sophisticated than the dichotomous debates (population/no population, growth/no growth) that characterized previous eras. The debates are also more diverse among IEP scholars than they are in the general public and in most policy debates among governments. In the future, it would be interesting to consider to what extent such growing sophistication within the field of IEP has been able to reconfigure policy debates and popular understandings of environmental issues.

We also see a resolute move towards a political economy of the environment with few if any environmental issues seen as separate from the overall political economy. Important research areas, such as sustainable development or ecological modernization, can hardly avoid taking a political economy approach. There is increasing attention to the challenge of simultaneously addressing poverty and environmental degradation in an era of globalization (Bruyninckx, Kütting and Rose, Parks and Roberts). Moreover, IEP scholars increasingly recognize the importance of global financial and economic institutions in environmental governance (Betsill, Biermann, Clapp). Related to the move towards political economy, questions about production and consumption, the interface of the environment and health, commodity and policy chains and the built environment are likely to become more important in the future. The centrality of North–South relations is in large part tied to scholarly recognition of the importance of the political economy of the environment.

With respect to broad empirical research agendas we see a strong move toward considering normative issues, albeit increasingly based on solid empirical research. These include issues of equity, democracy, participatory processes, and so on. This trend complements the move to political economy and a focus on North–South relations. Questions of governance are likely to remain high on the agenda but will be increasingly affected by discussions over levels of governance and the changing nature and variety of actors involved in governance processes (Betsill, Biermann). Procedural issues and the ability to assert and claim authority will continue to be central dimensions of governance.

In our view IEP scholars ought to pay attention to the formation of environmental problems with the same empirical sensitivity that we have been using to understand their impacts and efforts at solving them. Tracking international environmental problems even before they are viewed as such (for example, China's energy and transportation policies) will enrich our understanding of other aspects of the various issues and may also allow us to offer more proactive policy recommendations.

In all these new areas of study, the field is likely to be influenced by the substantive shifts of a number of actors in IEP over the last decade. The US move towards unilateral international policies and frequent anti-environmentalism is especially notable. The European Union is often counterpoised to the United States as the world's new environmental leader, but those accomplishments could be exaggerated, especially with the challenges of its recent enlargement. On the other side of the global North–South divide, Southern countries now also sometimes find themselves as the drivers of environmental initiatives, such as the Johannesburg Conference in 2002 or the desertification negotiations (Bruyninckx). Private actors, once assumed to be universally opposed to environmental regulation, now develop policy tools and techniques for environmental protection. Complex emerging issues like the global trade and management of transgenic organisms have already rearranged global negotiating coalitions and positions altogether, in ways that seem likely to become more common.

methodological issues

In general, methodological concerns have not been central in studies of IEP, with the regime effectiveness and environmental security literatures being notable exceptions (Hochstetler and Laituri, Swatuk, Wettestad). We detect signs that this is slowly changing and expect that the methodological sophistication of IEP scholars will grow in the future. For example, we anticipate developments in the use of quantitative techniques that complement the in-depth qualitative case studies that have characterized much of IEP scholarship. IEP is also well situated to integrate important methodological insights and techniques from other disciplines, such as geography (for example, geospatial information technologies) and ecology. The prominence of global change models requires, in our view, a renewed debate on the use of models and forecasting, with particular emphasis on the integration of human practices and choices into what are largely naturalistic assumptions. Debates in archaeology over the impact of climatic changes on past civilizations could provide us with a useful introduction as would a revisiting of the literature from the 1970s. In sum, we believe the field would greatly benefit from more explicit consideration of methodological issues and the development of a mixed methods approach and hope that trends in this direction continue.

theoretical issues

All of the substantive debates and developments outlined above seem likely to contribute to theoretical developments as well. Such changes provoke theoretical debates as they point out the inconsistencies between theoretical assumptions and concrete unfolding events. When the global hegemon has clearly become an environmental laggard rather than leader, for example, certain theoretical explanations of global environmental change appear less tenable. Such changes often provoke rethinking that goes beyond the particular event.

With respect to broader theoretical perspectives we expect that the substantive attention to questions of equity is also likely to offer a strong challenge to the theoretical perspectives of liberal institutionalism and liberal constructivism which have been largely inattentive to these questions. This will likely accelerate emergent debates within these perspectives on how to integrate questions of equity. Emphasis on the ecological foundations of IEP is similarly likely to influence radical social perspectives and force them to address ecological issues more seriously than they have done in the past. As a result we also expect to see an acceleration of the emergent debates within these perspectives.

In general, we believe that the theoretical diversity of IEP research is salutary and appropriate for such a young field with so many concerns and disciplinary foundations. We have been careful in this book to sample widely across different approaches to the study of IEP in the English speaking world. A next step would be to more consistently include theories and understandings from the global South, especially given the growing importance of North–South relations and normative concerns in IEP. One drawback of such a pluralist orientation is the field's resulting inability to give clear and consistent guidance to those who govern the environment. Without this, the impact of scholarship in IEP on the practice of IEP may be limited.

The discrepancy between the study and practice of IEP has been and ought to be a central theoretical and normative issue. Why is there divergence during some periods or with respect to some issues, but not in others? Are there variations across countries and what explains them? While we think that the study of IEP ought to address policy issues and offer specific proposals this should not be at the expense of speaking truth to power. Tailoring IEP research to specific political or economic exigencies would be tantamount to taking theoretical pursuits out of physics or ecology or any other systematic field of study.

linkage to other disciplines

Our focus in this book is on IEP and how the field of international relations (IR) approaches the environment in particular. The IR field is uniquely situated to take the lead in dealing with important aspects of IEP, whether conflict or governance. At the same time, however, the study of IEP clearly benefits from drawing on other disciplines in the social and natural/ecological sciences. The question, therefore, is how and when IEP/IR scholars should create bridges with other disciplines. In the chapters' presentations of the history of particular areas of study, we have noted that the study of IEP has often followed the lead of the ecological and physical sciences while economics has also played a central role (Stevis). There is no doubt that the study of IEP has to pay close attention to these disciplines and even recognize their leading role in dealing with particular issues. It seems to us, however, that additional social sciences, and IR in particular, also have much to offer to the study of international environmental problems. Which topics are invisible or trivial from the perspective of IR and how might other disciplines make those more visible? To what extent does the study of IEP require theories and approaches distinct from other areas of research within IR?

Several examples of contributions from other social science disciplines are evident in the chapters. Geography can offer some important methodological tools enabling IEP scholars to better link to the natural and physical sciences (Hochstetler and Laituri). On the issue of governance, geography, with its emphasis on scale, will likely contribute to debates about multilevel governance in IEP. Urbanists and regional planners have long been at the forefront of ecological thinking. The human habitat, for instance, was one of the major themes of the Stockholm Conference. The organization of space is central to the creation and solution of environmental problems and there is much that IEP scholars can learn from those disciplines. Sociological perspectives have informed IEP scholarship and often have been the source of important and now mainstream concepts, such as risk analysis and ecological modernization. For example, studies of transnational actors rely on concepts of framing and social movements borrowed from sociology (Betsill). Finally, world systems approaches inform studies of IEP, particularly related to issues of globalization and equity (Kütting and Rose, Parks and Roberts).

Archaeology, history and cultural anthropology have much to offer us, as well. Archaeology and history can provide us with long term perspective as well as important cases that will help us better evaluate some of the assumptions behind key environmental issues of the day, such as climate change, biodiversity, deforestation and desertification. Cultural anthropologists have produced some of the most intriguing studies of the impacts of human practices over time but, also, are particularly sensitive to practices and relations – for example, the importance of non-state actors and the 'powerless' – that IR scholars and others focusing on large systems and modernity may easily miss.

One important question that we must address is that of the economization of IR and IEP. The adoption of microeconomic thinking by important strands of IR has been the subject of serious debate going back a few decades. The move towards economic thinking and instruments in environmental affairs is certainly evident in IEP. While many economists and IR scholars may argue that economic instruments are just that, others have made a convincing argument that every social policy instruments carries with it key political and social assumptions (Bruyninckx, Clapp, Wettestad). Assigning prices to the environment or life is not a simple technical device. Rather, it indicates that everything can be subjected to the same common denominator and that alternative criteria, whether moral or political, cannot be used to guide policy choices.

How much and what kind of borrowing from other disciplines the study of IEP needs to do depends on what we wish to understand. If we are looking at what explains the formation of international environmental politics and policies, one could make the argument that IEP does not need theories beyond those of IR. Other IR issue areas have contributed a great deal to emerging understandings of the multiple actors, processes, and levels of governance now important in IR generally and IEP in particular. IR already offers significant tools for understanding such governance issues. At the same time, however, the centrality of natural and physical sciences in this process does distinguish IEP from other areas of IR scholarship. As the literature on effectiveness has already noted, there may be large gaps between successful *political* outcomes and successful *environmental* ones (Wettestad). Ecological and physical sciences will continue to be important for understanding the underlying causes of environmental degradation. Socially sustainable outcomes may also require the insights of disciplines like sociology and anthropology, as suggested above (Parks and Roberts). In short, while IEP has found a hospitable and productive home in IR, IR has never been the only source of its insights and should not be expected to be in the future.

In any case, the contributions of IEP to IR should also be briefly noted. IEP has exemplified certain new global developments – the rise of non-state actors and private governance, the changing meaning of security, the fungibility of nation-state boundaries, the importance of North–South and global equity concerns – in ways that have been clarifying and suggestive for IR as a whole. The willingness of IEP scholars to reach across disciplines for the necessary tools to understand these phenomena is one of its most important contributions.

index

Acción Ecológia, 346
acid rain, 23, 311
  see also Convention on Long-Range Transboundary Air Pollution
adaptation, 215, 279, 286, 343
Africa, 129–30, 132–5, 162, 179, 204, 220–7, 254, 282, 287, 289, 339, 343
  see also individual countries
African Union, 229
Agenda 21, 193, 268, 273, 276, 281–2, 288–9
  see also United Nations Conference on Environment and Development
agenda-setting, 182, 349, 363
agriculture, 129–35
  see also trade
Alliance for a Corporate-Free UN, 156
Alliance of Small Island States (AOSIS), 191
Ambiente e Sociedade, 35
Ambio, 24
American Journal of International Law, 20
American Society for International Law, 27
Americas (The), 131, 179, 205, 282, 339, 344
  see also individual countries and Latin America
anarchy, 55–9, 71–2, 204, 207, 217, 220, 229n1
Angola, 223
anthropocentrism, 70
anti-globalization movement, 180
apartheid, 214–15, 223–5
Article XX, *see* General Agreement on Tariffs and Trade

Asia, 151, 162, 179, 205, 227, 254
  see also individual countries
Asia Pacific Economic Cooperation (APEC), 151, 229, 165n3
Asia Pacific People's Environmental Network, 177
Australia, 36, 40, 151, 161, 165n3, 191
Australian Export Finance and Insurance Corporation (EFIC), 161
Austria, 314

balance of power, 204
Bangladesh, 331, 343
Bariloche Foundation, 19, 24, 35
Basel Convention on the Transboundary Movement of Hazardous Wastes and their Disposal (1989), 149, 150, 162–3
Belgium, 280
bilateral agreements, 150, 165
biodiversity, 65, 158–9, 221–2, 225
  see also biosafety *and* Cartagena Protocol *and* Convention on Biological Diversity
biosafety, 154, 155
  see also Cartagena Protocol
Bhopal, India, 164, 334
Botswana, 223–4
Brazil, 35, 86, 157–8, 159, 173, 305, 323n9, 334, 342
  see also United Nations Conference on Environment and Development *and* World Social Forum
British International Studies Association, 34–5
British Petroleum, 273

Brundtland Report (1987), 23, 33, 265–6, 270, 333
　see also World Commission on Environment and Development;
　and global solutions, 267
　and liberalism, 123–4
　North–South, 272–3
Bush, George H.W., 274
Bush, George W., 274
business, *see* industry
Business Action for Sustainable Development, 154, 177
Business Council for a Sustainable Energy Future, 191
Business Council on Sustainable Development, 68, 154, 177

Cameroon, 159, 395, 323n9
Canada, 25, 30, 151,165n3, 280
Canadian Export Development Bank, 160–1
capitalism, 114–15, 125, 174
　capital accumulation, 66–9, 74n16, 74n17
　carboniferous capitalism, 219–20
　and critical theory, 102
　modern, 117–19, 128, 208, 235
　and structuralist analysis 205–6
　transformation of, 72, 180, 181
Capitalism, Nature, Socialism, 32
Caribbean, 129–30
Cartagena Protocol, 149
Carter, Jimmy, 266
carrying capacity, 20
Centre for Science and the Environment, 31
chemicals, 42n11, 161, 337
　see also Convention on Long-Range Transboundary Air Pollution *and* Rotterdam Convention *and* Stockholm Convention
　agriculture, 130, 132
　trade, 147
Chile, 151, 165n3
China, 151, 165n3, 305, 323n9, 342
　and climate change, 343, 345
　Three Gorges Dam, 161
Chinese Taipei, 165n3
Christian Aid, 156

Cities for Climate Protection, 177
civil society, 121, 135, 173, 174, 182, 341
　see also global civil society
　and industry, 121, 178, 363
　organizations 115, 190
　participation in IEP, 248–50, 276, 289, 291
Climate Action Network (CAN), 172, 177, 188–92, 194n10
climate change, 34, 67, 274, 360, 366
　see also Kyoto Protocol *and* United Nations Framework Convention on Climate Change
　climate justice, 331, 342–7, 350n7
　EU policy, 150
　Global Environment Facility (GEF), 158
　governance 73n4, 246–7, 310
　industry response, 155–6, 202, 339
　Intergovernmental Panel on Climate Change (IPCC), 29, 61, 248–50, 285, 344–6
　negotiations, 188–92, 335–6
　North–South, 65, 362–3
　and trade, 149
Climate Policy, 34
Clinton, Bill, 340
Club of Rome, 23–4, 26–7, 116, 127, 266, 291n1
　Limits to Growth 127, 291n2
Colorado Journal of International Environmental Law and Policy, 30
common pool resources, 25, 37, 92
commons, *see* global commons
community-based natural resource management (CBNRM), 226
comparative advantage, 144, 146
compensatory justice, 331
Conservation International, 221
constructivism, 15, 72, 125–6, 361
　see also critical perspectives
　and anarchy 55, 58, 59
　and knowledge, 59, 60
　liberal 15, 366–7
　methods, 84, 207–8, 218
　and sustainability, 69
　and transnational actors 180, 339, 341

consumerism, 127, 208
consumption, 37, 67, 124, 208, 211, 279, 339
  see also Club of Rome
  and agriculture, 129–34
  and equity, 268–9
  Northern 126–7, 214, 227, 274
  regional level, 280
  structuralists, 277–8
  transnational networks, 282
Convention on Biological Diversity, 68, 277, 286, 287
Convention on Long-Range Transboundary Air Pollution (CLRTAP), 299–300, 305, 309–22
  EMEP system, 311, 313
  Executive Body, 311, 317–20
  Geneva Protocol on Volatile Organic Compounds (VOCs), 312, 317, 319
  Implementation Committee, 310, 318–20
  Protocol for the Reduction of Sulphur Emmissions, 312
  Protocol to Abate Acidification, Eutrophication and Ground-Level Ozone (Gothenburg Protocol), 279, 292n4, 313, 315, 317–18
  Second Sulphur Protocol, 312, 317, 319, 323n21
  Sofia Protocol on Nitrogen Oxide, 312
  Trust Fund for Countries in Transition, 317
  Working Group Strategies, 318
Convention on the International Trade in Endangered Species of Wild Fauna and Flora (CITES) (1973), 150, 308
cooperation, 216, 222, 229, 306, 310
  and ecoauthoritarianism, 58
  environmental conflict, 213
  game theory, 335–6
  and liberal institutionalism, 57–8, 303, 305
  public–private, 238, 242
  and realism, 56–7, 58
  scale, 71

scientific information, 60
transnational networks, 282, 283
critical environmental security studies, 204, 216–20, 226, 235
critical perspectives, 54, 72, 206–7, 227
  see also critical environmental security studies
  and IPE, 144, 164, 165n2, 170
  and methodology, 97–103, 207–9
  postmodernism, 125–6, 204
  and science, 61–2
  and TNCs, 152–3, 154
  and trade 146–7
currency devaluation, 159–60
Czech Republic, 292n5

dams, 161, 335
  Narmada Dam Scheme, 157–8
  Three Gorges Dam, 161
Darussalam, 165n3
debt forgiveness, 214
deep ecology, 333
deforestation, 65, 157, 159
democracy, 89, 221, 223, 365
  see also global governance
  and environment, 70, 86, 90, 155, 214
democratic peace, 214, 221
Denmark, 314
dependency theory, 64, 205–6, 340
desertification, 65, 130, 290, 292n7, 331, 366, 368
  see also United Nations Convention to Combat Desertification
  soil degradation, 103–2, 135, 160
developing countries
  see also North–South
  agriculture, 129–35
  climate change, 191, 248–50, 342–6
  decentralization, 289
  digital divide, 97
  ecological modernization, 125
  environmental security, 211–12
  finance, 156, 162
  in the global economy, 65, 336
  global governance, 250–4, 305, 327
  globalization and 116, 120
  hazardous waste trade, 162–4

developing countries *continued*
    ozone depletion, 309, 349n3
    sustainable development, 272–3
    TNCs, 152, 155
    transnational actors, 121, 176, 248, 281
diplomacy, 151, 204, 252
distributive issues, 25–9, 32, 36, 38
    distributional justice, 65, 330
    ecological distributional conflicts, 334–5
distributive justice, 65, 330
double standard, 152, 164
Duke University Press, 24, 34

Earthscan, 24, 29, 31, 34
ecoauthoritarianism, 27–8, 55–8, 69–70, 72, 144
ecodevelopment, 33
ecocentrism, 70, 122, 333
ecofeminism, 333
    see also feminism
eco-labeling, 63, 283
ecological debt, 346–7
ecological distributional conflicts, 334–5
    see also distributive justice
ecological economics, 20, 26, 118, 135, 144–5, 146
ecological justice, 332–3, 362
ecological marginalization, 211, 228
ecological modernization, 26, 69, 72, 124–5, 127–8, 135, 187, 365, 368
ecological footprint, 70, 219, 285
ecological shadow, 219
Ecologist (The), 24
Economic Commission for Latin America and the Caribbean (ECLAC), 278
economic growth, 42n14, 133, 142
    see also capitalism *and* Club of Rome
    ecological economists, 144, 146
    and environmental justice, 333
    institutionalists, 145
    neoclassical economists, 144
    North–South, 330
    steady-state economic approach, 118, 135

    and trade, 146, 147
    World Bank, 122, 157, 159
ecopolitics, 14, 15, 20–1, 25–6, 36, 42n10
effectiveness, *see* international regimes
efficiency, 118, 124, 127, 131, 146, 147, 160, 181, 253–4
Enlightenment, 177, 207
Environment Liaison Centre, 177
environmental agreements, *see* international environmental agreements
Environmental Conservation, 24
Environmental Defense Fund (Environmental Defense), 157–8
environmental degradation, 55, 57
    see also environmental security studies *and* tragedy of the commons
    agriculture, 130–5
    capitalism, 67–8
    ecological economists, 146
    globalization, 114, 116–19, 126, 127–9, 179, 365
    and growth, 122–4
    health, 272
    justice, 329
    liberalism, 336, 351n13
    power, 70
    and science, 61, 369
    structural inequality, 64–6, 74n16, 118
environmental equity, *see* environmental justice *and* inequality
environmental impact assessments, 158, 161
environmental justice, 65, 329–49
    agriculture, 132
    climate justice, 331, 342–7, 350n7
    ecological economics and, 118
    ecological modernization, 125
    global governance, 135, 237
    intergenerational, 33
    movement, 66, 330
    North–South, 330, 362
    in the study of IEP, 15, 29, 32, 38–9, 42n15, 361, 362

Environmental Kuznets Curve (EKC), 144
environmental performance standards, 153, 155, 161–2, 190, 304
Environmental Politics, 17
environmental racism, 329–30
environmental security, 73n5, 157, 203–29, 362–3, 366
  see also environmental security studies, scarcity, *and* security
environmental security studies, 28, 33, 96, 210–16, 227–9
  syndrome approach, 211, 229n2
Environmental Studies Section (International Studies Association), 17, 25, 30, 34–6
Environment, Development and Sustainability, 35
epistemic communities, 33, 37, 60, 176, 225, 285, 305, 322n8, 339, 341
  see also knowledge
Europe, 18, 205, 249, 270, 272, 279
  see also European Union *and individual countries*
  agriculture, 133
  climate change, 344
  NGOs, 187
  in the study of IEP, 24, 32, 36, 240
  transboundary air pollution, 311–13, 317
European Consortium for Political Research, 35
European Union (EU), 150–1, 229, 238, 292n71, 310, 323n9, 365
  climate change, 190–1, 339
  sustainable development, 278–80
  transboundary air pollution, 314
  tuna–dolphin dispute, 148–9
expertise, 59, 216, 218, 226
  see also epistemic communities *and* knowledge
  and global governance, 251, 254, 320
  postpositivism, 85
  transnational actors, 182, 189–90, 248–51

export credit agencies (ECAs), 157, 160–2, 165
  see also individual agencies
export crops, 160
  see also agriculture
export elites, 338
Export-Import Bank of the United States (Ex-Im Bank), 160–1
export subsidies, 146, 161
Exxon Mobile, 339

feminism, 54, 64, 65, 98–9, 128, 206, 217
  ecofeminism, 333
finance, 114, 142–5, 156–65, 273, 365
  see also investment
Finland, 271, 312, 314, 323n11
fisheries, 208, 211
foreign direct investment (FDI), 151–3, 160
  see also investment
Forest Stewardship Council, 63, 245
forestry, 153, 211, 283, 314
Foro Internacional, 35
frames, 2, 34, 194n7
  environmental justice, 329, 347
  North–South, 15
  sustainable development, 271, 275
  transnational actors, 183, 330, 334, 342, 368
France, 272, 280, 323n11
Freeport McMoRan's Grasburg Mine, 335
Friends of the Earth International, 63, 156, 160, 176
fuel, 116, 160, 190–1, 337

game theory, 56–7, 91–3, 104n3, 335–6, 363
Gandhi, Indira, 329
Gandhi, Mahatma, 345
gas, 161, 314, 346
Gaza/Kruger/Gonarezhou (GKG) transfrontier conservation area, 223–4
genetically modified organisms, 131
General Agreement on Tariffs and Trade (GATT), 115, 123, 143, 147–50, 163
  Article XX, 148–9

Georgetown International Law Review, 30
geopolitics, 14, 20–1, 25, 33–4, 36, 40, 65
geospatial informational technologies, 83–4, 94–7, 104, 366
Germany, 161, 189, 250, 280, 323n11, 339
  climate change, 345
  ecological modernization, 125
  global governance, 240
  liberal institutionalism, 15, 33
  study of IEP in, 35, 36
  transboundary air pollution, 314, 316–17
Ghana, 159
global change, 22, 27, 29–34, 37, 40, 366
  see also global environmental change
global civil society, 38, 174
  see also civil society
  and globalization, 37, 113, 115, 121–3
  and pluralism, 63
  transnational actors, 183–5
Global Climate Coalition, 177, 191
global commons, 22, 26, 27
Global Compact, 156, 244
global culture, 128
  see also globalization
Global Environmental Change, 17
global environmental change, 17, 29, 32, 128–9, 136, 366
  see also environmental degradation *and* global change
Global Environmental Politics, 17
Global Environment Facility (GEF), 158–9, 249, 253–4, 273–4, 350n5
global governance, 27, 32–3, 36–7, 63–4, 73n11, 119–23, 173–4, 217–18, 237–56, 363–6, 368–9
  see also international regimes
  and conflict, 214–16
  democratization of, 183, 185, 188, 193, 202, 288, 289, 291
  economic institutions, 132–4, 143, 145–6, 147–51, 153–6
  legitimacy, 245–6, 256

  post-Westphalian, 215–16, 243
  and sustainable development, 265–91
  and transnational actors, 72–93, 243–5
globalists, 14, 26–7
globalization 37, 113
  beginning of, 114–17, 125
  ecological, 34, 113–36, 237, 240–1, 247, 272, 362, 368
  economic, 68, 115, 120, 123–5, 127, 135, 142, 154, 164, 205–6, 220, 237, 240–1, 247, 276, 283, 329
  political, 119–23, *see also* global governance
  social, 237
  socio-cultural, 116, 125–9, 135, 247
  structural origins of, 114–16
global political economy, *see* international political economy
global warming, *see* climate change
Gore, Al, 191
Gothenburg Protocol, *see* Convention on Long-Range Transboundary Air Pollution
governance, *see* global governance
Group of 8 (G8), 278
Greece, 150
Greenpeace, 176
  Bhopal Principles, 156
growth, *see* economic growth
Guidelines on Multinational Enterprises, 156

hazardous waste, 162–4
  see also Basel Convention
Heavily Indebted Poor Countries (HIPC) initiative, 134
heavy metals, 312–13, 322
Hermes Kreditversicherung-AG (Hermes), 161
Homer-Dixon, T., 28, 87, 210–15, 219, 227, 228, 285
Hong Kong, 165n3
Human Dimensions of Global Environmental Change Programme (The), 285
Hungary, 292n5, 323n9

implementation, 182, 225, 269, 287, 350n11
see also individual agreements
domestic, 289, 304–5, 314, 322
effectiveness, 302, 310, 318, 320, 322n2
participation, 269, 287, 288
sustainable development, 270–1, 273, 275, 277, 279, 280
transnational actors, 182, 245, 321, 349
World Environment Organization, 254
incremental costs, 158
India, 24, 31, 102, 157–8, 283, 305, 323n9, 338–9, 342–3, 345
Indiana University Press, 24, 34
Indonesia, 151, 165n3, 335, 342
industrial agriculture, 157–8
industry (as international actor), 124, 154–6, 181–5, 190, 243–4, 283–4, 322n2, 337
see also nongovernmental organizations
accountability, 153, 156, 165
lobbying, 154, 163
structural power, 154, 181, 225
voluntary conduct, 155, 156
industrial flight, 152–3, 163–4
inefficiency, see efficiency
inequality, 64–7, 145, 188, 206, 207, 211, 222, 225, 266–7, 274, 277
see also environmental justice and North–South
infrastructure, 125, 157–8, 292n6, 343, 346
ingenuity, 90, 214, 219, 304
institutional effectiveness, 237, 247
see also international regimes
institutional interlinkages, 315, 362–3
institutionalism, see liberalism
institutionalized global politics, 119, 129, 172–93, 242, 246, 248–50, 265–91, 291–2n3
Institutions for the Earth project, 302, 306
integration theory, 173
intergenerational justice, 33–4, 38–9, 332–3, 350n8

intergovernmental organizations (IGOs), 23–4, 37, 172, 180–1, 187, 193–4n1 242, 244
see also global governance and individual organizations
Intergovernmental Panel on Climate Change (IPCC), 61, 248–50, 285, 344–6
International Affairs, 20, 23
International Association for Ecological Economics, 35
International Association for the Study of Common Property, 30
International Chamber of Commerce, 177
International Climate Change Partnership, 191
International Convention for the Prevention of Pollution from Ships (MARPOL), 304
International Council for Local Environment Initiatives (ICLEI), 281
International Council for the Exploration of the Sea, 244
International Council of Scientific Unions (ICSU), 18–9
International Court of Justice (ICJ), 244–5, 348
International Criminal Court, 245
International Environmental Affairs, 30
International Environmental Agreements, 34, 36
international environmental agreements, 121–2, 129, 147, 158, 225, 251–2, 304–6, 345, 348
see also individual agreements
bilateral agreements, 150, 165
implementation, 154, 159
multilateral environmental agreements, 149, 162–5, 308
International Financial Corporation (JEXIM), 160–1
International Friends of Nature, 178
International Geosphere-Biosphere Programme (IGBP), 29–30
International Human Dimensions Programme on Global Environmental Change (IHDP), 30, 35

International Institute for Applied
    Systems Analysis (IIASA), 306
International Institute for
    Environment and Development
    (IIED), 23–4, 28–9
International Labour Organization
    (ILO), 249–51, 252–3, 284
International Monetary Fund (IMF),
    115, 122–3, 159–60, 250, 278,
    289
international negotiations
    see also individual agreements
    air pollution, 322, 314, 317–18,
        323n29
    environmental justice, 331,
        350n12
    and geographic information
        technologies, 96
    and the global economy, 122
    industry, 153–4, 244
    NGOs, 100, 182, 185, 189–92, 243,
        284
    liberalism, 335
    North–South, 363
    ocean politics, 22
    ozone depletion, 31, 310
    science, 244
    South, 330
    sustainable development, 283
    two-level games, 93
    and WEO, 252
International Organization, 20, 23
International Organisation for
    Standardization, 68
international political economy (IPE),
    14, 21–3, 29, 33, 68, 142–65, 365
international regimes, 21, 27, 33, 69,
    71, 238
    see also global governance
    climate, 146–7
    desertification, 287–90
    effectiveness, 238, 299–322, 322n2,
        322n3, 323n9, 362–3, 366
    liberalism, 14, 37, 58, 59
    NGOs 63–4
    South, 140, 253
    study of, 37, 247, 255
International Security, 23

International Studies Association
    (ISA): Environmental Studies
    Section (ESS), 17, 25, 30, 34–6
International Studies Quarterly, 23
International Sociological Association,
    35
International Tribunal for the Law of
    the Sea, 245
International Union for the
    Conservation of Nature and
    Natural Resources (IUCN), 18–9,
    21–2, 194n1, 267–8
International Union for the
    Protection of Nature (IUPN) *see*
    International Union for the
    Conservation of Nature and
    Natural Resources
International Union of Forestry
    Research Organizations, 178
investment, 115, 122, 124–5, 142,
    143, 145, 151–6, 162–5
    see also finance
    guarantees, 160
    liberalization, 159–60
ISO 14000, 155
Italy, 150, 323n11

Jakarta Declaration for Reform of
    Official Export Credit and
    Investment Insurance
    Agencies (2000), 162, 165n4
Japan, 151, 165n3, 191, 205, 323n9
Japan Bank for International
    Cooperation (JBIC)/International
    Financial Corporation
    (JEXIM), 160–1
Johannesburg Conference, *see* World
    Summit on Sustainable
    Development
*Journal of Agricultural and
    Environmental Ethics*, 31
Journal of Conflict Resolution, 20, 23
*Journal of Environment and
    Development*, 30, 31
Journal of Peace Research, 19, 20, 23

Kenya, 102, 282
Kgalagadi Transfrontier Park, 223–4

knowledge, 28, 33, 37, 59–62, 72, 102, 285–7, 305
  indigenous/local, 62, 95, 128–9, 286–8
Kyoto Protocol, 149, 189, 190, 217–18, 247, 331, 345, 347

land degradation, 159, 210–1, 272, 346
  see also agriculture *and* soil erosion
Latin America, 31, 162, 227, 254
  see also individual countries
liberalism, 15, 54, 68, 123–4, 127, 133, 146
  liberal environmentalism, 13, 39, 59
  liberal institutionalism, 3–4, 15, 32, 55, 57, 70, 72, 73n6, 145, 147, 153, 164–5, 165n2, 204–5, 210, 217, 275–9, 299, 335, 339–40, 364, 367
  neoliberalism, 122–3, 127, 132–3, 159, 173, 180, 221, 241, 279
  pluralism, 36–9, 54, 63–4, 71, 73n11, 103, 173, 180, 367
liberalization, *see* globalization *and* trade
lifestyle emissions, 344
Limits to Growth, 127, 291n2
limits to growth, 26, 32, 59, 67, 70, 144–5
livelihood emissions, 344
logging, 161

Malaysia, 165n3, 331
maldevelopment, 212–14
Malthusian, 20, 26, 42n13
  neo-Malthusian, 209, 212, 278
maquiladora firms, 164
Marine Policy, 22
Marine Pollution Bulletin, 22
Marine Stewardship Council, 245
Marxism, 64–7, 72, 74n16, 244
  see also structuralism
  neo-Marxism, 241
mercury, 318
methodology 172
  see also qualitative methodology *and* quantitative methodology

Mexico, 148, 151, 164, 165n3, 215
migration schemes, 157–8
military and environment, 28, 33
  see also security
mining, 153, 159–61, 330–1, 340
modernity, 36, 65, 114, 117, 125, 207, 218–20, 239, 368
modernization, 213–14, 219, 240
Mohamed, Mahathir, 331
Montreal Protocol on Substances that Deplete the Ozone Layer (1987), 150, 218, 246, 249, 254, 308–10, 331, 349n4
  Copenhagen Amendment, 310
  London Amendment, 310
Mozambique, 223–5
multilateral environmental agreements (MEAs), 149, 162–5, 308
multinational corporations (MNCs), 37, 115, 173
  see also industry; agriculture, 134–5
  and globalization, 126, 187
  governance, 242, 244–5
  as transnational actor, 177–8
  transnational corporations (TNCs), 42, 124, 151–6, 332, 337–40

Namibia, 223–4
Narmada Dam scheme, 157–8
National Association for the Advancement of Colored People (NAACP), 329
natural resources, 129, 153
  conflict, 215–16
  economic growth, 125, 144
  environmental justice, 334, 346–7
  finance, 160; 210–12
  in the study of IEP, 16, 19
  sustainability, 69–70
  trade, 146, 148
  transboundary management, 220–7
  world-systems theory, 118, 340
Natural Resources Forum, 24
Natural Resources Journal, 19, 20
neoclassical economics, 143–5, 146, 152, 153, 164–5, 165n2
neorealism, 173, 204–5, 217, 335
  see also realism

Netherlands, 125, 272, 280, 306, 314
New International Economic Order (NIEO), 24, 267, 330–1, 349n3
New Partnership for Africa's Development (NEPAD), 278
New Zealand, 36, 40, 165n3, 191
Nigeria, 334
nitrogen dioxides, 313–14, 318
  see also Convention on Long-Range Transboundary Air Pollution
nongovernmental organizations (NGOs), 21, 38, 121, 134–5, 156, 165, 174–6, 185, 265, 339, 344
  see also transnational actors
  business/industry, 163, 176, 181, 193, 250, 265, 321
  environmental (ENGOs), 66, 121, 155, 157–9, 161–3, 176–9, 184, 187, 189, 191–3, 221, 225, 243, 255, 271, 280, 282–4, 321, 340
  international (INGOs), 175, 178–9, 221, 223, 226, 305
  labour organizations, 284, 340
  legitimacy, 182, 184
  national, 175, 177, 305
  representation, 243–5, 248–50, 280–1, 288, 313
  strategies, 126, 181, 189–90, 348–9
  umbrella organizations, 283–4
non-state actors, 40, 174, 245, 248, 341, 363, 365, 368–9
  see also nongovernmental organizations
North American Agreement on Environmental Cooperation (NAAEC), 150
North American Commission on Environmental Cooperation (CEC), 150
North American Free Trade Agreement (NAFTA), 143, 150, 151, 153, 180
North Atlantic Treaty Organization (NATO), 23, 211–12
North Sea Cooperation regime, 310
North–South (issues of) 253–4, 298, 300, 309, 331
  see also developing countries *and* inequality

environmental justice, 342–7, 349n4
  in the study of IEP, 14–15, 19, 21, 24, 27, 29, 33–6, 38–40, 41n8, 362–3, 365, 367, 369
  sustainable development, 330
  transnational actors, 188
Norway, 19, 23, 33, 304, 306, 314–16, 323n11
nuclear power, 161
Nueva Sociedad, 35

Ocean Development and International Law, 22
oceans, 18–19, 21–3, 32, 34
  see also water
oil, 161, 273, 284, 330, 338, 340
OPEC, 191
OPIC, *see* US Overseas Private Investment Corporation
Organization for Economic Cooperation and Development (OECD), 23, 121, 144, 161–3, 331
  Guidelines on Multinational Enterprises, 156
Our Common Future, *see* Brundtland Report
ozone depletion, 32, 155, 158–9, 309, 310
  see also Montreal Protocol

Pakistan, 31, 215
Papua New Guinea, 165n3
Paterson, Matthew, 5, 143–5
peace-building, 28, 214–16, 221, 226–8
peace dividends, 229
peace parks, 221–6
persistent organic pollutants (POPs), 154, 155, 159, 312–13, 318, 322
  see also Stockholm Convention
Peru, 165n3
Pesticide Action Network, 177
pesticides, 42n11, 130, 160, 164
  see also Rotterdam Convention
Philippines, 151, 159–60, 165n3, 215
pluralism, 36–9, 54, 63–4, 71, 73n11, 103, 173, 180, 367
  see also liberalism
Poland, 312

policy formulation, 182, 269, 306
  see also individual agreements
political economy, *see* international political economy
pollution, 14, 17, 19, 27, 34, 40, 146–7, 152, 160, 203, 266, 330, 334, 337
  acid rain, 23, 311
  air, 23, 150, 310–20
  oil, 303–4
  havens, 33, 152–3, 163–4
  water, 130
Polonoreste road project, 157–8
population, 40, 330–1, 337, 342–3
  and conflict, 212–13, 228–9
  in the study of IEP, 17–21, 26–7
populationists, 20, 27
  see also Malthusian
Portugal, 150
positivism, 218
  positivist methodology, 83–97, 208
  rational choice, 58, 71, 85, 87, 91–4, 102–4
  rationalism, 60–1, 125, 337, 351n13
postmaterialism, 128
postmodernism, 125–6, 204
  see also critical perspectives
postpositivism, 84, 97–103
  see also critical perspectives
poststructuralism, 54, 98–9, 204, 206–8, 210, 218–20
  see also structuralism
power, 134, 335, 342, 347–8, 364
  asymmetries, 224, 336
  balance of, 204
  business 67, 68, 338, 340
  critical perspectives, 98–9, 101–2, 206–7, 210, 217, 218, 277–8
  environmental degradation, 70
  feminism, 65–6
  institutionalism, 27, 58
  knowledge, 33, 62, 227
  state, 120, 204, 226, 289
  transnational networks, 341–2
prior informed consent, 149, 162–3
production and processing methods (PPMs), 148

qualitative methodology, 83–4, 87, 96, 99–101, 103
  case studies, 2, 83, 87–8, 92, 129–35, 165, 172, 178, 185–6, 188–92, 220–8, 229–300, 306, 322, 323n9, 366
  critical, 83, 97–101
  positivist, 85
quantitative methodology, 83–90, 96, 102–3, 206, 306–7, 320, 366
  statistical analysis, 83, 86–7, 89, 152
quotas, 146

race to the bottom, 146–7, 152
rationalism, 60–1, 125, 337, 351n13
  see also positivism
Reagan, Ronald, 32, 291n2
realism, 54–9, 72, 73n5, 172, 204–5, 208, 335, 337–40
  neorealism, 173, 204–5, 217, 335
recycling, 163
regime, *see* international regimes
regional trade agreements, 150, 165
  see also individual agreements
regulatory chill, 153
Republic of Korea, 165n3
Research Group on Development Strategies, 23–4
Responsible Care, 155
resource capture, 211, 228, 347
resource scarcity, *see* scarcity
Resources for the Future, 19, 21
Resources Policy, 23
Review of European Community and International Environmental Law, 31
Ricardo, David, 144
Rio Earth Summit, *see* United Nations Conference on Environment and Development
road building, 157–8, 161
Rotterdam Convention, 149
Russia, 86, 151, 165n3, 304–6, 315, 323n9
Rwanda, 215, 228

Scandinavia, 15, 36–7, 310–11
scarcity, 26, 32, 131, 146, 160
  and conflict, 28, 212
  environmental scarcity, 17, 20, 25–9
  and violence, 209, 211–14, 221

science, 24, 29, 33, 37, 61–2
  see also expertise and knowledge
  technology, 21, 26, 28
security, 172
  see also environmental security
  and development, 205
  discourses of, 203–20
  and gender, 218
  global, 203
  human, 28, 33, 218, 222
  redefinition, 38, 218–20, 369
  and regionalism, 229
security studies, 203–29
  see also environmental security studies
Shell Oil, 273–4, 284, 334
Sierra Leone, 213
Singapore, 165n3
Slovenia, 292n5
social ecology, 332–3
social movements, 134–5, 176–7, 334–5, 348
  environmental justice, 66–7, 330
  in the study of IEP, 368
societal politics, see transnational relations
soil degradation, 103–2, 135, 160
  see also land degradation
South Africa, 156, 215, 221, 223–5, 228
  Group for Environmental Monitoring, 221
South Africa Peace Parks Foundation, 221–2
Southern African Development Community (SADC), 221–2, 224
sovereignty, 226, 236
  and anarchy, 55
  constructivism, 59
  critical perspectives, 71–2, 115
  decline, 120, 241
  transnational actors, 180, 183, 291
Soviet Union, see Russia
spatial development initiatives (SDIs), 222
spatial relationships, 217, 219–20
Spain, 150, 323n11
Specialization, see comparative advantage

Stakeholder Forum, 177
state, 204, 205, 207–10, 213, 214–15, 278, 335
  agriculture 130–5
  critical perspectives, 217–20
  global governance, 42n19, 119–20
  globalization, 115, 120
  institutionalism, 37
  environmental justice, 332, 347
  state failure, 214
  structuralism, 205–6, 337–8, 340–1
  in the study of IEP, 31, 363, 369
  sustainable development 280–1
  transnational actors, 38, 121, 173–4, 179, 180–3, 185, 187
statistical analysis, 83, 86–7, 89, 152
  see also quantitative methodology
steady-state economics, 26, 117–18
Stockholm Conference, see United Nations Conference on Human Development
Strong, Maurice, 274
structural adjustment programmes (SAPs), 122–3, 133, 159–60, 164
  see also World Bank
structuralism, 64–6, 71–2, 74n16, 135, 361
  see also Marxism and world-systems theory
  environmental security, 204, 205–6, 218–20
  environmental justice, 337, 340
  methodology, 98, 101–2
  poststructuralism, 98–9, 126, 208
  sustainable development, 266–7, 272, 275–8
subsidies, 159–60
  see also export subsidies
sulphur dioxide, 311, 313, 318
  see also Convention on Long-Range Transboundary Air Pollution
sustainability, 69–71
  see also limits to growth and sustainable development
sustainable development, 214, 265–91, 333, 348
  see also Brundtland report
  discourse, 59, 120
  environmental justice, 330–1

intergovernmental organizations, 122, 147, 151
  in the study of IEP, 17, 26, 28–35, 38–9, 362, 365
  transnational actors, 179
  transnational corporations, 154, 157
  World Conservation Strategy, 22–3
Sweden, 161, 271, 316, 323n11
Switzerland, 161, 314, 323n11
syndrome approach, 211–12

Tanzania, 228
tariffs, 146
  see also trade
Tata Energy Research Institute (TERI), 24, 35
technology, 21, 26, 28
technology transfer, 149, 152, 251, 253, 274, 300, 309, 362
Thatcher, Margaret, 32
Thailand, 160, 165n3
Third World Network, 177
Third World Quarterly, 31
Three Gorges Dam, 161
toxic substances, 146, 147
  see also Basel Convention, Stockholm Convention, *and* Rotterdam Convention
trade, 28, 142, 145–51, 162–5, 172, 245, 273–4
  agreements, 122, 147–51, 153, 165
  and agriculture, 129–35
  comparative advantage, 144, 146
  competitiveness, 132, 134, 146–7, 152, 153
  impact on the environment, 118, 130, 132, 146–7, 150, 162, 216
  liberalization, 114, 123, 144, 146–7, 159–60, 164, 242–3
  restrictions, 134, 146–8, 155, 163, 242, 246
tragedy of the commons, 56, 92
transboundary natural resource management (TBNRM), 204, 220–7
transfrontier conservation areas (TFCAs), 222–6

transnational actors, 113, 172–93, 368
  see also nongovernmental organizations *and* industry
  accountability, 178, 183, 186–8
  pluralism, 63–4
  global governance, 121–3, 135, 282–4
  legitimacy, 178, 186–8
  representation, 178, 183, 186–8, 243–5, 269, 288
  strategies, 188, 194n5
transnational advocacy networks, 172, 176–7, 188–92, 339
transnational corporations, *see* multinational corporations
transnational networks, 17, 35, 121, 175–7, 181, 282, 305, 342, 347
transnational relations, 15, 172–93
transportation, 131–2
Tuna-Dolphin debate, 143, 148–9, 179–80

Underal, Arlid, 299–303, 306, 314
Unitary Rational Actor (URA) model, 305–6
United Kingdom (UK), 161, 306, 312, 314–15, 323n11, 323n21, 323n22
  study of IEP in, 14, 18–19, 23, 32
UK Export Credit Guarantee Department (ECGD), 161
United Nations (UN), 21, 115, 122, 291
  institutionalism 276–8
  NGOs, 175, 179, 183, 187
United Nations Commission for Sustainable Development (UNCSD), 276, 285, 291
United Nations Commission on Global Governance, 239–40
United Nations Conference on Environment and Development (UNCED) (1992), 287, 329, 331
  Agenda 21, 193, 268, 273, 276, 281–2, 288–9
  nongovernmental organizations, 179, 283
  Rio Declaration, 268, 349–50n5
  and the study of IEP, 17–18, 173
  sustainable development, 265–6; 268–70, 276

United Nations Conference on
   Human Development (1972), 24,
   127, 267, 329–30, 368
  Environment Fund, 330
  nongovernmental organizations,
   179, 283
  Resolution on Institutional and
   Financial Arrangements, 330
  and the study of IEP, 17, 22, 238
United Nations Conference on Trade
   and Development (UNCTAD), 24
United Nations Convention to
   Combat Desertification
   (UNCCD), 266, 277, 287–90
United Nations Development
   Programme (UNDP), 30, 158–9,
   254
United Nations Economic
   Commission for Europe
   (UNECE), 311
United Nations Educational, Scientific
   and Cultural Organization
   (UNESCO), 18–19, 22, 24
United Nations Environment
   Programme (UNEP), 22, 149, 248,
   348
  in the study of IEP, 24, 32
  Global Environment Facility, 158–9
  peace-building, 227–8
  World Environment Organization,
   250–4
United Nations Framework
   Convention on Climate Change,
   287, 345
United Nations Human Rights
   Commission, 348
United States (as international actor),
   18, 217–18, 221, 241, 266,
   292n4, 323n9, 365
  agriculture, 131, 133–4
  APEC, 151
  Central Intelligence Agency (CIA),
   214
  civil rights movement, 329–30,
   334–5
  climate change, 191, 339–40, 342–3
  environmental justice, 329–30, 334,
   348–9

  Environmental Protection Agency
   (EPA), 158, 250
  export credit agencies, 161
  global conferences, 274
  Homeland Security Department, 96
  nongovernmental organizations,
   187
  ocean politics, 22
  study of IEP in, 14–15, 19, 23–6, 30,
   33, 361
  transboundary air pollution, 312
  tuna–dolphin dispute, 148–9
  World Bank, 157–8
United States Agency for International
   Development (USAID), 161, 223,
   226
United States Marine Mammal
   Protection Act, 148–9
Union Carbide, 164, 334
urban politics, 219
Uruguay, 31

Vienna Convention, 309
Vietnam, 165n3
volatile organic compounds (VOCs),
   318, 322
  see also Convention on Long-Range
   Transboundary Air Pollution

war on terror, 96, 217–18
waste, 65, 117–18, 150, 154, 155, 272,
   338, 346, 363
  see also hazardous waste
water, 130, 292n6, 292n7, 329
  dams, 157–8, 161, 335
  environmental security, 211, 215,
   220–1
  Global Environment Facility, 158–9
  oceans, 18–9, 21–3, 32, 34
  sustainable development, 272
W. H. Freeman, 24, 34
World Bank, 23, 30, 156–62
  globalization, 115
  global governance, 122–3, 250,
   253–4
  greening of, 158–60
  nongovernmental organizations,
   179, 245
  sustainable development, 278, 289

World Business Council on
    Sustainable Development
  see World Summit on Sustainable
    Development
World Climate Research Programme,
  29
World Commission on Dams, 246
World Commission on Environment
  and Development (WCED), 23,
  31, 266–7
  see also Brundtland Report
World Conservation Union, 221
World Development, 31
World Economic Forum (WEF), 278
World Environment Organization
  (WEO), 150, 244, 250–4
World Health Organization (WHO),
  251, 253, 315
World Meteorological Organization
  (WMO), 248, 251
World Order Models Project (WOMP),
  24, 27
World Resources, 30
World Resources Institute, 30
World Social Forum, 283
World Summit on Sustainable
  Development (WSSD) (2002),
  269, 273, 329, 366
  corporate accountability, 156
  economic globalization, 127

institutionalism, 165
Johannesburg Declaration, 277
nongovernmental organizations,
  154, 179, 283
trade, 149
world-systems theory, 64–5, 337,
  351n5
  ecological, 118
  environmental justice, 332, 338–41,
  346
  methodology, 102
  statism, 205–6
  in the study of IEP, 368
World Trade Organization (WTO),
  147, 153, 163, 185, 278
  agriculture, 132–4
  global governance, 68, 122–4, 150,
  155, 250
  globalization, 115, 123–4
  and multilateral environmental
  agreements, 308, 165, 245
Worldwatch Institute, 23
World Wide Fund for Nature, 22–3,
  63, 176

*Yearbook of International Cooperation on
  Environment and Development*, 31

Zambia, 159, 223
Zimbabwe, 223